Modern Circuit Placement
Best Practices and Results

Series on Integrated Circuits and Systems

Series Editor: Anantha Chandrakasan
Massachusetts Institute of Technology
Cambridge, Massachusetts

Modern Circuit Placement: Best Practices and Results
Gi-Joon Nam and Jason Cong
ISBN 978-0-387-36837-5

CMOS Biotechnology
Hakho Lee, Donhee Ham and Robert M. Westervelt
ISBN 978-0-387-36836-8

SAT-Based Scalable Formal Verification Solutions
Malay Ganai and Aarti Gupta
ISBN 978-0-387-69166-4, 2007

Ultra-Low Voltage Nano-Scale Memories
Kiyoo Itoh, Masashi Horiguchi and Hitoshi Tanaka
ISBN 978-0-387-33398-4, 2007

Routing Congestion in VLSI Circuits: Estimation and Optimization
Prashant Saxena, Rupesh S. Shelar, Sachin Sapatnekar
ISBN 978-0-387-30037-5, 2007

Ultra-Low Power Wireless Technologies for Sensor Networks
Brian Otis and Jan Rabaey
ISBN 978-0-387-30930-9, 2007

Sub-Threshold Design for Ultra Low-Power Systems
Alice Wang, Benton H. Calhoun and Anantha Chandrakasan
ISBN 978-0-387-33515-5, 2006

High Performance Energy Efficient Microprocessor Design
Vojin Oklibdzija and Ram Krishnamurthy (Eds.)
ISBN 978-0-387-28594-8, 2006

Abstraction Refinement for Large Scale Model Checking
Chao Wang, Gary D. Hachtel, and Fabio Somenzi
ISBN 978-0-387-28594-2, 2006

A Practical Introduction to PSL
Cindy Eisner and Dana Fisman
ISBN 978-0-387-35313-5, 2006

Thermal and Power Management of Integrated Systems
Arman Vassighi and Manoj Sachdev
ISBN 978-0-387-25762-4, 2006

Leakage in Nanometer CMOS Technologies
Siva G. Narendra and Anantha Chandrakasan
ISBN 978-0-387-25737-2, 2005

Statistical Analysis and Optimization for VLSI: Timing and Power
Ashish Srivastava, Dennis Sylvester, and David Blaauw
ISBN 978-0-387-26049-9, 2005

Gi-Joon Nam
Jason Cong

Modern Circuit Placement

Best Practices and Results

Springer

Editors:
Gi-Joon Nam
IBM Austin Research Laboratory
Austin, TX
USA

Jason Cong
University of California, Los Angeles
Los Angeles, CA
USA

Series Editor:
Anantha Chandrakasan
Department of Electrical Engineering and Computer Science
Massachusetts Institute of Technology
Cambridge, MA 02139
USA

ISBN 978-1-4419-4231-9 e-ISBN 978-0-387-68739-1

Printed on acid-free paper.

9 8 7 6 5 4 3 2 1

springer.com

Dedicated to VLSI circuit placement researchers and practitioners whose creative and persistent efforts made it possible for us to handle the exponential increase of circuit placement complexity in the past four decades. –Gi-Joon & Jason

Contents

Foreword

I have a very clear memory of the first time I ever read anything about placement algorithms. I was a graduate student, and the research community was crackling with the excitement and challenges of the early days of the "VLSI" revolution. I went to my university's library to track down a copy of the book chapter "Placement Techniques" by Maurice Hanan and Jerome M. Kurtzberg, in Mel Breuer's book on *Design Automation of Digital Systems*. This was an eye-opening experience for a young student. The seminal Hanan-Kurtzberg material was a wonderfully clear review of what was known at the time about placement problems; it was a beautiful mix of geometry, algorithms, heuristics, optimization, and real experiments on real (and by today's standards, really *small*) designs. Reading this paper was a significant "Aha!" moment in my own career in the physical design area.

The intervening decades have dramatically broadened the portfolio of successful placer strategies, beyond the simple iterative improvement and partitioning approaches of those early days. We have more powerful iterative paradigms like annealing; we have vastly improved partitioning technologies; we have large-scale analytical solutions that formulate and solve enormous numerical optimization problems; we have clustering and multi-scale methods; and we have a hierarchy of geometric models, from coarse initial placement to final legalization. In this vastly more complex landscape of challenges and solutions, where does the enterprising student of placement look to figure out what's what in the placement business today?

This book, I hope.

Based on a set of placers that competed in recent contests sponsored by the ACM International Symposium on Physical Design (ISPD), and using the tremendously important sets of common placer benchmarks associated with ISPD, this volume offers an excellent overview of what we know about placement today. We owe a great debt to its editors, Gi-Joon Nam of IBM and Jason Cong of UCLA, for organizing all this material into one accessible and coherent volume. As our problems continue to grow in size, and we layer ever more constraints like timing, power and reliability on these tools, it's clear that people are no less excited about placement problems today than when I first read about the topic as a student.

Rob A. Rutenbar
Carnegie Mellon University
April 2007

Preface

Research in placement algorithms for VLSI circuits has enjoyed a renaissance in recent years. Today, there are a number of high quality academic placers that have been developed in universities. The amount of research on this topic clearly reflects the importance of the placement as the single most critical component for achieving timing/design closure in a modern physical synthesis tool. Placement algorithm itself has been researched for more than three decades. Yet, the problem is still very challenging for multiple reasons. First, the exponential increase of the circuit density according to Moore's Law has led to designs with tens of millions of placeable objects today. Although such complex designs are composed hierarchically based on the logic or function hierarchy, multiple studies (e.g. [3]) show that placement based on the logic hierarchy may lead to considerably inferior results. The preferred methodology is to place the entire design flat (with millions or tens of millions of placeable objects) to derive a good physical hierarchy and then use it to guide the subsequent physical synthesis process. Therefore, the modern placers have to handle extremely large problem sizes. Second, today's System-on-Chip (SoC) designs introduce complex constraints, such as routability and timing constraints, as well as the support of mixed size macros, area I/Os, multi-Vt and multi-Vdd islands for power optimization. Moreover, recent work on placement optimality studies ([1,2]) suggest that there exists significant room for improvement even for wire length optimization alone (details will be discussed in Chap. 2). All these reasons stimulated renewed interests in research in circuit placement problems, both in academia and industry, in the past a few years.

To help further stimulate advances in placement research, ISPD (International Symposium on Physical Design [7]) hosted two placement contests using new, large-scale benchmark suites based on real industrial designs ([5,6], see Chap.1 for more detailed discussion). The common goals of the two ISPD placement contests were:

- To provide new modern placement benchmarks to stimulate new development in placement research

- To provide a common basis for quantitative measurements of contemporary placement algorithms, and help the academic community to publicize their placement tools and results
- To provide an educational forum on a variety of state-of-art placement algorithms for future placement researchers

These two placement contests were huge success with participation from a number of academic placers and provided a common platform to evaluate various placement algorithms on the same set of realistic benchmarks. This book is the product of these academic efforts on placement contests and it can be considered as the year 2006 snapshot of state-of-the-art modern placement techniques employed in the field. The book provides in-depth description of the best practices of placement algorithms used in the research community today. Each book chapter provides detailed description of the underlying algorithm and implementation features of a placement tool that participated in the two contests, including the experimental results on ISPD placement benchmark circuits and the optimality analysis on PEKO-MS benchmarks.

This book is organized in four parts:

- Part I introduces placement benchmark suites. In Chap. 1, new industry design-driven ISPD 2005/2006 benchmark circuits are presented with contest results. Chapter 2 describes the details of PEKO-MS benchmarks that can be used for placement optimality analysis.
- Part II describes flat placement techniques, which formulate and solve the entire placement problem directly (although the numerical solvers used in these placers may use multilevel methods). Chapter 3 describes the most recent analytical placer DPlace that is an anchor cell-based quadratic placement engine. The Kraftwerk placement algorithm, the winner of ISPD 2006 placement contest, is presented in Chap. 4.
- Part III presents top-down partitioning-based placement techniques. It includes Capo, a congestion driven placer (Chap. 5) and the Dragon placer that combines simulated annealing optimization with a partitioning algorithm (Chap. 6).
- Part IV is about multilevel placement methods that have attracted significant attentions recently. It covers APlace (Chap. 7), which was the winner of the 2005 placement contest, the runtime efficient force-directed placer, FastPlace (Chap. 8), the mFAR fixed-point addition based placer (Chap.9), and the multilevel non-linear optimization placer mPL (Chap. 10) that produced the highest quality solutions in the 2006 placement contest. Also, NTUplace3 (Chap. 11), a new analytical placer for large scale mixed-size designs, is presented here.

The idea of this book emerged in April 2006, right after the ISPD 2006 placement contest, as a way of capturing a technology snapshot of dominant placement algorithms. We sent out invitations to all placement contest participants, and every team agreed to contribute to this book. By February 2007, all chapter manuscripts were submitted. In fact, some of them included the latest progress they made after the 2006 placement contest. Therefore, the results reported in some of the chapters are different (better) from the original placement contest results, which we provided at the end of Chap. 1 for reference.

The editors are well aware of the limitations of placement objectives used in the two contests. The 2005 contest uses wire length minimization as its sole objective function, while the 2006 contest uses a combination of wire length minimization, cell density control and runtime as its objective function (see Chap. 1 for more details). Real placement problems need to consider a number of other objectives, such as timing, power, and thermal optimization, as well as interaction with various physical synthesis operations, such as buffer insertion and gate sizing. A direct comparison of different placers under all these objectives and constraints may not be possible or meaningful, as each design has its own emphasis, and the final result is not determined by the placement algorithm alone. Many other steps, such as timing analysis, global and detailed routing, and various physical optimization operations can affect the final result. Therefore, we think that it is appropriate to use rather simple metrics in the two placement contests to measure the capability of the core wire length optimization engines employed in the different placers. As pointed in [4], a placer with good wire length minimization engine can be extended to handle other design objectives through weighted wire length minimization using various weighting functions.

This book is intended for graduate students, researchers, and CAD tool developers in the physical synthesis and physical design area. Each chapter is mostly self-contained and can be read independently. We hope that the readers can benefit from this collection of modern placement algorithms and potentially contribute to the field with new perspective. Please note this book is not intended to provide a comprehensive review of all available placement techniques, but to highlight the most successful techniques and practices used in modern placers. We refer the reader to [4] for a more comprehensive survey for the existing placement techniques.

We would like to thank the ISPD organizing committee for sponsoring the two placement contests, and IBM Corporation for providing the benchmark examples. We are indebted to the time and efforts of all the chapter authors who made this book possible. Finally, we would like to thank David Papa at the University of Michigan for thorough reviews of all chapters.

Gi-Joon Nam
IBM Research
Austin, Texas

Jason Cong
University of California
Los Angeles, California

March 2007

References

1. C.-C. Chang, J. Cong and M. Xie, "Optimality and Scalability Study of Existing Placement Algorithms," *Asia South Pacific Design Automation Conference*, 2003, pp. 621–627
2. C.-C. Chang, J. Cong, M. Romesis and M. Xie, "Optimality and Scalability Study of Existing Placement Algorithms," *IEEE Transactions on Computer-Aided Design of Integrated Circuits*, pp. 537–549, April 2004
3. J. Cong, "An Interconnect-Centric Design Flow for Nanometer Technologies", *Proceedings of the IEEE*, vol. 89, No. 4, pp. 505–528, April 2001
4. J. Cong, T. Kong, J. Shinnerl, M. Xie and X. Yuan, "Large Scale Circuit Placement," *ACM Transaction on Design Automation of Electronic Systems*, vol. 10, no. 2, pp. 389–430, April 2005
5. Gi-Joon Nam, "ISPD 2006 placement contest: Benchmark suite and results," *Proceedings of the International Symposium on Physical Design*, pages 167–167, 2006
6. G.-J. Nam, C.J. Alpert, P. Villarubbia, B. Winter, and M. Yildiz, "The ISPD2005 placement contest and benchmark suite," *Proceedings of the International Symposium on Physical Design*, pages 216–219, 2005
7. `http://www.ispd.cc`

Part I

Benchmarks

1

ISPD 2005/2006 Placement Benchmarks

Gi-Joon Nam[1], Charles J. Alpert[1], and Paul G. Villarrubia[2]
[1]IBM Austin Research Lab
[2]IBM EDA
{gnam, alpert, pgvillar}@us.ibm.com

1.1 Introduction

Benchmarks can contribute significantly to algorithm development of many fields by providing a common basis for quantitative measurement and comparison. The early MCNC benchmarks and ISPD98 benchmarks [1] helped the academic community significantly to measure the advances in physical design in 1990s. While still being used extensively in placement and floorplanning research, those benchmarks can no longer be considered representative of today's physical design challenges. To further aid future advances in placement, new benchmark suites, dubbed as *ISPD 2005/2006 Placement Benchmarks*, have been released in conjunction with ISPD placement contests. There are total 16 benchmark circuits that are directly derived from modern industrial ASIC designs. These benchmarks include rather unique features to represent modern physical design challenges:

- Design size
- I/O objects
- Macros
- Floorplans
- White space
- Density target

Due to scaled technology and requirements of more complex functionality, modern designs contain more circuit elements than ever. In today's environment, several million gate designs are considered a norm and this number will increase year after year. This situation renders more scalable placement algorithm to design size preferred in the field. ISPD placement benchmarks provide five circuits with more than a million placeable objects. These large circuits serve to test the scalability of placement algorithms.

In older technologies, I/O pins usually preside around the chip boundary, and they are called perimeter I/Os. Modern technologies, however, allow I/O pins to be populated within chip area to reduce signal propagation delay through I/O pins. This new technology is called the area-array I/O technology. The new ISPD benchmark suites have multiple placement instances using the area-array I/O technology.

In today's design environment, hierarchical design methodology is extremely popular to reduce the turn-around-time of taping out chips. Due to this, more and more macro blocks (either fixed or movable) are introduced into modern ASIC designs. Indeed, these macro blocks can cause serious problems during placement. For example, legalization becomes extremely tricky with large movable macros. All ISPD placement benchmarks contain either movable or fixed macros or both, and serve as a good test set for mixed-size placement algorithms.

Another interesting aspect with macros is that they produce much more variety of floorplans for modern placement tools. Early decisions of where to put these macros during global placement have significant impact on the quality of placement solution. Therefore, modern placement tools have to be more reliable and robust to various floorplans. ISPD placement benchmarks provide a variety of floorplans with many macro blocks.

Modern high-performance ASIC designs have abundant white space primarily due to the needs of potential gate sizing, buffering, and other optimizations that follow the placement process. In global placement algorithm, white space management has become one of the most important considerations because it could affect the wire length of placement solution significantly. ISPD 2005/2006 benchmarks have wide range of white space to emphasize the importance of white space management.

If a placement solution is not routable, there is no point of discussing whether it is a good or bad solution. In other words, placement solution has to be routable. Unfortunately, it is extremely difficult to model the routability accurately or measure congestion during placement without actually running global/detailed router. Density target forces global/detailed placement algorithm to preserve some amounts of white space in arbitrary neighborhood of placement regions. These additional white space can serve to mitigate local congestion. All ISPD 2006 benchmarks have associated *density targets* (ranging from 50% to 90%) to represent different congestion constraints.

Next we briefly review the benchmark statistics and quality measurements of two placement contests.

1.2 ISPD 2005 Placement Contest and Benchmark[1]

As an inaugural placement contest, the quality of placement solutions was solely measured by half-perimeter bounding box wire length (HPWL) with pin locations respected. HPWL is widely used when comparing results in placement and

[1] Portions reprinted from the reference [11] with © [2005] ACM.

Table 1.1. ISPD 2005 benchmark characteristics.

Circuit	#Obj	#Mov	#Fixed	#Net	Density (%)	Util.(%)	# Peri.I/Os
adaptec1	211447	210904	543	221142	76	57	480
adaptec2	255023	254457	566	266009	79	44	407
adaptec3	451650	450927	723	466758	75	34	0
adaptec4	496045	494716	1329	515951	63	27	0
bigblue1	278164	277604	560	284479	54	45	528
bigblue2	557866	534782	23084	577235	62	38	0
bigblue3	1096812	1095519	1293	1123170	86	57	0
bigblue4	2177353	2169183	8170	2229886	65	44	0

floorplanning research. HPWL is easy to measure and still a reasonable first-order estimation for the routed wire length for small nets. For larger nets, steiner-tree wire length is necessary to accurately estimate the routed wire length.

Table 1.1 summarizes the characteristics of circuits in ISPD 2005 placement benchmark suite. The reported statistics are:

- *#Obj*. The total number of objects of design
- *#Mov*. The number of movable objects
- *#Fixed*. The number of fixed objects
- *#Net*. The number of nets
- *Density(%)*. The total design density defined as the ratio of the area sum of all objects divided by placement area
- *Util.(%)*. The design utilization defined as the ratio of the area sum of only movable objects divided by available free space. The available free space is defined as difference between entire placement area and the area sum of fixed objects
- *#Peri. I/Os*. The number of perimeter I/O objects. If this number is 0, a design contains only area-array I/Os

Figure 1.1 presents layout figures of benchmarks. To produce these layout figures, the Capo [14] placer's utility plotter is used that can be obtained from GSRC Bookshelf [3]. These circuits include many fixed blocks demonstrating current SoC-style VLSI design methodology. Big fixed macro blocks basically dictate the overall placement footprints while placing movable *dust logic* around them becomes the main task of the placement algorithm. It is still a challenging problem because placement algorithm must be able to cope with large number of objects (scalability), satisfy timing constraints (timing closure) and produce routable solutions (routability/congestion). Each design presents slightly different styles of placement problems. For example, *adaptec2* and *adaptec3* have large fixed blocks in the center of placement region, which may cause larger variations in wire lengths depending on which sides of fixed blocks movable cells are placed. In *bigblue1*, placing movable objects at the center region in more compact manner seems to be a more critical task. The benchmark *bigblue2* have relatively large number of pins from regularly placed small fixed blocks. The design density of *bigblue3* is over 85% primarily due to several large fixed blocks whereas its design utilization is around 55% with abundant free

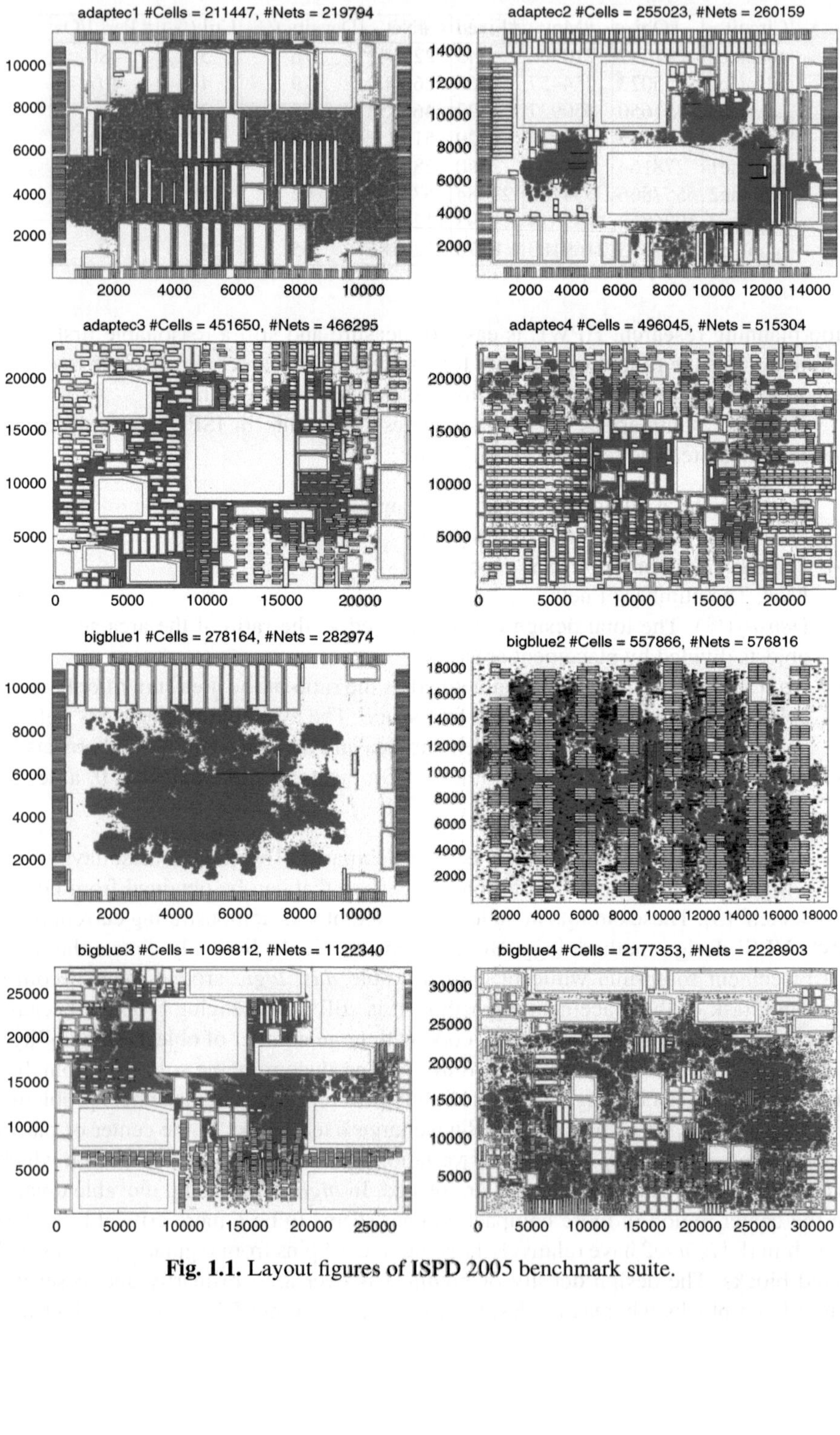

Fig. 1.1. Layout figures of ISPD 2005 benchmark suite.

space available. Two benchmarks *bigblue3* and *bigblue4*, with more than one million placeable objects, are good test cases for testing scalability of placement algorithms.

1.3 ISPD 2006 Placement Contest and Benchmark

To address the routability and congestion mitigation, ISPD 2006 placement contest benchmarks were released with associated density targets that the placer should obey. Density target is a constraint that forces a placer to reserve specified white space in any subregions of placement area. The density target is a floating number that is larger than or equal to *design density* that is defined as the total movable cell area divided by the total available area in the placement region. For example, if density target is 0.7, any local region should be less than or equal to 70% occupied. The lower the density target is, the more spreading is required. The idea behind this constraint is to improve routability and allow space for buffering, gate sizing, and clock tree that will be inserted later. In other words, this spacing helps make the placement instance more realistic to the practice in the field.

In 2006 placement contest, the quality of placement solution was measured by the adjusted HPWL function:

$$HPWL \times (1 + scaled\_overflow\_factor + cpu\_factor). \tag{1.1}$$

To measure *scaled_overflow_factor*, the placement region is divided into a set of equal-sized bins (10 circuit row heights). For each bin b, *bin_overflow* is calculated as follows

$$bin\_overflow(b) = \sum_{movable v \in b} [area(v) - free\_space(b) \times density\_target] \tag{1.2}$$

where $free\_space(b)$ is the available white space, i.e., the difference between the single bin area and the sum of fixed object areas that belong to the corresponding bin. The *total_overflow* of a design is simply the sum of $bin\_overflow(b)$ over all bins. Then, *scaled_overflow_factor* is calculated by

$$scaled\_overflow\_factor = \frac{total\_overflow \times single\_bin\_area \times density\_target}{(\sum_{movable v \in design} area(v)) \times c} \tag{1.3}$$

where c is a constant. Note that the term

$$\frac{\sum_{movable v \in design} area(v)}{single\_bin\_area \times density\_target} \tag{1.4}$$

is the minimum number of bins that can accommodate all movable objects in a design with a given density target. This term is independent of placement region area. Rather it is a unique design related factor. Therefore, the *scaled_overflow_flow* gives the (scaled) average overflow over the minimum number of bins required. For ISPD 2006

Table 1.2. ISPD 2006 benchmark characteristics.

Circuit	#Obj	#Mov	#Fixed	#Net	Density (%)	Util. (%)	Density target
adaptec5	843128	842482	646	867798	79	50	0.5
newblue1	330474	330137	337	228901	86	83	0.8
newblue2	441516	330239	1277	465219	86	62	0.9
newblue3	494011	482833	11178	552199	85	26	0.8
newblue4	646139	642717	3422	637051	66	46	0.5
newblue5	1233058	1228177	4881	1284251	75	50	0.5
newblue6	1255039	1248150	6889	1288443	59	39	0.8
newblue7	2507954	2481372	26582	2636820	76	49	0.8

contest, the constant c is set to 400, and to emphasize the importance of overflow factor, the squared *scaled_overflow_flow* term is used.

The purpose of *cpu_factor* is designed to gently encourage CPU performance. The intention is to encourage placers to get faster but not at the cost of significant solution quality. The *cpu_factor* is calculated as

$$cpu\_factor = 0.04 \times \ln \frac{placer\_cputime}{median\_cputime} \tag{1.5}$$

Therefore, if placer A is 2 times slower (faster) than a median speed placer, it gets 4% wire length penalty (advantage). If it is 4 times slower (faster), 8% wire length penalty (advantage) is imposed on placer A. However, *cpu_factor*'s wire length penalty/advantage is limited to maximum 10%.

Table 1.2 summarizes the characteristics of circuits in ISPD 2006 placement benchmark suite with associated density targets, and Figure 1.2 shows the layout figures that were generated with the Capo [14] placer's utility plotter from GSRC Bookshelf [3]. Again, each circuit in ISPD 2006 benchmark suite presents different styles of placement challenges. The density target of *adaptec5* is set to 0.5 while its design utilization is a little bit below 50%. Thus, it forces a placer to spread movable objects uniformly over the entire placement area. *newblue1* has several large movable macro blocks, and test the floorplanning capability of placement algorithm. In *newblue2*, all standard cells are inflated by 2× for congestion mitigation. Since cells are already inflated, a rather high-density target 0.9 is enforced. *newblue3* presents an unique floorplan with large fixed macros on both sides of placement region. The density targets of *newblue4* and *newblue5* are both set to 0.5, almost similar value to design utilization. Therefore, uniform spreading is required for both test cases. Also, *newblue5*, *newblue6* and *newblue7* have more than a million movable objects and the scalability of placement algorithm is important for these large circuits.

1.4 ISPD Placement Contest Results

Table 1.3 presents the results of 2005 placement contest. Columns 2–7 show the actual half-perimeter bounding box wire lengths of all placer's on each circuit and column 8 is the average ratio of a placer's wire length over the best wire length of

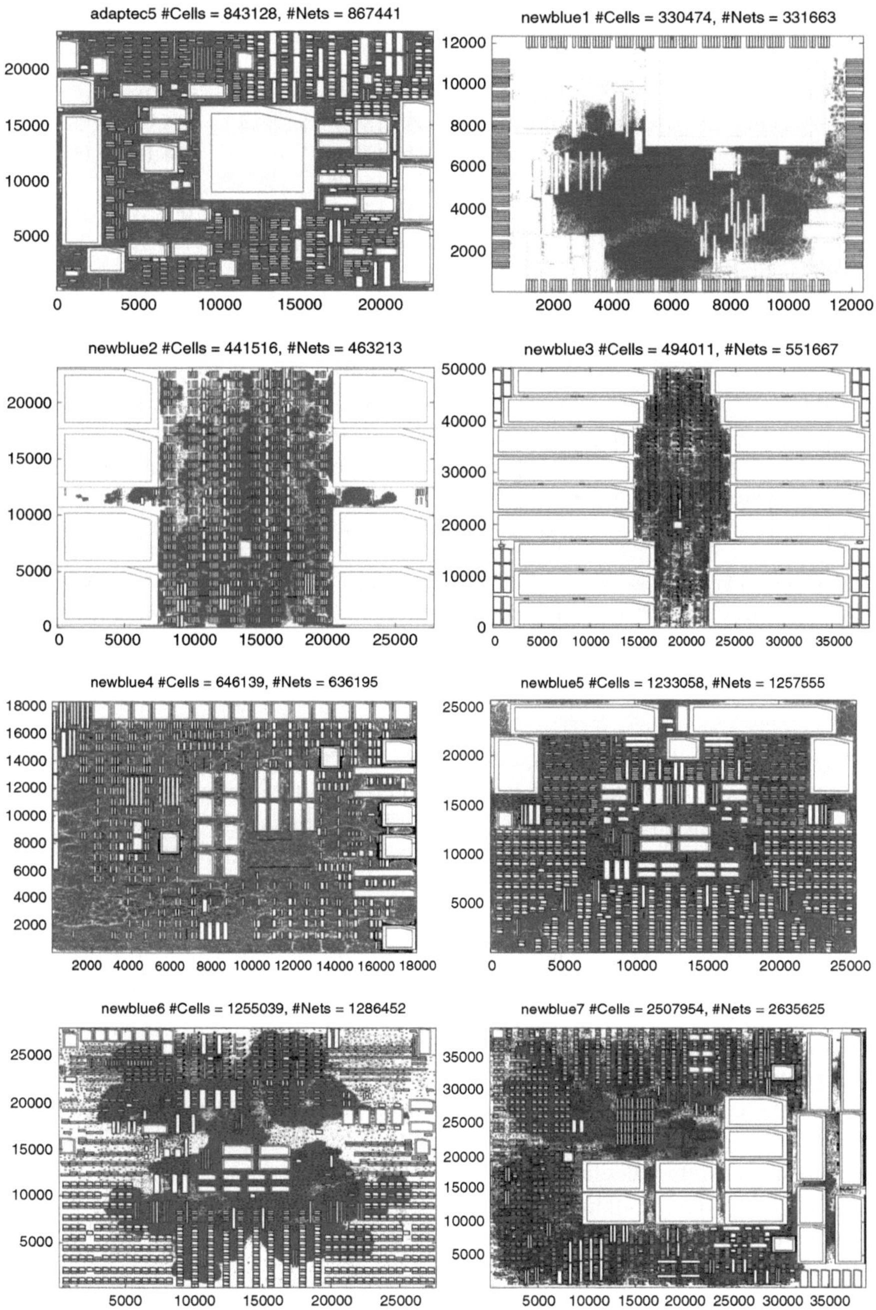

Fig. 1.2. Layout figures of ISPD 2006 benchmark suite.

Table 1.3. ISPD 2005 contest results.

Placer	adaptec2	adaptec4	bigblue1	bigblue2	bigblue3	bigblue4	Ratio
Aplace [9]	87.31	187.65	94.64	143.82	357.89	833.21	1.00
mFar [8]	91.53	190.84	97.70	168.70	379.95	876.28	1.06
Dragon [16]	94.72	200.88	102.39	159.71	380.45	903.96	1.08
mPL [4]	97.11	200.94	98.31	173.22	369.66	904.19	1.09
FastPlace [18]	107.86	204.48	101.56	169.89	458.49	889.87	1.16
Capo [14]	99.71	211.25	108.21	172.30	382.63	1098.76	1.17
NTUplace [6]	100.31	206.45	106.54	190.66	411.81	1154.15	1.21
FengShui [2]	122.99	337.22	114.57	285.43	471.15	1040.05	1.50
Kraftwerk [13]	157.65	352.01	149.44	322.22	656.19	1403.79	1.84

Table 1.4. ISPD 2006 contest results: Scaled HPWL.

Placer	ad5	nb1	nb2	nb3	nb4	nb5	nb6	nb7
Kraftwerk	457.92	78.60	208.41	280.93	315.53	569.36	545.94	1170.85
mPL6 [5]	431.14	67.02	200.93	287.05	299.66	540.67	518.70	1082.92
NTUplace3 [7]	432.58	63.49	203.68	291.15	305.79	517.63	532.79	1181.30
mFAR	476.28	77.54	212.90	303.91	324.40	601.27	535.96	1153.76
Aplace3 [10]	520.97	73.31	198.24	273.64	384.12	613.86	522.73	1098.88
Dragon [17]	500.74	80.77	260.83	524.58	341.16	614.23	572.53	1410.54
FastPlace	805.64	84.55	212.30	362.99	429.79	962.06	574.18	1236.34
DPlace	572.98	102.75	329.92	380.14	364.45	752.08	682.87	1438.99
Capo [15]	494.64	98.48	309.53	361.25	362.40	659.57	668.66	1518.75

each circuit. Note that the sole quality metric of 2005 placement contest was the half-perimeter bounding box wire length, and no CPU time usage limit was applied.

Aplace won the first contest convincingly by producing the best wire length on all circuits. Generally, analytical placers such as APlace, mFAR, mPL, and FastPlace produced better placement solutions than other algorithms. However, Kraftwerk, another analytical placement tool ended up the last place. Placeable objects in Kraftwerk solutions were spread uniformly over the entire placement regions leading to worse wire lengths. This turns out to be a critical mistake in low-utilization circuits. FengShui is a recursive partitioning-based placement tool that participated only in the first contest. Also, note that NTUplace was implemented as a partitioning-based placement tool in 2005.

The detailed data of ISPD 2006 placement contest are presented in Tables 1.4 and 1.5 [12]. Note that the scaled HPWLs in Table 1.4 include the bin overflow penalty factor in wire length calculation while CPU time factor is not included. Thus, the scaled HPWL is a representative metric for quality of solutions.

In Table 1.5, the reported statistics are:

- *Avg. HPWL Ratio*. The average ratio of a placer's HPWL over the best HPWL of each circuit. The closer to 1.0 this value is, the better a placer's HPWL is.
- *Avg. OV Factor %*. The average scaled bin overflow penalty factor to scaled HPWL calculation in percentage. For example, Kraftwerk got 1.68% increase in

Table 1.5. ISPD 2006 contest results: HPWL, Bin overflow, and CPU time factor.

Placer	Avg. HPWL Ratio	Avg. OV Factor (%)	Avg. CPU Factor (%)	Score Ratio
Kraftwerk	1.09	1.68	−5.04	1.03
mPL6	1.03	1.36	1.58	1.04
NTUplace3	1.02	4.10	1.66	1.05
mFAR	1.11	2.71	−0.12	1.11
Aplace3	1.10	3.82	5.31	1.16
Dragon	1.33	0.12	−5.90	1.24
FastPlace	1.18	22.09	−5.62	1.33
DPlace	1.34	9.32	−4.54	1.36
Capo	1.38	0.32	2.69	1.39

wire length due to its bin overflow penalty. FastPlace, however, was imposed over 22% wire length penalty implying that the density target was not strictly honored. Note that this penalty was already applied in the scaled HPWLs reported in Table 1.4.

- *Avg. CPU Factor %.* The average CPU time factor to wire length calculation in percentage. Negative value implies that a placer is faster than others and the contest scoring wire length is reduced by that amount. For example, Kraftwerk, mFAR, Dragon, FastPlace, and DPlace were rather fast placement tools and got some wire length advantage from 0.12% to 5.90%.
- *Score Ratio.* The average ratio of contest's scoring metric of a placer over the best one on individual circuit. This value represents the final ranking of the contest.

Overall, the Kraftwerk placer won the 2006 placement contest. The applied contest scoring metric with the consideration of HPWL, bin overflow, and runtime was successfully able to identify three best placers among all participants. The discrepancy between top three teams turned out to be negligible. If only HPWL and bin scaled factor (i.e., scaled HPWLs) are considered, mPL6, NTUplace3, and Kraftwerk again remain as top three teams.

References

1. C. Alpert, "The ISPD98 Circuit Benchmark Suite," in *Proc. ACM/IEEE International Symposium on Physical Design*, 1998, pp. 80–85
2. A. Agnihotri, S. Ono and P. Madden, "Recursive Bisection Placement: Feng Shui 5.0 Implementation Details," in *Proc. ACM/IEEE International Symposium on Physical Design*, 2005, pp. 230–232
3. A.E. Caldwell, A.B. Kahng, I.L. Markov, VLSI cad bookshelf. `http://vlsicad.eecs.umich.edu/BK/`. See also Caldwell AE, Kahng AB, Markov IL (2002) Toward cad-ip reuse: the marco gsrc bookshelf of fundamental cad algorithms. IEEE Design and Test 72–81
4. T.F. Chan, J. Cong, M. Romesis, J.R. Shinnerl, K. Sze and M. Xie, "mPL6: A Robust Multilevel Mixed-Size Placement Engine," in *Proc. ACM/IEEE International Symposium on Physical Design*, 2005, pp. 227–229

5. T.F. Chan, J. Cong, J.R. Shinnerl, K. Sze and M. Xie, "mPL6: Enhanced Multilevel Mixed-Size Placement," in *Proc. ACM/IEEE International Symposium on Physical Design*, 2006, pp. 212–214
6. T.-C. Chen, T.-C. Hsu, Z.-W. Jiang and Y.-W. Chang, "NTUplace: A Ratio Partitioning Based Placement Algorithm for Large-Scale Mixed-Size Designs," in *Proc. ACM/IEEE International Symposium on Physical Design*, 2005, pp. 236–238
7. Z.-W. Jiang, T.-C. Chen, T.-C. Hsu, H.-C. Chen and Y.-W. Chang, "NTUplace2: A Hybrid Placer Using Partitioning and Analytical Techniques," in *Proc. ACM/IEEE International Symposium on Physical Design*, 2006, pp. 215–217
8. B. Hu, Y. Zeng and M. Marek-Sadowska, "mFAR: Fixed-Points-Addtion-Based VLSI Placement Algorithm," *Proc. ACM/IEEE International Symposium on Physical Design*, 2005, pp. 239–241
9. A.B. Kahng, S. Reda, and Q. Wang, "APlace: A General Analytic Placement Framework," in *Proc. ACM/IEEE International Symposium on Physical Design*, 2005, pp. 233-235
10. A.B. Kahng, and Q. Wang, "A Faster Implementation of APlace," in *Proc. ACM/IEEE International Symposium on Physical Design*, 2006, pp. 218–220
11. G.-J. Nam, C.J. Alpert, P. Villarrubia, B. Winter and M. Yildiz, "The ISPD2005 Placement Contest and Benchmark Suite,", in *Proc. ACM/IEEE International Symposium on Physical Design*, 2005, pp. 216–220
12. Gi-Joon Nam, "ISPD 2006 placement contest: Benchmark suite and results," *Proceedings of the International Symposium on Physical Design*, pages 167–167, 2006
13. B. Obermeier, H. Ranke and F. M. Johannes, "Kraftwerk - A Versatile Placement Approach," in *Proc. ACM/IEEE International Symposium on Physical Design*, 2005, pp. 242–244
14. J.A. Roy, D.A. Papa, S.N. Adya, H.H. Chan A.N. Ng, J.F. Lu and I.L. Markov, "Capo: Robust and Scalable Open-Source Min-Cut Floorplacer," in *Proc. ACM/IEEE International Symposium on Physical Design*, 2005, pp. 224–226
15. J.A. Roy, D.A. Papa, A.N. Ng and I.L. Markov, "Satisfying Whitespace Requirements in Top-down Placement," in *Proc. ACM/IEEE International Symposium on Physical Design*, 2006, pp. 206–28
16. T. Taghavi, X. Yang, B.K. Choi, M. Wang and M. Sarrafzadeh, "DRAGON2005: Large-Scale Mixed-Size Placement Tool," in *Proc. ACM/IEEE International Symposium on Physical Design*, 2005, pp. 245–247
17. T. Taghavi, X. Yang, B.K. Choi, M. Wang and M. Sarrafzadeh, "DRAGON2006: Blockage-Aware Congestion-Controlling Mixed-Size Placer," in *Proc. ACM/IEEE International Symposium on Physical Design*, 2006, pp. 209–211
18. N. Viswanathan, M. Pan and C.C. -N. Chu, "FastPlace: An Analytical Placer for Mixed-Mode Designs," in *Proc. ACM/IEEE International Symposium on Physical Design*, 2005, pp. 221–223

2

Locality and Utilization in Placement Suboptimality

Jason Cong,[1] Michalis Romesis[2], Joseph R. Shinnerl[3], Kenton Sze[4], and Min Xie[1]

[1]UCLA Computer Science
[2]Magma Design Automation, Inc.
[3]Tabula, Inc.
[4]UCLA Mathematics
{cong,xie}@cs.ucla.edu
michalis@magma-da.com
jshinnerl@tabula.com
nksze@math.ucla.edu

2.1 Introduction

Placement is a critical step in VLSI design. Interconnect delay dominates system performance, and placement determines the interconnect more than any other step in physical design. The complexity of modern designs, however, makes estimation of suboptimality difficult [14, 16, 28]. Studies on simplified, synthetic benchmarks with known optimal-wire length placements (PEKO [7]) initially suggested that many leading tools may produce solutions with excess wire length from 60% up to 150% or more. These results have generated wide interest in both industry [13] and academia [19, 22, 28]. Recent progress in placement [1, 5, 6, 17] has reduced the wire length gap on PEKO to about 12–40%.

The PEKO benchmarks, however, have well-known limitations. Although their cell counts, net counts, and net-degree statistics match corresponding quantities in standard industrial benchmarks [2], the PEKO circuits are simplified in three key ways, in order to guarantee known optimal solutions. First, all cells are squares of the same size. Second, the known optimal placements for the PEKO circuits are packed layouts with zero white space. Third, all nets in an optimal PEKO placement are local – the netlist of a PEKO circuit is defined over cells arranged in a regular array, with adjacent cells grouped into local nets of minimum HPWL.

Subsequent studies [9, 16] derived useful lower bounds on the HPWL suboptimality of placements of circuits with more realistic netlists. The PEKU circuits [9] add nonlocal nets to packed, uniform-grid PEKO layouts but sacrifice any assurance of optimality. Zero-change netlist transformations [16] preserve both module shapes and core utilization, but they quantify the sensitivity of a placement tool to netlist changes, not the suboptimality of a given placement on a given netlist. It is not known how close the lower bounds on suboptimality are to the true suboptimality gaps for either the PEKU circuits or the zero-change netlists.

The benchmarks described in this chapter directly address several of the shortcomings in existing suboptimality benchmarks. Two new sets of placement examples are constructed, one targeting the role of nonlocal nets in suboptimality, and another targeting the role of white space and large variations in module sizes. The first set, PEKO-MC, is a set of standard-cell circuits with nonlocal nets in known optimal placements. A given netlist is modified so as to render a given placement for it optimal for the new netlist. Cell dimensions and locations are not changed, net-degree statistics are matched exactly, and over 60% of the original netlist is left unchanged. The second set, PEKO-MS, incorporates a parametrized percentage of white space into a mixed-size placement which precisely matches given macro dimensions and locations as well as the net-degree distributions of the ISPD 2005 benchmark suite [21]. HPWL for the placements generated for the PEKO-MS circuits are proven to be less than 3% above optimal for most cases and within 8% of optimal on all cases.

The concept of a monotone path for a circuit signal has been used in performance-driven logic synthesis [23, 26], coupled timing-driven placement and logic synthesis [27], performance-driven multilevel partitioning [15], and the analysis of wire length models in timing-driven placement [24]. The concept is also employed in two of the three netlist transformations used by Kahng and Reda [16]. To our knowledge, however, the work described here is the first to employ monotone chains in the construction of netlists with known optimal-wire length placements.

Typically, mixed size placement proceeds in three stages: global placement (GP), legalization, and detailed placement (DP). The goal of *GP* is to position each cell within some relatively small neighborhood of its final position, while eventually obtaining a sufficiently uniform distribution of cell area over the entire chip. Typically, large sets of cells are moved simultaneously under some relaxed or incremental formulation of area density control – scalable algorithms do not strictly enforce pairwise nonoverlap constraints during this stage. The goal of *legalization* is, given a sufficiently good GP P_g, determine positions of all cells so that (1) no two cells overlap and (2) a given objective, e.g., approximate total wire length or total displacement from P_g, is minimized. During *DP*, all constraints are strictly enforced. Typically, DP proceeds by a sequence of refinements made one at a time on small, contiguous subregions [4, 12] or on individual rows [18].

In practice, GP is terminated when iterations are observed to make little or no reduction in the objective and the module-area distribution is sufficiently uniform.

How much of the optimality gap left by contemporary methods should be attributed to deficiencies in global-placement algorithms, and how much to legalization and DP? On a real circuit, there is no way of knowing how far a cell is from its nearest optimal location, at any stage. On circuits with known optimal or near-optimal placements, however, it is possible to evaluate precisely the quality of any of the three engines in isolation from its counterparts. Thus, the benchmark circuits described here provide a more precise means of quantifying the relative effectiveness of the methods used in the three stages. Results estimating the separate suboptimality contributions of GP and legalization and DP are described in Sect. 2.4.

2.2 Peko-MC Benchmark Construction

Each PEKO-MC example has an optimal-wirelength placement in which over 50% of the nets are non-local. Module shapes, core utilization, and net-degree statistics match corresponding quantities in a given benchmark exactly. The PEKO-MC construction is described in this section.

2.2.1 Monotone Chains

The definition of a monotone chain in a netlist uses simple ideas from both graphs and hypergraphs. First, consider a path P in a *graph* G whose vertices lie in the plane. Let P consist of n consecutive edges $(e_1, \ldots, e_n)$ connecting $n+1$ vertices $(v_0, \ldots, v_n)$, vertex v_i with coordinates (x_i, y_i) and edge e_i connecting vertices v_{i-1} and v_i. Then P is *monotone* if and only if, for every $i \in \{1, \ldots, n\}$, $|x_n - x_i| \leq |x_n - x_{i-1}|$ and $|y_n - y_i| \leq |y_n - y_{i-1}|$. Hence, a path in a graph embedded in the plane is monotone if and only if the Manhattan distance between its two terminal vertices equals the sum of the Manhattan lengths of its edges.

In a hypergraph, a *hyperpath* is a finite sequence of hyperedges in which each hyperedge intersects with its predecessor and successor. We say that a hypergraph lies in the plane or, equivalently, is placed in the plane, if the nodes (modules) of the hypergraph (netlist) have been assigned specific locations in the plane. In this case, the total length of a hyperpath is the sum of the HPWLs of its hyperedges (HPWL denotes minimum bounding-box half-perimeter). An edge $e = (v, w)$ is called the *equivalent edge* of a hyperedge h of a hypergraph in the plane, if (1) its vertices v and w are in h and (2) e's minimum-HPWL bounding box is the same as h's minimum-HPWL bounding box. A hyperedge in the plane may have zero, one, or two equivalent edge(s).

A path P is called the *equivalent path* of a hyperpath H in a hypergraph in the plane, if there is a one-to-one correspondence between the edges of P and the hyperedges of H, such that every edge of P is the equivalent edge of its corresponding hyperedge in H. A *monotone chain* is a hyperpath which has an equivalent path that is monotone.

Assuming that no two vertices can occupy the same location, neighboring hyperedges in a monotone chain have exactly one vertex in common. These common vertices form the equivalent monotone path. The two terminal vertices of a monotone chain are the terminal vertices of its equivalent path. Hence, the length of a monotone chain equals the HPWL of the edge defined by the chain's two terminals.

Observation 2.2.1 *If the terminal vertices of a monotone chain $P = (h_1, h_2, \ldots, h_n)$ of a hypergraph G are fixed in the plane, there is no other planar embedding of hypergraph G which reduces the length of P.*

Given a placement of the hypergraph in the plane, a *local net* is a hyperedge the HPWL of which is the minimum possible, subject to some spacing constraints between vertices. From Observation 2.2.1, it is evident that a placement has optimal HPWL if all its nonlocal nets can be partitioned into netwise-disjoint monotone chains with fixed endpoints.

2.2.2 The Peko-MC Algorithm

Starting from the placement of the real benchmark, sets of nets are identified that can be grouped together into netwise-disjoint monotone chains between well-separated fixed terminals. Initially, these chains are not complete and have gaps called intervening regions. These are later filled by other nets that are modified from the original netlist. Local nets in the given placement are not modified. The main steps of the PEKO-MC algorithm are sketched below.

Placement generation. The PEKO-MC generator requires a placement of the original netlist. This placement is held fixed, while the netlist is changed so that the given placement attains the optimal HPWL for the modified netlist. Starting from a random placement is possible, but experiments show that starting from a placement computed by a tool increases the final similarity of the original and the derived circuits, because a real placement has many more locally optimal nets than a random placement.

Net categorization. The nets of the original hypergraph are divided into three different categories depending on the placement of their pins:

(1) Locally optimal-HPWL nets
(2) Nets that do not have equivalent edges
(3) Nets with equivalent edges

Nets of Type (2) cannot be members of monotone chains and are therefore modified. Nets of Type (3) are labeled according to the directions of the monotone chains of which they can be members: from lower left toward upper right, or from lower right to upper left. Some of these nets can be members of chains in either direction.

Chain generation. As illustrated in Figure 2.1, sets of nets that can be members of the same chain are identified along with sets of intervening regions that must later be filled by nets in order to complete the monotone chains. All nets of Type (3) are assigned to chains during this step.

Chain removal. In our experiments, the number of intervening regions between pairs of nets created during chain generation is higher than the number of nets of Type (2). Hence, to preserve netlist statistics, some of the chains generated are removed in order to reduce the number of such intervening regions and increase the number of

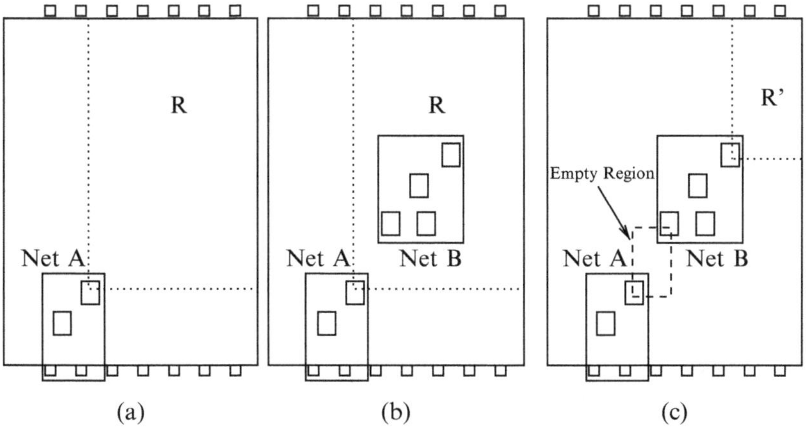

Fig. 2.1. Example of chain generation. (**a**) Net A, containing two cells, has already been added to the chain. A search for a new net takes place in region R. (**b**) Net B is selected to be added to the chain. (**c**) An intervening region is inserted between nets A and B that will be covered later by a new net. A new search is initiated for nets in region R'.

available nets. Chains with the highest ratios of intervening regions to contained nets are removed until the number of available nets equals or exceeds the number of gaps between terminals in chains.

Gap covering. In the final step, empty regions between nets in chains are filled by new nets. Each new net replaces some available net in the original netlist. The new net includes the two pins defining the equivalent edge of the bounding box of its intervening region R as well as additional pins selected from within R in order to match the degree of the replaced net. The cells whose degree in the current netlist are smallest compared to their original degree are given priority. In this way, the cell-degree distribution of the new netlist closely follows the corresponding distribution of the original circuit. Most intervening regions are covered by one net, but a few are covered by two nets, when the number of available nets exceeds the number of empty regions.

Experiments reported in Sect. 2.4 suggest that, on the 2004 FastPlace-IBM standard-cell circuits with 20% white space, nonlocal nets probably do not represent a significant source of suboptimality for these tools. In order to amplify the suboptimality observed on mixed-size cases as much as possible, the PEKO-MS benchmarks described next include only local nets by default.

2.3 Peko-MS Benchmark Construction

We refer to our placement suboptimality benchmarks with parametrized white space as PEKO-MS. As shown in Figure 2.2, the PEKO-MS generator produces a benchmark closely approximating the following four targets (1) net-degree histogram $N_{\#}$, (2) given placement P_{mac} of all macros, (3) number of standard cells N_{sc}, and (4) white

Set grid-resolution limit $\overline{N_G}$.

input

$N_{\#}$	target net-degree histogram
P_{mac}	macro placement in core region $\mathcal{R}$
N_{sc}	target number of standard cells
ϕ_{ws}	target white-space fraction

$\phi_{mac} := \left(\sum_{v_i \in P_{mac}} a(v_i)\right) / a(\mathcal{R})$.
$\phi_{ws} := \min\{\phi_{ws}, 1 - \phi_{mac} - N_{sc}/\overline{N_G}\}$.
$\phi_{sc} := 1 - \phi_{mac} - \phi_{ws}$.
$N_G := N_{sc}/\phi_{sc}$.
if ($N_G > \overline{N_G}$) **then**
  $N_G := \overline{N_G}$; $\phi_{sc} := N_{sc}/N_G$; $\phi_{ws} := 1 - \phi_{mac} - \phi_{sc}$.
end if
Snap P_{mac} into G, truncating macros as necessary;
  mark grid cells assigned to macros.
$N_{ws} := \phi_{ws} \cdot N_G$.
repeat
  Randomly select unvisited non-macro grid cell c.
  if (the spatial neighbors of c remain spatially
    connected in G when c is removed) **then**
      Mark c as white space and decrement N_{ws}.
  end if
until (N_{ws} == 0 **or**
      every non-macro grid cell has been examined)
if ($N_{ws} > 0$) **report failure and exit** **end if**
Mark all unmarked grid cells as standard cells;
  $\mathcal{V}$:= {macros} ∪ {standard cells}.
Following Figure 2.3, generate a minimal netlist
  "backbone" E_B, a connected set of local nets
  consistent with $N_{\#}$ which covers $\mathcal{V}$.
while ($N_{\#}$ still has nonzero entries **and**
    available locations for local nets still exist)
  Randomly select an available location p for a local net
  if (no new local net can be generated at p) **then**
    remove p from list of available local-net locations.
  else
    generate a local net of maximum possible degree k
    still represented in $N_{\#}$. Decrement $N_{\#}[k]$.
  end if
end while
output the placement suboptimality benchmark netlist

Fig. 2.2. The Peko-MS benchmark generator.

space fraction ϕ_{ws}. The ith component of vector $N_{\#}$ is the target number of nets of cardinality i. A macro is any module, fixed or movable, with height greater than the standard-cell row height. The generator places N_{sc} standard cells between macros and defines nets locally such that the total HPWL of the given placement is no more than a small, explicitly computed factor (1.00–1.08) above optimal for the final benchmark. Connectivity of the constructed netlist is ensured by inserting white space in such a way that all remaining cells and macros form a spatially connected set in the placement region.

As described in Figure 2.2 and later, the PEKO-MS generator proceeds in four stages:

1. Input target statistics; definition of uniform grid G; definition of mapping f_G which snaps a given macro placement P_{mac} into G.
2. Designation of white-space grid-cells, leaving cells, and macros spatially connected.
3. Construction of the netlist backbone (Figure 2.3), a minimal connected set of local, near-optimal-HPWL nets connecting all cells and macros.
4. Construction of additional, optimal-HPWL local nets to match target netlist statistics as closely as possible.

An optional additional stage for the addition of optimal-HPWL *nonlocal* nets is described in Sect. 2.4.

Every legal mixed-size placement induces a complicated partition $\mathcal{R} = \mathcal{R}_{mac} \cup \mathcal{R}_{sc} \cup \mathcal{R}_{ws}$ of its placement region $\mathcal{R}$ into three disconnected subregions: $\mathcal{R}_{mac}$ occupied by macros, $\mathcal{R}_{sc}$ by standard cells, and $\mathcal{R}_{ws}$ left as white space. The PEKO-MS generator preserves a given macro placement P_{mac} precisely with respect to a fixed core region $\mathcal{R}$. Let $a(S)$ denote the area of subregion S. Region $\mathcal{R}$ is neither shrunk nor expanded relative to the macros – both $a(\mathcal{R})$ and $a(\mathcal{R}_{mac})$ are held fixed. Instead, standard cells are uniformly shrunk or inflated to attain a higher or lower white-space targets, respectively. With this fixed-outline and fixed-macro-layout strategy, $\phi_{mac} \equiv a(\mathcal{R}_{mac})/a(\mathcal{R})$ is fixed, and it is evident that white space cannot be increased beyond the space left to it by the macros and standard cells:

$$\phi_{ws} \leq 1 - \phi_{mac} - \phi_{sc}^{\min}$$

where $\phi_{sc}^{\min}$ denotes the minimum fraction of $\mathcal{R}$ which can be left for standard cells. The exact value of $\phi_{sc}^{\min}$ is determined by storage and run-time considerations, as described next.

A tight lower bound on the optimal HPWL of each PEKO-MS benchmark is obtained by mapping the given macro layout P_{mac} into a uniform rectangular integer grid G of square cells over which all nets are defined. The mapping is denoted by $f_G : P_{mac} \rightarrow \text{Rect}(2^G)$, where $\text{Rect}(2^G)$ denotes the set of all contiguous rectangular subsets of grid cells in G. Each macro is identified by the mapping f_G with a distinct rectangular subset of grid cells in G. A nonoverlapping macro placement ensures that the grid-cell subsets associated with distinct macros are disjoint. Each center of each grid cell represents a candidate pin location. Pin locations on macros

```
input   C := ∅ = set of vertices contained in nets
B := ∅ = set of vertices not yet in C but spatially
        adjacent (in G) to at least one vertex in C.
Create a local net e at a random location.
Insert all v ∈ e into C and all G-neighbors of e into B.
while (B is not empty)
    Select an as yet unconnected grid cell b ∈ B and a
        connected grid cell c ∈ C such that b and c are
        adjacent in G. Cell b may be either a standard cell
        or a grid cell assigned to the boundary of an as yet
        unconnected macro. Cell c may be either a
        standard cell or an as yet unconnected grid cell
        assigned to the boundary of a connected macro.
    Create a net e containing b and c and containing as
        many other standard cells as possible, up to the
        maximum target net degree remaining in N#.
    if (N# [|e|] > 0) then decrement N# [|e|]
    else
        k := min{j | j > |e| and N# [j] > 0}.
        decrement N# [k] and increment N# [k − |e|],
    end if
    for ( all v ∈ e )
        remove v from B and insert it into C.
        for ( each grid neighbor w of v )
            if (w ∉ e and w ∉ C) insert w into B end if
        end for
    end for
end while
output minimal connected netlist E_B covering all v ∈ V
```

Fig. 2.3. Peko-MS Netlist backbone generator.

are restricted to grid-cells on macro boundaries and kept distinct. I.e., the center of each grid-cell along any macro's boundary can serve as a pin for at most one net. For simplicity, however, all pins on each standard cell are located at the same point at the center of that cell; i.e., the center of each standard cell may represent several pins for several different nets.

With all t pins of a given net placed at distinct grid-cell centers, the minimum HPWL of a t-pin net in such a grid is $r+s-2$, where $r = \lceil \sqrt{t} \rceil$ and $s = \lceil t/r \rceil$. This result is easily derived by packing the t square grid cells of the net into a rectangle of least possible perimeter. However, as shown in Figure 2.4, the optimal HPWL for a t-pin net may be attained by pin configurations with bounding boxes of different shapes.

In order to construct a local net of optimal or near optimal HPWL containing a small subset of rectilinearly connected seed pin locations, rectangles of gradually increasing sizes containing the seeds are recursively examined. Each such rectangle is a rectangular subset of grid-cells containing the seed locations and representing

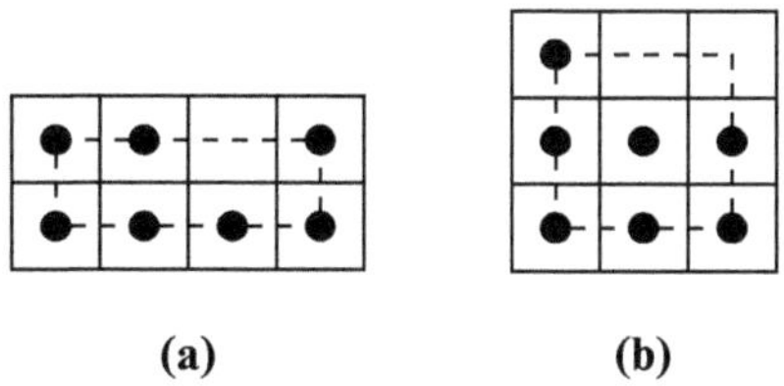

Fig. 2.4. On a uniform square grid, the optimal HPWLof a 7-pin net (4 grid units) can be attained by pin configurations with either of the two bounding boxes shown as *dashed line* segments.

a bounding box for a candidate net. In addition to the seeds, it may contain white space, standard cells, or grid-cells on the boundaries or interiors of macros. Of these, the available pin locations are the centers of the standard cells and the centers of the grid-cells on macro boundaries which have not yet been used as pins in other nets. As long as the number of available, rectilinearly connected pin locations in each such rectangle R is high enough to ensure optimal HPWL of the corresponding net, four larger rectangles containing R may also be considered. As shown in Figure 2.5, a rectangle is enlarged by adding to it a row or column of grid-cells along one of its four edges. Hence, the candidate rectangles for a given set of seeds form a quad-tree, the rectangles increasing in size along any path from root to leaf. Rectangles are enlarged until either optimal-HPWL cannot be obtained or the maximum-degree net remaining in $N_{\#}$ can be formed.[1]

At each seed location, the highest-degree optimal-HPWL net possible is formed, subject to the constraint that the number of nets of that degree in $N_{\#}$ has not yet been attained in the benchmark. The reason to form high-degree nets first is simply that they are the most difficult to construct. As pin locations along macros are gradually taken, high-degree nets become ever harder to construct. As not all high-degree targets in $N_{\#}$ may be attained during the construction, a compromise is made in the backbone-construction phase. When the degree d_b of a large backbone net b is no longer available in $N_{\#}$ but a larger target degree $d_t > d_b$ in $N_{\#}$ exists (i.e., $N_{\#}[d_t] > 0$ for some (net-degree) index $d_t > d_b$), then (1) net b is retained in the constructed netlist, (2) the maximum net-degree target remaining in $N_{\#}$ is decremented, and (3) the difference degree target entry $N_{\#}[d_t - d_b]$ is incremented. In this way, the total pin count of the constructed netlist is typically assured of matching the total pin count in the original benchmark.

As is suggested by the labeling in Figure 2.5, incremental enumeration of distinct candidate optimal-HPWL bounding boxes amounts to the enumeration of distinct finite sequences $\{d_i\}_1^N$, where each $d_i \in \{n, s, e, w\}$ represents the direction of enlargement at the ith step, and $N = 1, 2, \ldots$ is the total number of enlargements for a given box. Two sequences of the same length N are distinct if and only if the numbers of occurrences of all the symbols $\{n, s, e, w\}$ are not the same for both.

[1] To reduce search time, rectangles after a certain level in the quad-tree are enlarged in only one of the most promising directions, i.e., a direction containing the most available pin locations.

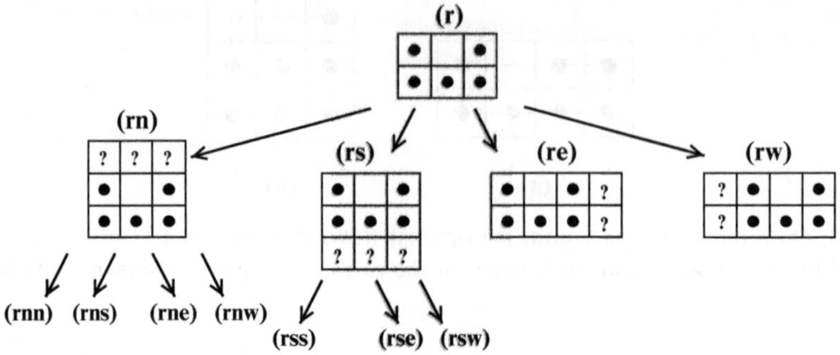

Fig. 2.5. The first level of local search for the largest optimal-HPWL net containing a given 5-pin seed. After the first level, many duplicate (e.g., *rsn*) and suboptimal (e.g., *ree*) cases at the subsequent levels can be pruned.

E.g., ns and sn are equivalent and lead to the same bounding box containing the initial seed box, but nse and nsn are distinct. The number of distinct sequences of length N is the number of ways $p_4(N)$ that the integer N can be expressed as the sum of four non-negative integers; asymptotically, $p_4(N)$ grows with order $N^3/6$.[2] However, sequences for suboptimal bounding boxes (such as *ree* and *rew* in Figure 2.5) and their descendants can be easily avoided.

Ideally, the resolution of grid G should be high enough to capture all macro and cell dimensions exactly. Our implementation simplifies the definition of f_G in two ways. First, P_{mac} is represented in floating point; macro positions and dimensions are expressed as fractions of chip dimensions prior to their conversion to integer grid units. Macro dimensions are truncated in G as needed to snap macros into the grid.[3] Second, each standard cell is represented by just one of the square grid cells of G – variations in standard-cell width are ignored. These two assumptions significantly reduce the size of G necessary to accurately represent P_{mac}. However, the resolution of G must still be large enough that:
(1) Each macro has nonzero height and width.
(2) The number of grid cells not used for macros is large enough to form both the requested number of standard cells N_{sc} and the requested fraction of white space ϕ_{ws}.

2.4 Experiments

Four sets of experiments with leading academic placement tools are reported. The first is on standard-cell PEKO-MC circuits generated from the 2004 FastPlace-IBM benchmarks. The second is on mixed-size PEKO-MS circuits derived from the ISPD

[2] The precise expression is $p_4(N) = (N^3 + 6N^2 + 11N + 6)/6$, which is the coefficient of x^N in the Taylor series for $(1-x)^{-4} = (1 + x + x^2 + x^3 + \cdots)^4$, assuming $|x| < 1$ [3].

[3] A small fraction of the transformed macros in G may be discarded due to error incurred in the truncation, e.g., macros mapped to zero-width rectangles in G, or one of a pair abutting macros in P_{mac} which overlap in G.

2005 suite. The third considers the impact of introducing chains of optimal-HPWL nets into a PEKO-MS benchmark. The fourth examines the suboptimality of legalization and detailed-placement engines in isolation from their global-placement counterparts on a parametrized adaptation of the PEKO-MS circuits.

2.4.1 Nonlocal Nets (Peko-MC)

All PEKO-MC benchmarks used in our experiments are generated from the FastPlace [8] versions of the 2002 IBM/ISPD benchmarks [2]. The white space in these test cases is approximately 20%. The FastPlace-IBM benchmarks modify the original IBM benchmarks by replacing macros with standard cells. However, the PEKO-MC algorithm can also be applied to examples with macros for the generation of mixed-size circuits with known optimal placements. Although no new pads are explicitly inserted, most existing pads are connected to several nets each to allow for more chains.

The PEKO-MC benchmark generator described in Sect. 2.2 requires as input both a netlist and an initial "seed" placement of that netlist. Dragon 3.01 [25] and mPL4 [10] were used to seed separate suites of PEKO-MC benchmarks. Using other placers as seeds was observed to have negligible impact on final results, even when the placer used to create the seed placements was run on the resulting PEKO-MC netlists.

The PEKO-MC suite matches the FastPlace-IBM benchmarks exactly in number of cells, cell areas, number of nets, and net-degree distribution. Roughly 60–70% of the nets in the original and synthetic benchmarks are identical, and the distributions of net lengths in the optimal placement of the synthetic benchmarks are nearly identical to those of their seed placements on the original netlists. Moreover, the cell-degree distributions of the original and synthetic benchmarks are very similar (the *degree* of a cell is the number of nets containing the cell). Almost 80% of the cells in an PEKO-MC netlist have a cell-degree difference at most 1 from their corresponding cells in the original netlist. Detailed statistics are shown in Figures 2.6 and 2.7.

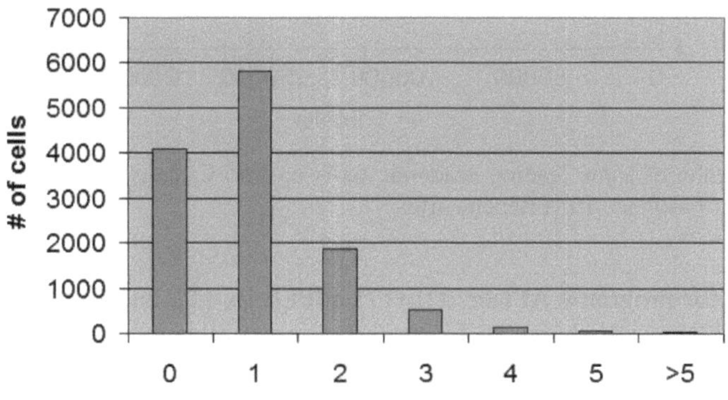

Fig. 2.6. The cell-degree difference (in absolute values) distribution between the cells of mPL-MC01 and their corresponding cells in FastPlace-ibm01.

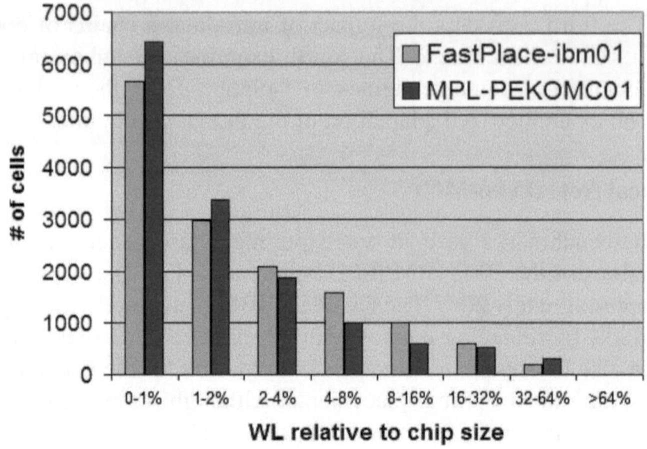

Fig. 2.7. The wire length distribution (relative to the chip half-perimeter) of the nets in FastPlace-ibm01 (as placed by mPL4) and the nets in mPL-MC01 (in their optimal placements).

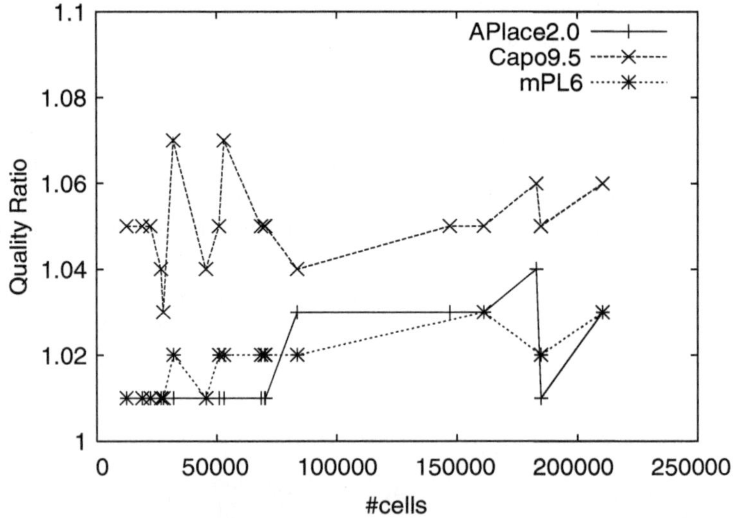

Fig. 2.8. Results of some leading academic tools on MC Circuits seeded by Dragon 3.01 placements of FastPlace-IBM Benchmarks.

Results for programs APlace 2.0 [17], mPL6 [6, 11], and Capo 9.5 [1] on the Dragon-MC suite are shown in Figure 2.8. Very similar results (not shown) were obtained for all the tools on the mPL4-MC suite. The overall results show very good performance by all tools on all the benchmarks, regardless of which tool generates the initial placement used to seed the benchmark construction. The worst reported quality ratio by any of the placers on any benchmark is 1.07. We attribute this result

Table 2.1. Peko-MS benchmark circuit statistics, with notation.

circuit	$\overline{N_{sc}}$	N_{mac}	$\overline{N_{nets}}$	ϕ_{mac}	ϕ_{ws}^{0}	ϕ_{ws}^{max}	ρ_{max}
PWS-A1	216180	63	233982	0.43	0.24	0.30	1.01
PWS-A2	264793	159	299358	0.62	0.21	0.29	1.03
PWS-A3	474287	723	531843	0.62	0.25	0.28	1.03
PWS-A4	531245	1329	563521	0.49	0.37	0.38	1.03
PWS-B1	280141	32	301577	0.17	0.46	0.46	1.01
PWS-B2	583514	23084	624625	0.38	0.38	0.40	1.03
PWS-B3	1137839	3778	1265913	0.67	0.14	0.24	1.08
PWS-B4	2237605	8170	2469988	0.38	0.35	0.40	1.03

$\overline{N_{sc}}$	average number of standard cells
N_{mac}	number of macros
$\overline{N_{nets}}$	average number of nets
ϕ_{mac}	macro-area utilization
ϕ_{ws}^{0}	original benchmark's white-space fraction
ϕ_{ws}^{max}	maximum white-space fraction attained by the generator
ρ_{max}	maximum ratio of generated HPWL to its lower bound

Averages and maxima are taken over the 4 different white-space values by which each circuit is parametrized. Standard deviations of N_{sc} range from 0.5% to 4.2% of N_{sc}; standard deviations of $\overline{N_{nets}}$ range from 10% to 16% of $\overline{N_{nets}}$.

to the increased range of optimal locations available to modules in multipin nets of monotone chains.

2.4.2 Parametrized White Space (Peko-MS)

The PEKO-MS approach gives the user control over the layout of the macros. In the ISPD 2005 benchmarks, all macro locations are prespecified for all circuits anyway, except `bigblue3`. For our construction based on `bigblue3`, we extracted movable macro locations from the placement generated for it by APlace [17] for the ISPD 2005 placement contest [29].

Each PEKO-MS local net's construction proceeds by depth-limited local search from a given subset of adjacent grid cells. A small amount of HPWL suboptimality is tolerated in some nets to simplify the implementation.[4] The optimal and attained HPWLs of the individual nets are simply added up to determine the limit on the total HPWL suboptimality in the final benchmark. These limits are shown in Table 2.1. On some circuits, nets e in the source netlist with more than a few hundred pins are represented by small subsets of high-degree nets whose pin counts sum to $|e|$.

Quality ratios of mPL6, APlace 2.0, and Capo 9.5 are listed in Table 2.2. The results show substantial variation both between tools and across different white-space values.

2.4.3 Suboptimality Under Both Parametrized White Space and Nonlocal Nets

The preceding results separate the impact of white space and mixed-size modules from that of nonlocal nets. However, the PEKO-MC and PEKO-MS techniques can

[4] However, we still refer to the placements as optimal, because the set of modules in each net is rectilinearly connected and hence supports an optimal *routed* wire length of the net.

Table 2.2. Results for mPL6, APlace 2.0, and Capo 9.5 on Peko-MS-ISPD2005 suboptimality benchmarks parametrized by white-space fraction. Displayed are quality ratios of total computed HPWL to near-optimal HPWL upper bounds. Results with uniformly distributed white space are shown for 5%, 10%, 20%, and the maximum possible white space values. For Peko-MS-adaptec1–4, quality ratios are also shown for benchmarks with optimal zero-white-space layouts ("pack") on the left side of the core region and 10% white space on the right. "mem" denotes an out-of-memory error. Capo 9.5 was run with option-noHMetis on Peko-MS-a3, Peko-MS-a4, and all four of the packed benchmarks; otherwise, all tools are in default mode in all cases.

	mPL6					APlace 2.0					Capo 9.5				
ckt\ ws	pack	5%	10%	20%	max	pack	5%	10%	20%	max	pack	5%	10%	20%	max
PWS-A1	1.80	1.35	1.48	1.70	1.80	1.33	1.50	1.22	1.15	1.54	6.17	3.33	3.14	3.05	2.67
PWS-A2	2.11	1.48	1.48	1.36	1.54	3.46	fail	fail	3.65	2.29	8.06	4.01	3.85	3.53	3.12
PWS-A3	4.32	2.14	1.52	1.41	1.33	2.27	1.23	1.14	1.13	1.10	4.10	2.10	1.93	1.53	1.33
PWS-A4	4.39	1.50	1.32	1.51	1.24	1.70	1.29	1.23	1.34	1.44	3.09	2.08	1.92	1.62	1.35
PWS-B1	–	1.30	1.34	1.30	1.24	–	1.44	1.32	1.17	1.33	–	2.50	2.42	2.08	1.77
PWS-B2	–	2.10	2.16	1.64	1.39	–	1.25	1.26	1.58	1.44	–	2.42	2.13	1.83	1.51
PWS-B3	–	1.54	1.62	1.99	2.02	–	2.14	1.59	2.02	2.23	–	2.49	2.10	1.89	1.91
PWS-B4	–	1.51	1.46	1.71	mem	–	1.26	1.21	1.16	1.33	–	mem	mem	mem	mem
Averages	3.16	1.61	1.55	1.58	1.51	2.19	1.45	1.28	1.65	1.59	5.35	2.70	2.50	2.22	1.96

be combined into a single set of suboptimality benchmarks supporting parametrized percentages of both nonlocal nets and white space. A combination derived from the Peko-MS construction (Figure 2.2) was tested on the mixed-size IBM01 benchmark from the ICCAD2004 test suite [1], as follows. Following the construction of the Peko-MS netlist backbone (Figure 2.3), monotone chains of nonlocal nets are constructed as follows:

1. The set of all boundary pads and candidate pin locations of fixed macros is partitioned by a simple heuristic into pairs of fixed terminals, such that the terminals in each pair are relatively far apart.
2. For each pair of terminals, designate one terminal in the pair as the start, and another as the end. A chain of nonlocal nets is iteratively constructed for the pair of terminals by the following sequence of steps (compare to Figure 2.1):
   (a) Randomly select an available pin location in the bounding box of the end terminal and the net corner pin most recently added to the chain. The selected location is the next net corner pin.
   (b) Randomly select additional pins in the resulting bounding box of that new net-corner pin location and the preceding net-box corner pin to populate the net.

   Net-box corner-pin locations are selected at randomized distances from one another approximately $1/10$ of the width or height of the placement region, until the end terminal of the chain is reached.

Results of APlace 2.0, Capo 10, and mPL6, all run in default mode, are shown for the combined PEKO-MSPEKO-MC IBM01 benchmark in Table 2.3, both without and with nonlocal nets. Macros larger than ten cell rows high were treated as fixed,

Table 2.3. HPWL Suboptimality of APlace 2.0, Capo 10, and mPL6, compared on 10% and 40% white-space versions of a PWS circuit derived from the ICCAD 2004 IBM01 mixed-size benchmark, both without (*top*) and with (*bottom*) the addition of optimal-HPWL nonlocal nets. Approximately 13,15% of the nets in the second set are nonlocal, accounting for 57%, 68% of total HPWL.

With Local Nets Only			
	APlace	Capo	mPL
ibm01-10WS	1.20	1.88	1.31
ibm01-40WS	1.40	1.96	1.27
Averages	1.30	1.92	1.29

With Chains of Optimal-HPWL Non-local Nets					
	$\frac{\#nln}{\#nets}$	$\frac{WL_{nonloc}}{WL_{total}}$	APlace	Capo	mPL
ibm01-10WS	0.15	0.57	1.11	1.49	1.16
ibm01-40WS	0.13	0.68	1.08	1.67	1.10
		Averages	1.10	1.58	1.13

their boundaries thus supplying some additional terminal locations. As expected, the presence of monotone chains of nonlocal nets decreases all placers' suboptimality ratios.

2.4.4 Suboptimality of Detailed Placement

Optimal GPs (OGP) parametrized by bin size were generated from the optimal PEKO-MS placements as follows. Uniform rectangular bin grids of user-specified dimensions were superimposed. Cells and macros centered in the same bin were moved to the bin center, where they were placed concentrically. These OGP placements were then used as benchmarks for the DP engines of mPL6 [6, 11] and APlace2.0 [17]. Each PEKO-MS circuit can generate several different OGP circuits, one for each bin size. The DP engines were run on a set of these OGP circuits, and the rate of degradation in their quality with respect to bin size and white-space value was observed. For each of the different white-space values, the quality ratios obtained by the DP engines were averaged over the eight different circuits. The result is illustrated in Figure 2.9. The benchmarks reveal opposite trends in these engines with respect to increasing white space. For these test cases, mPL's performance degrades as white space increases, while APlace's improves. APlace's cell-swapping strategy may have some advantage on these benchmarks, because the standard cells in these test cases are all of uniform size and shape. Under higher white space, the size of the set of candidate swaps is reduced, making successful swaps more likely to be found. On the other hand, mPL's local-window-based refinement is apparently a drawback on the higher-white-space cases, where larger scale moves are apparently needed.

Results on the OGP benchmark derived from the PEKO-MS-`adaptec2` benchmark with 10% uniformly distributed white space are summarized in Figure 2.10. Results are shown for two scenarios: one in which all macros are held fixed, and

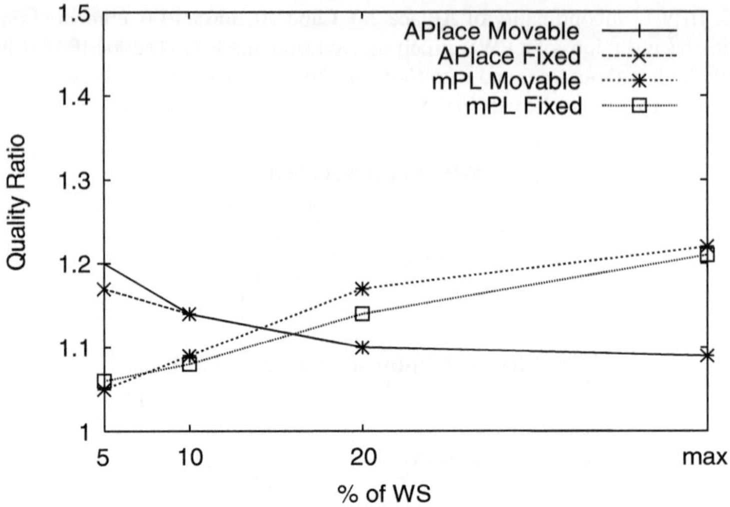

Fig. 2.9. Average quality ratios of APlace2.0-DP and mPL6-DP over the eight different netlists of OGP DP benchmarks.

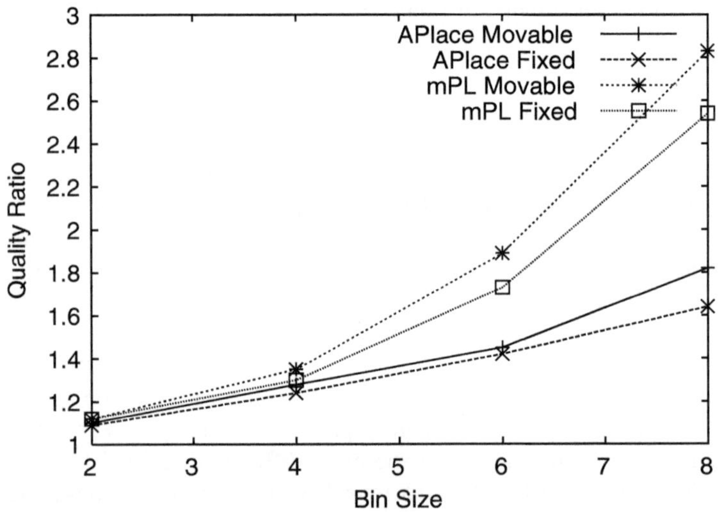

Fig. 2.10. Suboptimality of APlace2.0-DP and mPL6-DP on the OGP DP benchmarks generated from Peko-MS-adaptec2 with 10% white space.

hence only standard cells are aggregated into bin centers, and another in which both kinds of objects are moved from their locations in the known near-optimal placement to the nearest bin center. The results of these experiments show that the quality of DP deteriorates fairly rapidly as the bin size increases, even for these uniformly accurate GPs. For bin sizes up to 4×4, macro legalization is not a primary source of suboptimality, but for larger bin sizes, it is.

Table 2.4. Estimated suboptimality of mPL6 global (GP) and detailed placement (DP) engines. The GP estimate is obtained simply by subtracting the observed DP quality ratio obtained on the 2×2 OGP benchmark from the overall quality ratio observed for the corresponding Peko-MS benchmark.

%WS	GP	DP	Total
5	51%	10%	61%
10	46%	9%	55%
20	39%	17%	56%
40	29%	22%	51%

The OGP benchmarks provide a means of estimating how much of a placer's suboptimality is attributable to its GP, and how much to legalization and DP. Table 2.4 compares the suboptimality observed for mPL on 2×2 OGP cases to that observed for mPL6, including both GP and DP, on the corresponding PEKO-MS source circuits from which the OGP cases are derived. Subtracting the observed DP suboptimality on the 2×2 OGP benchmark from the total mPL6 GP+DP suboptimality on the corresponding PEKO-MS benchmark gives an estimate of the mPL6 GP suboptimality. It should be noted, however, that these suboptimality values are not truly additive, for at least two reasons. First, the starting configuration for DP on the OGP benchmark is very different from the DP starting configuration on the corresponding PEKO-MS benchmark. Second, relative module positions in a GP below the resolution of the 2×2 OGP grid will typically be used as hints during legalization and DP to improve results.

Figure 2.11 displays line segments between modules' placed locations and their optimal locations in a small PEKO-MS testcase (IBM02) constructed with 5% white space from the ICCAD2004 mixed-size suite [1]. Results for both global and detailed placements of APlace 2.0 and mPL6 are shown. From these plots, it is clear that displacement errors are not at all randomly distributed, and, on the contrary, display large-scale systematic bias. We observe similar trends in displacement plots for other tools on other Peko-MS circuits and at other white-space fractions. We conclude that, even when each cell in a GP is very close to (one of) its optimal location(s), further reduction in the objective can often only be achieved by moving large subsets of cells simultaneously by small amounts. Iterative, local, window-based refinement will not remove the systematic error.

2.4.5 HPWL Suboptimality Comparison of Leading Academic Tools on Peko-MS 2005

Recent (early 2007) versions of the placers entered in the 2006 ISPD Placement Contest were run on PEKO-MS adaptations of the ISPD 2005 and ISPD 2006 benchmark suites. The ratios of attained HPWL to the known near-optimal HPWL on the 2005 PEKO-MS benchmarks at 80% free-space utilization are shown in Table 2.5. Summary statistics for these benchmarks are shown in Table 2.6. Notational abbreviations are summarized in Table 2.7.

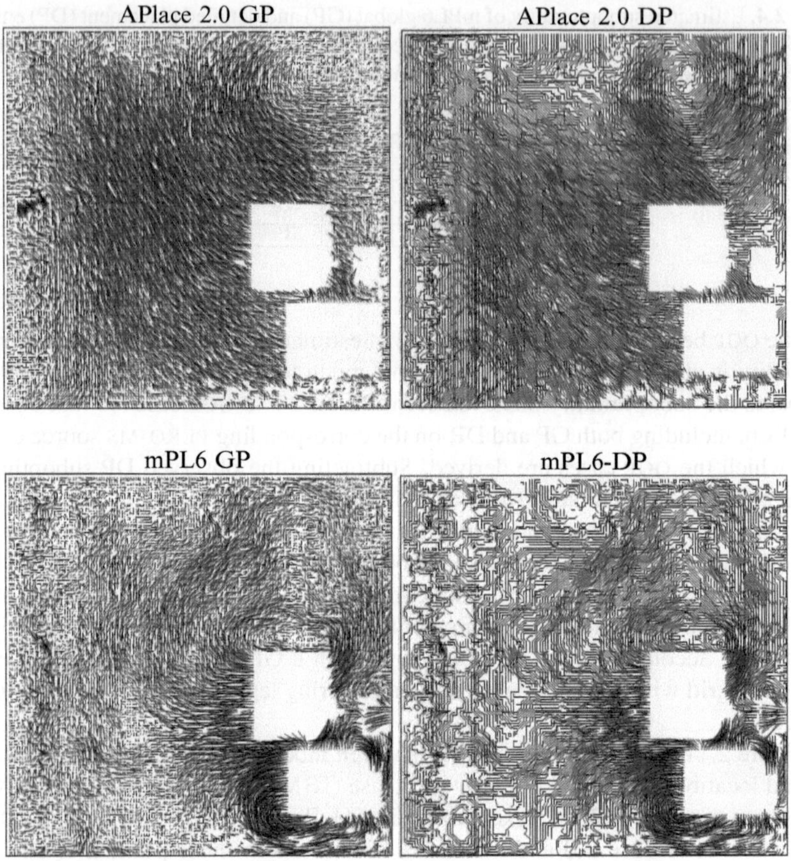

Fig. 2.11. Individual module displacements from optimal on the Peko-MS-ICCAD04-IBM02 benchmark with 5% white space. Displacements of both global and detailed placements are shown for APlace (*top*) and mPL6-DP (*bottom*). The HPWL quality ratios observed on this benchmark are 1.45 for APlace and 1.23 for mPL.

There are at least two ways in which these results may be useful in identifying weaknesses of the tools. First, the PEKO-MS benchmarks tend to amplify the suboptimality associated with local nets. Hence, a relatively high-average suboptimality gap on these test cases by a given tool (e.g., DPlace, mFar) suggests that tool might benefit from enhancements designed to help it better identify such local nets and reduce their lengths. Second, detailed investigation of a given tool's computation on a particular test case where it exhibits a relatively large gap (e.g., Kraftwerk on bigblue2, APlace on bigblue1, mPL6 on bigblue3) compared to its own results on other test cases may be useful in improving the tool's robustness. Such anomalous gaps may be particularly useful when the circuits on which they are observed have some distinguishing features, e.g., the relatively large number of movable macros in bigblue2, or the relatively low-area-fraction of fixed objects in bigblue1.

Table 2.5. HPWL suboptimality ratios of leading academic placers on Peko-MS ISPD 2005 benchmarks. For notation, see Table 2.7.

PEKO-MS-05 (80% free-space utilization) suboptimality ratios										
Circuit	DPlace	Kraft	Capo	APlace	FastP	Mfar	NTUPlace3	mPL6	Dragon	Averages
adaptec1	1.47	1.19	1.50	1.13	1.58	2.44	1.30	1.27	2.63	1.61
adaptec2	1.62	1.17	1.61	1.12	1.61	2.61	1.61	1.32	3.07	1.75
adaptec3	1.77	1.21	1.60	1.13	1.79	2.50	1.58	1.43	2.95	1.77
adaptec4	1.73	1.23	1.47	1.13	1.71	2.21	1.37	1.29	2.21	1.59
bigblue1	1.52	1.22	1.42	1.31	1.63	2.67	1.24	1.21	2.74	1.66
bigblue2	1.61	1.45	1.57	1.29	1.65	–	1.34	1.34	4.75	1.87
bigblue3	2.49	1.27	2.01	1.19	1.84	3.04	3.01	1.56	6.73	2.57
bigblue4	2.20	1.29	1.48	1.29	2.00	3.61	1.39	1.31	3.03	1.96
Averages	1.80	1.25	1.58	1.20	1.73	2.72	1.60	1.34	3.51	1.85

Table 2.6. Peko-MS ISPD 2005 benchmarks statistics. For notation, see Table 2.7.

PEKO-MS-ISPD2005 statistics (80% free-space utilization)										
Circuit	#obj	#mac	a(mac)	a(fix)	#term	# net	# pin	#pad	util	HPWL
adaptec1	211977	63	0.431	0.430	542	243402	939918	480	0.80	20056216
adaptec2	261153	159	0.615	0.613	543	290584	1061117	407	0.80	24969764
adaptec3	466560	723	0.615	0.615	723	515720	1864504	0	0.80	40954784
adaptec4	511448	1129	0.486	0.486	1329	558781	1903080	0	0.80	39391712
bigblue1	273708	32	0.172	0.172	559	307892	1136899	528	0.80	20858240
bigblue2	578938	22984	0.384	0.346	3313	634210	2117165	0	0.80	42256768
bigblue3	1122682	3778	0.680	0.668	675	1207133	3791237	0	0.80	94399040
bigblue4	2244173	8169	0.376	0.358	664	2494307	8792369	0	0.80	171477120

2.4.6 Suboptimality of Routability-Aware Placement

The Peko-MS algorithm was adapted to model the 2006 ISPD placement-contest scaled-HPWL objective, excluding run time, as follows. The scaled HPWL objective is computed as HPWL $\times$ $(1 + 0.01 \times \sigma)$, where σ is the "scaled overflow factor," defined by summation of bin-overflow penalties over each bin in a uniform grid consisting of square bins ten standard-cell rows high. We refer to this uniform grid as the "utilization evaluation grid." The utilization penalty for a given bin B is [20],

$$\left(\sum_{\{\text{modules } m\}} \text{area of } m \text{ overlapping with B}\right) - \text{utilization\_target} \times \text{free\_area(B)},$$

where the free area in B is simply the total area in B not occupied by fixed objects. Although this measure may vary somewhat with the size and position of the utilization-evaluation grid relative to fixed objects, it does test a placer's ability to target area congestion in a specified subregion.

The PEKO-MS generator of Figure 2.2 in Sect. 2.3 was adapted to target user-specified bin utilizations simply by iterating its random white-space insertion separately over all bins in the utilization-evaluation grid, terminating when either the

Table 2.7. Notation for column labels in this section.

Notation	
#obj	number of objects, movable and fixed
#mac	number of macros, i.e., objects of height greater than 1 row
a(mac)	fraction of core area occupied by macros, movable, or fixed
a(fix)	fraction of core area occupied by fixed objects
#term	number of fixed terminals
# net	number of nets
# pin	number of pins
#pad	number of perimeter I/O objects
util	target utilization in the part of core not occupied by fixed objects
HPWL	total HPWL of the given near-optimal placement
Hratio	ratio of attained HPWL to near-optimal HPWL
SOV/bin	average scaled overflow per bin
SHPWL	HPWL, scaled by $1 + 0.01 \times$SOV/bin
Sratio	ratio of attained SHPWL to near-optimal SHPWL

Table 2.8. Peko-MS ISPD 2006 benchmarks statistics. For notation, see Table 2.7.

PEKO-MS-06 statistics							
Circuit	# objs	# macs	a(mac)	a(fix)	#term	# nets	# pins
adaptec5	872276	646	0.572	0.572	646	1029763	3394072
newblue1	385625	64	0.379	0.000	337	334324	1237412
newblue2	461252	5000	0.652	0.636	1171	463382	1772264
newblue3	511413	8756	0.792	0.792	8845	559874	1940730
newblue4	671548	3422	0.359	0.359	3422	711993	2418450
newblue5	1282550	4881	0.495	0.495	4881	1520814	4805020
newblue6	1318990	6505	0.334	0.334	6505	1301252	5305156
newblue7	2641754	25065	0.535	0.535	25065	2651867	10098844

Circuit	#pads	util (%)	hpwl	sov/bin	shpwl
adaptec5	0	50	81893792	9.99	9.01E+07
newblue1	337	80	20500032	1.73	2.09E+07
newblue2	0	90	32869280	10.29	3.63E+07
newblue3	0	80	73514272	9.55	8.05E+07
newblue4	0	50	49143584	9.26	5.37E+07
newblue5	0	50	102083104	9.58	1.12E+08
newblue6	0	80	90657856	8.36	9.82E+07
newblue7	0	80	206175072	7.07	2.21E+08

utilization target is reached or when a given limit on BFS iterations (from 200 to 800) is reached. Complex fixed-macro geometries often make the precise target difficult for this simple approach to attain; hence, the final bin-utilization Peko-MS benchmarks have a low but nonzero and hence suboptimal level of overflow in most evaluation bins. Characteristics of these benchmarks are described in Table 2.8.

Table 2.9. HPWL and SHPWL suboptimality ratios of leading academic placers on Peko-MS ISPD-2006 benchmarks. For notation, see Table 2.7.

PEKO-MS-06 suboptimality ratios									
	DPlace			Kraftwerk			Capo		
Circuit	Hratio	SOV/bin	Sratio	Hratio	SOV/bin	Sratio	Hratio	SOV/bin	Sratio
adaptec5	2.53	369.8	10.80	1.17	36.37	1.45	1.58	4.97	1.51
newblue1	2.42	116.1	5.14	1.49	11.6	1.63	2.75	1.53	2.74
newblue2	2.18	124.2	4.42	1.29	50.87	1.76	2.77	1.17	2.54
newblue3	2.19	141.8	4.82	1.10	53.27	1.55	1.65	1.72	1.53
newblue4	2.22	282.5	7.77	1.23	36.26	1.54	1.61	6.43	1.57
newblue5	2.96	360.7	12.45	1.37	66.94	2.09	1.64	6.33	1.60
newblue6	2.28	168.5	5.66	1.23	51.88	1.72	2.15	2.3	2.03
newblue7	3.70	261.0	12.48	1.25	46.07	1.71	2.04	2.11	1.94
Averages	2.56	228.1	7.94	1.27	44.16	1.68	2.02	3.32	1.93
	APlace			FastPlace			MFar		
Circuit	Hratio	SOV/bin	Sratio	Hratio	SOV/bin	Sratio	Hratio	SOV/bin	Sratio
adaptec5	1.13	117.2	2.23	2.09	100.3	3.80	3.30	213.1	9.40
newblue1	1.35	89.7	2.51	2.18	13.3	2.43	5.98	39.1	8.18
newblue2	1.43	162.5	3.41	1.61	45.4	2.12	3.72	186.8	9.66
newblue3	1.20	132.6	2.55	1.11	81.3	1.84	2.15	162.8	5.15
newblue4	1.12	75.1	1.80	1.54	98.5	2.80	3.08	219.2	9.00
newblue5	2.61	217.6	7.55	2.05	84.8	3.46	3.16	207.7	8.87
newblue6	1.15	48.4	1.57	1.39	45.6	1.87	2.98	176.5	7.59
newblue7	1.31	119.8	2.68	1.32	45.1	1.79	2.74	177.1	7.10
Averages	1.41	120.34	3.04	1.66	64.28	2.51	3.39	172.78	8.12
	NTUPlace3			mPL6			Dragon		
Circuit	Hratio	SOV/bin	Sratio	Hratio	SOV/bin	Sratio	Hratio	SOV/bin	Sratio
adaptec5	1.31	7.4	1.28	1.35	17.7	1.44	2.96	0.29	2.69
newblue1	1.26	14.2	1.42	1.50	11.5	1.65	3.11	0.01	3.06
newblue2	1.45	2.6	1.35	1.35	31.5	1.61	4.20	0.13	3.81
newblue3	1.28	5.3	1.23	1.36	20.6	1.50	3.49	0.14	3.19
newblue4	1.26	3.3	1.19	1.37	15.8	1.45	2.64	0.29	2.43
newblue5	1.29	4.7	1.23	1.29	17.0	1.38	2.94	0.30	2.69
newblue6	1.23	2.0	1.16	1.40	18.0	1.53	2.80	0.17	2.59
newblue7	1.36	8.3	1.38	1.54	26.7	1.83	3.76	0.11	3.52
Averages	1.30	6.0	1.28	1.40	19.8	1.55	3.24	0.18	3.00***

Results of the most recent available implementations of the placers entered in the ISPD 2006 Placement Contest on the 2006 PEKO-MS test cases are shown in Table 2.9; median results over all the placers are listed in Table 2.10. Overall, the high-scaled HPWL suboptimality values obtained by most tools on most of the these benchmarks reveals considerable room for improvement of these tools in the presence of congestion metrics. As with the other PEKO-MS benchmarks, the utility of the 2006 PEKO-MS test cases lies primarily in helping to identify particular test cases where investigation of a given tool's performance may reveal weaknesses.

Table 2.10. Median HPWL and SHPWL suboptimality ratios of all leading academic placers listed in Table 2.9 on Peko-MS ISPD-2006 benchmarks. For notation see Table 2.7.

Median PEKO-MS-06 Suboptimality Ratios			
Circuit	Hratio	SOV/bin	Sratio
adaptec5	1.58	36.37	2.23
newblue1	2.18	13.27	2.51
newblue2	1.61	45.40	2.54
newblue3	1.36	53.27	1.84
newblue4	1.54	36.26	1.80
newblue5	2.05	66.94	2.69
newblue6	1.40	45.61	1.87
newblue7	1.54	45.06	1.94
Medians	1.56	45.23	2.08

E.g., consider (1) Capo on newblue1 and newblue2, and (2) APlace, DPlace, and FastPlace on newblue5, etc. While such hints might also be obtained simply by comparing to results of several tools on the original ISPD 2006 benchmarks, use of the PEKO-MS test cases may reduce the time needed to identify deficiencies by providing an absolute measure of suboptimality. In particular, the PEKO-MS test cases facilitate analysis of the trade-off between routability optimization and HPWL optimization without the need for comparisons to results of other tools. E.g., on the PEKO-MS test cases, results suggest that the superior area-congestion reduction of Capo and Dragon comes at a significant cost in increased HPWL. APlace, on the other hand, typically attains excellent HPWL reduction but relatively high-bin-overflow values; this result suggest that its placements may sometimes be difficult to route.

2.5 Conclusions

Two new sets of synthetic benchmark circuits with known optimal-HPWL or near-optimal-HPWL placements have been presented. The PEKO-MC set quantifies the role of nonlocal nets in suboptimality; the PEKO-MS set quantifies the role of white space and modules of mixed size. Experiments with leading academic placement tools support four main conclusions. First, as shown in Table 2.2, different tools produce widely varying results on some of the mixed-size PEKO-MS benchmarks. Hence, these benchmarks can be used to identify deficiencies in tools producing relative poor results. Second, the presence of netwise-disjoint chains of nets linking pairs of numerous, well distributed, fixed terminals appears to make wire length-driven placement by contemporary methods considerably less difficult. Circuits designed to ensure the existence of monotone paths for all signals [23, 24, 26, 27] might reasonably be expected to have wire lengths far closer to optimal than what leading placement tools are able to achieve on other circuits. Third, the accumulation of small but systematic errors in the placement of local nets appears to be a greater source of suboptimality than the total error in identifying and placing nonlocal nets.

The corrective action needed to further reduce that suboptimality, whether taken during global placement, legalization, or detailed placement, must consider simultaneous motion of large subsets of objects in order to be effective. Restriction to subsets localized in an arbitrary way is, in general, insufficient to improve on existing results. Fourth, the high-scaled HPWL suboptimality values obtained by most tools on most of the bin-utilization-controlled PEKO-MS benchmarks suggest that considerable room for improvement of these tools remains, particularly on large complex test cases, and particularly when bin-area congestion is factored into the quality evaluation.

2.6 Acknowledgments

Financial support for this work was provided by Semiconductor Research Consortium Contract 2003-TJ-1091 and National Science Foundation Contracts CCF 0430077 and CCF-0528583. The authors thank Editor Gi-Joon Nam for his assistance with final preparation of the PEKO-MS benchmarks and collection of placement results.

References

1. S.N. Adya, S. Chaturvedi, J.A. Roy, D.A. Papa, and I.L. Markov. Unification of partitioning, placement and floorplanning. In *Proc. Int. Conf. on Comp.-Aided Design*, pages 12–17, 2004
2. C.J. Alpert. The ISPD98 circuit benchmark suite. In *Proc. Int. Symp. on Phys. Design*, pages 80–85, 1998
3. G. Andrews and K. Eriksson. *Integer Partitions*. Cambridge University Press, 2004
4. U. Brenner, A. Pauli, and J. Vygen. Almost optimum placement legalization by minimum cost flow and dynamic programming. In *Proc. Int. Symp. on Phys. Design*, pages 2–8, 2004
5. U. Brenner and M. Struzyna. Faster and better global placement by a new transportation algorithm. In *Proc. Design Automation Conf.*, pages 591–596, 2005
6. T.F. Chan, J. Cong, and K. Sze. Multilevel generalized force-directed method for circuit placement. In *Proc. Int. Symp. on Phys. Design*, pages 185–192, 2005
7. C. Chang, J. Cong, M. Romesis, and M. Xie. Optimality and scalability study of existing placement algorithms. *IEEE Trans. on Comp.-Aided Design of Integrated Circuits and Sys.*, pages 537–549, 2004
8. C. Chu and N. Viswanathan. FastPlace: Efficient analytical placement using cell shifting, iterative local refinement, and a hybrid net model. In *Proc. Int. Symp. on Phys. Design*, pages 26–33, April 2004
9. J. Cong, M. Romesis, and M. Xie. Optimality, scalability and stability study of partitioning and placement algorithms. In *Proc. Int. Symp. on Phys. Design*, pages 88–94, 2003
10. J. Cong, J.R. Shinnerl, M. Xie, T. Kong, and X. Yuan. Large-scale circuit placement. *ACM Trans. on Design Automation of Electronic Systems*, 10(2):389–430, 2005

11. J. Cong and M. Xie. A robust detailed placement for mixed-size IC designs. In *Proc. Asia South Pacific Design Automation Conf.*, pages 188–194, 2006
12. K. Doll, F.M. Johannes, and K.J. Antreich. Iterative placement improvement by network flow methods. *IEEE Trans. on Computer-Aided Design*, 13(10), October 1994
13. R. Goering. Placement tools criticized for hampering IC designs. *EE Times*, February 5, 2003 http://www.eedesign.com/story/OEG20030205S0014
14. L.W. Hagen, D.J.-H. Huang, and A.B. Kahng. Quantified suboptimality of VLSI layout heuristics. In *Proc. Design Automation Conf.*, pages 216–221, 1995
15. C.-S. Hwang and M. Pedram. PMP: Performance-driven multilevel partitioning by aggregating the preferred signal directions of i/o conduits. In *Proc. Asia South Pacific Design Automation Conf.*, pages 428–431, January 2005
16. A.B. Kahng and S. Reda. Evaluation of placer suboptimality via zero-change netlist transformations. In *Proc. Int. Symp. on Phys. Design*, pages 208–215, April 2005
17. A.B. Kahng, S. Reda, and Q. Wang. Architecture and details of a high quality, large-scale analytical placer. In *Proc. Int. Conf. on Comp.-Aided Design*, November 2005
18. A.B. Kahng, P. Tucker, and A. Zelikovsky. Optimization of linear placements for wirelength minimization with free sites. In *Proc. Asia South Pacific Design Automation Conf.*, pages 241–244, 1999
19. Q. Liu and M. Marek-Sadowska. A study of netlist structure and placement efficiency. In *Proc. Int. Symp. on Phys. Design*, pages 198–203, 2004
20. G.-J. Nam. The ISPD2006 placement contest and benchmark suite, April 2006 http://www.sigda.org/ispd2006/papers/7-3.pdf
21. G.-J. Nam, C.J. Alpert, P. Villarrubia, B. Winter, and M. Yildiz. The ISPD2005 placement contest and benchmark suite. In *Proc. Int. Symp. on Phys. Design*, pages 216–220, April 2005
22. S. Ono and P.H. Madden. On structure and suboptimality in placement. In *Proc. Asia South Pacific Design Automation Conf.*, January 2005
23. R. Otten, and R. Brayton. Planning for performance. In *Proc. Design Automation Conf.*, pages 122–127, 1998
24. S. Ramji, and N. Dhanwada. Design topology aware physical metrics for placement analysis. In *Proc. Great Lakes Symposium on VLSI*, pages 186–191, 2003
25. M. Sarrafzadeh, M. Wang, and X. Yang. *Modern Placement Techiques*. Kluwer, Boston, 2002
26. W. Gosti, A. Narayan, R. Brayton, and A. Sangiovanni-Vincentelli. Wireplanning in logic synthesis. In *Proc. Int. Conf. on Computer-Aided Design*, pages 26–33, 1998
27. W. Gosti, S. Khatri, and A. Sangiovanni-Vincentelli. Addressing the timing closure problem by integrating logic optimization and placement. In *Proc. Int. Conf. on Computer-Aided Design*, pages 224–231, 2001
28. Q. Wang, D. Jariwala, and J. Lillis. A study of tighter lower bounds in LP relaxation based placement. In *ACM Great Lakes Symp. on VLSI*, pages 498–502, 2005
29. http://www.sigda.org/ispd2005/contest.htm.

Part II

Flat Placement Techniques

3

DPlace: Anchor Cell-Based Quadratic Placement with Linear Objective

Tao Luo and David Z. Pan
The University of Texas at Austin
{tluo, dpan}@ece.utexas.edu

3.1 Introduction

Although circuit placement has been studied for decades, it continuously attracts research attentions. The placement problems grow rapidly in both problem size and complexity. Some industry placement problems contain multimillion gates and excessive number of blockages [1,2]. In this chapter, we introduce DPlace, an anchor cell and diffusion spreading-based quadratic placement engine that can handle large-scale placement problem.

Historically, existing circuit placement algorithms can be roughly classified into three major categories, i.e., simulated annealing [3], iterative partitioning-based approach [4–6], and analytical placement approach [7–14].

Among existing placement works, analytical placement has been successful in recent years and achieved impressive results on wire length, scalability, and the speed of convergence. A typical analytical placement formulates the wire length optimization into a mathematical problem, and minimizes a smooth, continuous, and derivable wire length formulation. According to the reported results of ISPD 2005 and 2006 placement contest [2, 15], most of the top ranked placers are analytical placers.

In placement, the Half-Perimeter Wire Length (HPWL) is a common estimation of the routed wire length. Since HPWL model is not smooth and derivable, quadratic placement optimizes the quadratic form of HPWL [7–12, 14], and nonlinear model placement [13, 16, 17] adopts a nonlinear estimation of HPWL model, such as the log–sum–exponential wire length approximation patented by Naylor et al. [18].

Three placers in the ISPD 2006 placement contest using the log–sum–exponential wire model have achieved impressive wire length results. It is agreed that placement uses log–sum–exponential wire model approximates the HPWL much closer than the quadratic estimation. However, although still controversial, some researchers believe that the quadratic placement potentially has advantages for timing driven placement, as the quadratic approximation of the HPWL gives larger penalty on longer wires.

Most of analytical placements are force directed placement. The initial placement solution generated in force directed placement has excessive overlap among cells. To push cells away from congestion, in subsequent iterations, force directed placer adds "spreading force" or density constraints into the original wire length formulation. In force directed quadratic placement, the density constraints are combined into the optimization objective either by adding the spreading forces as constant force terms or by adding fixed points to implement the spreading forces.

We present a new quadratic placement, DPlace, that does not explicitly add "force" or apply density constraint into the original wire length optimization framework [19]. Different from traditional force directed quadratic placement, we divide the wire length minimization and density control tasks in two steps. A concept of anchor cell is presented to split the overlap reduction and wire length optimization objectives into two problems. DPlace is based on, but not limited to, quadratic placement, and the new framework is applicable for other analytical placements. In brief, during every iteration in DPlace, we have two steps:

1. A preplacement step to spread cells for better density distribution. The wire length minimization are not explicitly considered in the preplacement step.
2. An unconstrained wire length minimization step to repair the wire length. In this step, anchor cells are inserted as the reference of the preplacement result and as the basis of the new wire length optimization formulation.

In traditional force directed placement, spreading forces are used to estimate where to push cells. There is no explicit control of the cell movements and a cell may be pushed to any placement region. However, explicit cell movement control is important to cope with some challenging placement tasks, such as the ECO placement and timing-driven placement. It is possible to control the cell movement in DPlace by specifying the cell movement explicitly in the preplacement stage. The following are a few characteristics of our approach, which differentiates DPlace from other analytical placement works.

- We propose a global placement framework that uses *anchor cells* to split a traditional placement/spreading iteration into two steps, the preplacement step to reduce the cell overlaping, and the unconstrained wire length minimization step to reduce the wire length.
- In order to reduce the gap between the quadratic wire length vs. linear wire length objective, we introduce a net weight linearization strategy that transforms the star model-based quadratic objective into HPWL objective exactly.
- The framework we propose in DPlace can be used for both the global placement and ECO placement. The preplacement step can be extended to control the cell movements explicitly, e.g, we can specify a certain group of cells be moved to a certain position of the chip. This capability has the advantage for ECO placement where the placement stability is crucial.
- Our quadratic formulation is efficient for large-scale placement. The Hessian matrix in our quadratic formulation has much lower dimension as well as extremely low density. The runtime to solve one iteration of the system of linear equations is improved by 24 times in our formulation.

In the following, we introduce the preliminaries and current status of the force directed quadratic placement in Sect. 3.2. The details of our global placement are described in Sect. 3.3. The legalization and detailed placement are presented in Sect. 3.4. We give the overall algorithm of DPlace in Sect. 3.5 and show the experimental results in Sect. 3.6. Finally, the conclusion is presented in Sect. 3.7.

3.2 Preliminaries and the Motivation

To motivate our proposed approach, Sect. 3.2.1 gives an overview of the force-directed quadratic placement and the analysis of the essential concept in some of the existing force directed quadratic placement approaches.

3.2.1 Quadratic Placement

In circuit placement, a netlist is normally modeled as a hypergraph with each node representing an object/cell and each edge representing a net. Let x_i and y_i denote the coordinates of each cell, HPWL is used as an estimation of the routed wire length. Because the equation of HPWL is difficult to optimize mathematically, quadratic placement minimizes the square of the length and width of the bounding box of a net, commonly referred as the quadratic wire length.

As multipin nets can not be processed in quadratic placement, each multipin net is transformed into multiple two pin connections with proper weights. Traditionally, clique model is used for multipin net transformation and one k-pin net will be transformed into C_k^2 connections in clique model. For the 4-pin nets in Figure 3.1(a), the clique model transformation is shown in Figure 3.1(b). The disadvantage of clique model is that it may increase the number of nonzero entries in the connectivity matrix significantly, as the example in Figure 3.6, which slows down the quadratic solver. Another type of transformation is the star model [11,20]. One k-pin net will be transformed into k connections in star model, as shown in Figure 3.1(c). The combination of the clique and star transformation is also referred as the hybrid model [11].

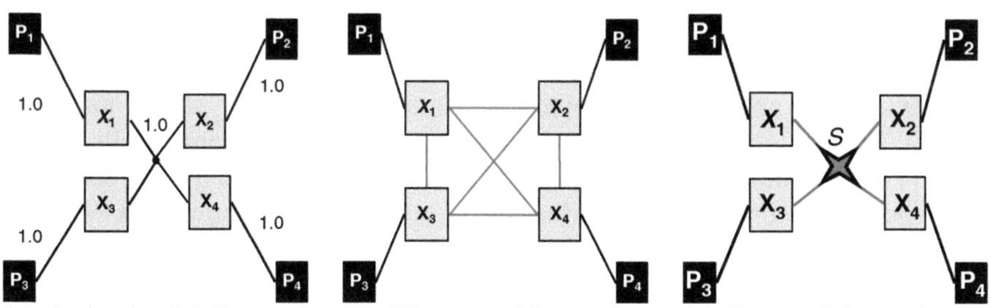

(a) A circuit with 5 nets, 4 cells, and 4 fixed pins, one of the nets is a 4-pin net

(b) Clique model transformation

(c) Star model transformation

Fig. 3.1. Transformations of the multipin net into multiple two-pin nets. Only the x coordinates are showed in these figures.

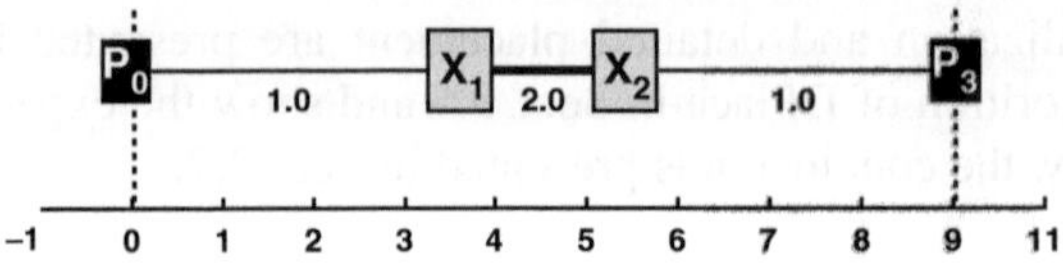

$$\begin{bmatrix} 3 & -2 \\ -2 & 3 \end{bmatrix}\begin{bmatrix} x_1 \\ x_2 \end{bmatrix} = \begin{bmatrix} p_0 \\ p_3 \end{bmatrix} = \begin{bmatrix} 0 \\ 9 \end{bmatrix}$$

$$x_1 = 3.6, \qquad x_2 = 5.4$$

Fig. 3.2. The quadratic placement formulation of a simple circuit in the x direction. p_0 and p_3 are the x coordinate of the fixed pins.

For a two pin net $e_{i,j}$ that connects cell i and j, the quadratic wire length is defined as $w_{i,j}((x_i - x_j)^2 + (y_i - y_j)^2)$, where $w_{i,j}$ denotes the weight of net $e_{i,j}$. The quadratic placement minimizes the sum of all quadratic wire lengths in the circuit. The optimization problems in x and y direction are separable and can be treated independently. Therefore, the cost function in x direction is given by

$$\Phi(x) = \frac{1}{2}\mathbf{x}^{\mathrm{T}}\mathbf{A}\mathbf{x} - \mathbf{b}^{\mathrm{T}}\mathbf{x} + \text{const.} \tag{3.1}$$

Assume there are n movable objects in the netlist. Let $\mathbf{A}$ denote the Hessian matrix of the quadratic system, which is essentially the n x n connectivity matrix of the netlist. $\mathbf{A}$ is symmetric and positive definite. $\mathbf{x}$ denotes the vector of x coordinates of all cells. $\mathbf{b}$ is the vector encoding all connectivity information between movable and fixed objects, and the pin offsets are captured in $\mathbf{b}$ as well. The minimizer of the cost function (3.1) can be obtained by taking the gradient of the cost function to zero, $\partial(\Phi(x))/\partial x = \mathbf{0}$, which is determined by the following system of linear equations

$$\mathbf{A}\mathbf{x} = \mathbf{b}. \tag{3.2}$$

Figure 3.2 shows a simple circuit with 2 movable cells and two fixed pins. The number associated with each net is the net weight. Cell 1 and 2 are in the force equilibrium status in Figure 3.2, i.e., the sum of the weighted wire length is the minimum.

3.2.2 Force-Directed Quadratic Placement

Solving the unconstrained minimization problem in (3.1) results in a placement with significant overlap among cells. A placer needs to push cells around to remove overlap. Some placers recursively partition the placement region to spread cells, such as Gordian [7]. The force-directed placers add spreading forces into the system in each solving process and reduce the overlap iteratively. Figure 3.3 shows that cell 1 and 2 are too close to each other, a force directed placer adds forces to push cells away from the center.

To apply spreading forces into the optimization framework, there are mainly two types of strategy to implement the force, the *constant force addition* and the *fixed point addition* approach. In each placement iteration, Kraftwork [8] and FDP [12] add a constant force vector **f** to the right-hand side of (3.2). The fixed point-based approach adds artificial pins and nets to move cells. mFar [9] uses multiple fixed virtual pins for each cell in every iteration, one is used to maintain a cell's force

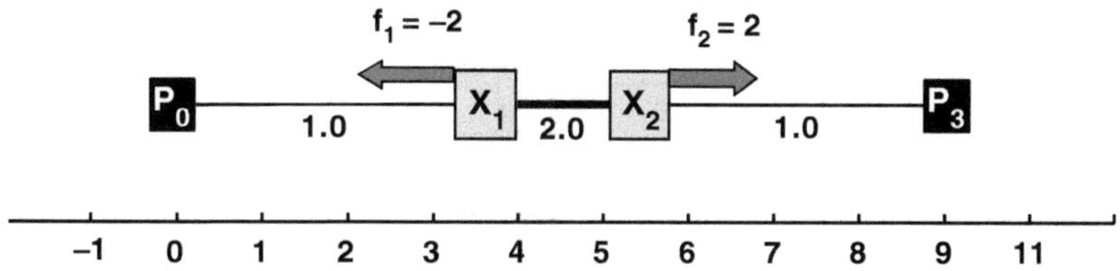

Fig. 3.3. Force directed placement: adding forces to push cells out of the region with congestion.

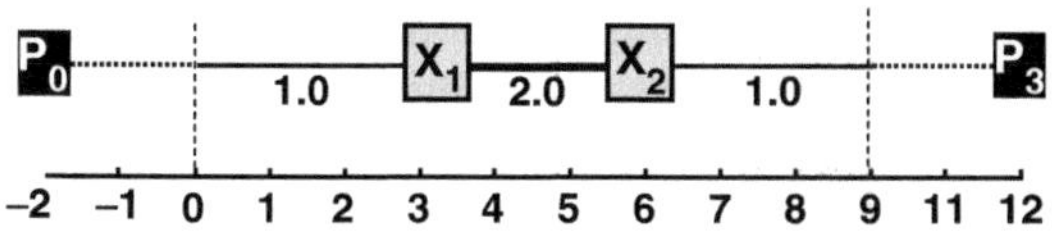

$$\begin{bmatrix} 3 & -2 \\ -2 & 3 \end{bmatrix} \begin{bmatrix} x_1 \\ x_2 \end{bmatrix} = \begin{bmatrix} p_0 + f_1 \\ p_3 + f_2 \end{bmatrix} = \begin{bmatrix} -2 \\ 11 \end{bmatrix}$$

$$x_1 = 3.2, \quad x_2 = 5.8$$

Fig. 3.4. Adding the constant force on a cell is equivalent to shifting its connected objects.

equilibrium state, and others are applied to perturb the cell. FastPlace [11] uses one fixed virtual pin for both purposes.

Constant Forces

In every iteration, the force for each cell is computed to reduce the overlap. In constant force-based approach, the force vector **f** is added to vector **b** in (3.2). The solution of the modified quadratic system generates a placement with less overlap among cells. In the ith iteration, the force vectors used in 1 to $(i-1)$th iterations are accumulated to prevent cells collapsing back. The modified equation with constant forces is given by

$$\mathbf{Ax} = \mathbf{b} + \sum_{k=1}^{i-1} \mathbf{f}^k + \mathbf{f}^i \tag{3.3}$$

In constant force-based approach, the Hessian (connectivity matrix) is not changed in each iteration unless the net reweighting is involved. In such case, the Hessian **A** only needs to be preconditioned once in the beginning, which will save runtime as the matrix preconditioning is runtime expensive.

The physical meaning of adding a spreading force to one cell is equivalent to shifting its connected pins and cells. To add the spreading force in Figure 3.3, a force vector is added into constant vector **b** in (3.2). To add a force vector is equivalent to shifting the connected objects of each cell, as shown in Figure 3.4. Pins are shifted outside of the chip, and cells may "jump" out the chip region if the magnitude and direction of the spreading forces are not properly adjusted. This tends to happen in the earlier placement iterations, where spreading forces are large and the force directions are not evenly distributed. Although such a scenario is not obvious in ISPD 2005 and 2006 benchmarks, where the initial density distributions are more even due to a large amount of fixed macros, the force scaling is tricky for placement with no fixed macros, such as the ISPD02 benchmarks [21].

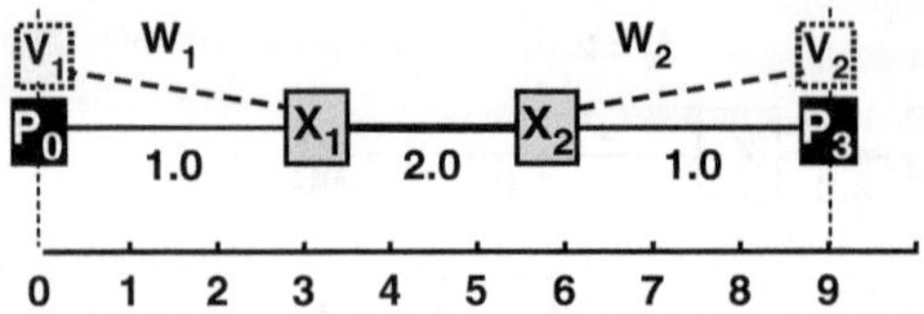

$$\begin{bmatrix} 3+w_1 & -2 \\ -2 & 3+w_2 \end{bmatrix} \begin{bmatrix} x_1 \\ x_2 \end{bmatrix} = \begin{bmatrix} p_0+w_1v_1 \\ p_3+w_2v_2 \end{bmatrix}$$

Fig. 3.5. An example of the fixed point addition formulation, p_0, p_3, v_1, and v_2 are the x coordinate of the fixed real and virtual pins, respectively.

Because the connectivity matrix is not strictly diagonal dominant, and often ill-conditioned, the solver of the linear system may have stability problem [12], i.e., cells may jump around when large forces are added. FDP adds a small weight to a portion of the diagonal terms of the Hessian and the new Kraftwerk [22] adds weight to all diagonal terms. Such a strategy is equivalent to adding a virtual fixed pin and net to a cell, as shown in Figure 3.5, which affects the quadratic objective and improves the stability of the quadratic solver.

Fixed Point Forces

In fixed point methods, the fixed points and nets are added to the original system of linear equations to perturb the placement. In fixed point-based methods, adding a virtual fixed point connection to a cell will add a diagonal term in the corresponding entry of the cell in the Hessian matrix **A** and the term in the constant vector **b**. In Figure 3.5, to add force to each cell, a virtual pin and connection are added to each cell with proper weight, and we can see the change in the Hessian and the constant vector in the figure. Therefore, adding a cell will make the corresponding row and column strictly diagonal dominant in Hessian **A**, and improve the condition number of the matrix. As a result, the fixed point addition-based method tends to be more stable.

The fixed point addition method guarantees cells moving inside the convex hull defined by the fixed points. If a large weight is used for the virtual nets, cells have less mobility and tend to move steadily toward force directions. However, the added large virtual net weights may dominate the actual net connections and affect the optimization objective. On the contrary, if using very small virtual net weights, fixed points will be off chip and cells may start to jump out of the boundary. In other words, the fixed point placement starts to behave similar as the constant force addition-based method. Furthermore, in fixed point-based approach, the connectivity weights will be updated in every iteration and the matrix needs to be preconditioned in every solving iteration.

3.2.3 The Proposed Approach

DPlace does not fall into the above categories. In each iteration in DPlace, the *cell anchoring* divides the constrained wire length minimization problem into two steps, the overlap reduction preplacement step and the unconstrained wire length minimization step. There is no "force" added to the quadratic system in DPlace, and there is no need to control the magnitude of forces, which is a nontrivial part in conventional force directed placements. As a result, no solver stability issue exists in DPlace.

Any smooth cell spreading techniques can be used for the density optimization preplacement step. We use the diffusion cell spreading [23] for preplacement step. Anchor cells are used to mark the preplacement result. We use the nets connecting anchor cells and real cells to formulate an unconstrained wire length minimization problem.

If a netlist is changed, without explicit cell movement control, the placement solutions before and after the changes could be completely different. In our approach, the explicit cell movement control can be naturally applied in preplacement step, which potentially provides flexibility for ECO placement. Furthermore, the fast growing of the problem sizes is a challenge to existing quadratic solvers. Our approach scales well to the problem size. The anchor cells used in quadratic framework significantly reduces the complexity of the problem.

3.3 Global Placement in DPlace

The global placement in DPlace is guided by a density driven preplacement method. We use the diffusion-based cell spreading technique [23] for the spreading smoothness.

3.3.1 Diffusion Preplacement

The global placement is guided by a diffusion-based cell spreading technique. Diffusion is the flow of particles from a region of highly concentration to a region with lower concentration, until the concentration on both regions is equal. The cell spreading in placement shares similar philosophy as the natural diffusion process, where cells are driven from high-density areas to low-density areas. Diffusion in placement is driven by the density gradient, i.e., the steepness of the density difference. Mathematically, the diffusion process is characterized by the following differential equation.

$$\frac{\partial d_{x,y}(t)}{\partial t} = D\nabla^2 d_{x,y}(t). \tag{3.4}$$

In the context of placement, $d_{x,y}(t)$ is the cell density at position (x, y) at time t. D is the diffusivity constant, which determines the speed of the diffusion process. The discrete approximation method in [23] can be used to solve the diffusion equation.

In diffusion-based preplacement, the placement region is cut into equal size bins. The bin density is computed as the total cell area enclosed in the bin divided the bin area. The discrete solver we use to solve the diffusion equation evens out the densities between neighboring bins as time proceeds.

In every global placement iteration of DPlace, cells are prediffused from high-density area to low-density area. The diffusion-based preplacement takes k sub-steps, where k is relatively small in earlier placement iterations and becomes larger in the later iterations. The cells will not be moved until we placed and locked all anchor cells.

3.3.2 Anchor Cells

Once a preplacement result is generated, we need to "memorize" the preplacement solution, in which cells have been spread out. Since the preplacement solution is often poor on wire length, we need to use the quadratic placement formulation to repair the wire length. To prevent cells collapsing back to the initial placement, we can fix a small percentage of cells in preplacement, and let the quadratic solver rearrange other cells. Another way is that we use virtual cells to mark the preplacement solution and replace some nets with virtual nets connecting the virtual cells and real cells. By updating the virtual connections into the wire length optimization objective and solving the unconstrained wire length minimization problem, cells will be "pulled" toward their anchors due to the wire tensions. In above scenarios, the fixed real or virtual cells are used as anchors to control the movement of real cells, and we name them "anchor cells."

We do not need to use one anchor per cell, which may over-restrict the movements of real cells. Instead, we can use one anchor for several cells, which gives more freedom for cells to move during the wire length optimization. We use star model to transform a portion of multipin nets into two-pin connections and use the star as the anchor of real cells. Compared with the method to use one anchor per cell, using stars as anchors will have much less impact to the original wire objective and imposes less constraint on cell movements.

In the hybrid model-based wire length transformation, the multipin nets are converted into star and clique model. All stars will be added back into the Hessian matrix **A** as moveable objects, which may increase the dimension of the matrix significantly. In ISPD 2005 benchmark, by using star model with a pin threshold as 5 will increase the dimension of the matrix up to 40%. For example, the dimension of the Hessian **A** for circuit *bigblue*4 in ISPD 2005 benchmark grows from 2.2 to 2.8 million. Under conventional formulation, solving one iteration of the system of linear equations with a dimension over 2 million will take several minutes.

Figure 3.6 shows the quadratic placement formulation of the circuit in Figure 3.1(a) by using the clique model. The dimension of the Hessian matrix is the same as the number of movable cells. Figure 3.7 is the formulation by using the star model. The dimension of the Hessian increases, but the matrix is more sparse compared with that by using the clique model. Unlike stars, anchor cells are fixed objects in our formulation. Therefore, anchor cells will not increase the dimension of the Hessian **A**. Furthermore, in the anchor cell-based quadratic formulation in Figure 3.8, we see that the Hessian matrix is extremely sparse compared with that by using both the star and clique models.

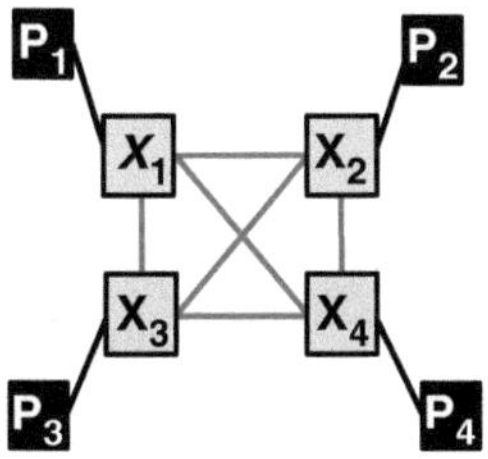

$$\Rightarrow \begin{bmatrix} 1.25 & -0.25 & -0.25 & -0.25 \\ -0.25 & 1.25 & -0.25 & -0.25 \\ -0.25 & -0.25 & 1.25 & -0.25 \\ -0.25 & -0.25 & -0.25 & 1.25 \end{bmatrix} \begin{bmatrix} x_1 \\ x_2 \\ x_3 \\ x_4 \end{bmatrix} = \begin{bmatrix} p_1 \\ p_2 \\ p_3 \\ p_4 \end{bmatrix}$$

Fig. 3.6. The quadratic placement formulation by using clique model. For simplicity, we assume the weight of each transformed two-pin net is weight 0.25.

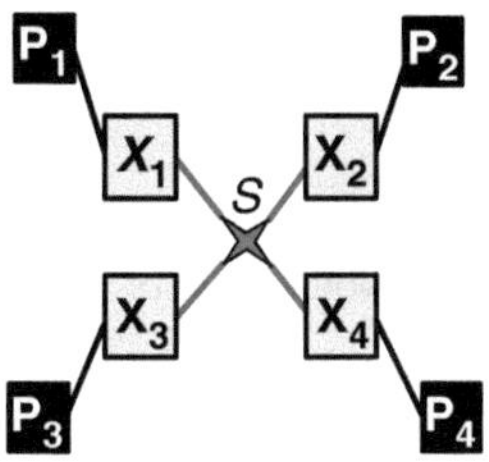

$$\Rightarrow \begin{bmatrix} 1.25 & & & & -0.25 \\ & 1.25 & & & -0.25 \\ & & 1.25 & & -0.25 \\ & & & 1.25 & -0.25 \\ -0.25 & -0.25 & -0.25 & -0.25 & 1 \end{bmatrix} \begin{bmatrix} x_1 \\ x_2 \\ x_3 \\ x_4 \\ s \end{bmatrix} = \begin{bmatrix} p_1 \\ p_2 \\ p_3 \\ p_4 \\ 0 \end{bmatrix}$$

Fig. 3.7. The quadratic placement formulation by using star model. S is the x coordinate of the star, which is a moveable object in the placement. For simplicity, we assume the weight of each transformed two-pin net is weight 0.25. The dimension of the Hessian matrix **A** is equal to the number of cells plus the number of stars.

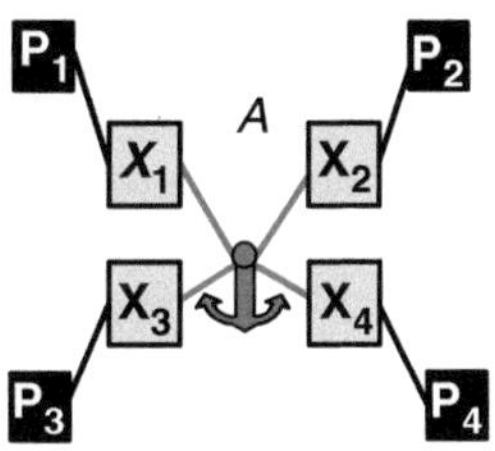

$$\Rightarrow \begin{bmatrix} 1.25 & & & \\ & 1.25 & & \\ & & 1.25 & \\ & & & 1.25 \end{bmatrix} \begin{bmatrix} x_1 \\ x_2 \\ x_3 \\ x_4 \end{bmatrix} = \begin{bmatrix} p_1+0.25C \\ p_2+0.25C \\ p_3+0.25C \\ p_4+0.25C \end{bmatrix}$$

Fig. 3.8. The quadratic placement formulation after the anchor cell insertion. C is the x coordinate of the anchor cell, which is a constant. The new Hessian matrix **A** is extremely sparse compared with that by using the star or clique formulation.

We assign anchor cells to the nets with a pin degree above th (e.g., 3 as in our implementation) only. With a small th, more anchor cells will be inserted, which may over-restrict the movement of cells. However, if the pin threshold th is too large, it will be more difficult to spread cells.

Let $\mathbf{A}'$ denotes the Hessian matrix in our new formulation. Anchor cells are not movable objects, thus do not appear in $\mathbf{A}'$. Matrix $\mathbf{A}'$ has the dimension as the number of movable objects in the netlist. We insert anchors after the completion of preplacement stage in each iteration. Once cells are preplaced, anchor cells are inserted at the gravity centers of their connected cells and locked. In such a way, anchor cells mark the preplacement result, and act as anchors to pull other cells around in

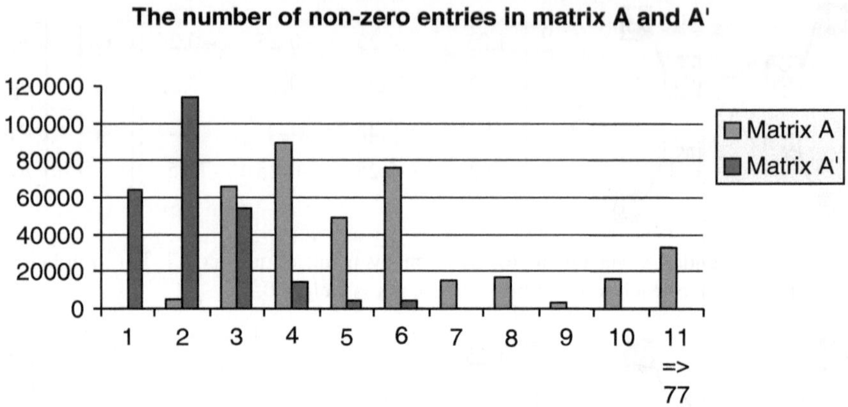

Fig. 3.9. The comparison of nonzero entries in all rows in the sparse matrix $\mathbf{A}$ and $\mathbf{A}'$. The x-axis is the number of nonzero entries, the y-axis is the row counts. Note: most of rows in matrix $\mathbf{A}'$ has only 2–3 nonzero entries.

the subsequent wire length minimization step. The new Hessian $\mathbf{A}'$ has a dimension much smaller than that in the conventional quadratic placement methods. Most importantly, the number of nonzero entries in each row of the new formulation $\mathbf{A}'$ is close to the number of pins on the cell, which is mostly around 2–4. A matrix with 2–4 nonzero entries is extremely sparse, and the linear system is trivial to solve using the anchor cell formulation.

Figure 3.9 shows the statistics of the number of nonzero entries in old Hessian $\mathbf{A}$ and new Hessian $\mathbf{A}'$ for circuit *adaptec*2 in ISPD 2005 benchmark. The dimension of the Hessian $\mathbf{A}$ is 354K, while only 254K for the new Hessian $\mathbf{A}'$. In most of rows, the number of nonzero entries in $\mathbf{A}$ are around 3–6, and 1–2 in new Hessian $\mathbf{A}'$. Circuit *bigblue*4 in ISPD 2005 benchmark contains 2 million objects. In our experiments for bigblue4, it takes 200 s for preconditioning and 75 s for solving using the conventional quadratic formulation, while only 11 s for preconditioning and 4 s for solving using our anchor cells-based formulation.

3.3.3 Unconstrained Wire Length Minimization

The initial placement seed is generated by solving a conventional quadratic formulation. In each following iteration, cells are diffused to obtain the desired density distribution, and anchor cells are inserted and locked. In the successive quadratic formulation, the locked anchor cells will be treated as fixed objects, which will be added to the constant vector $\mathbf{b}$ in (3.2). Therefore, in this step, the quadratic engine minimizes an unconstrained wire length objective. It is to be noted that anchor cells are used in hyper-nets decomposition, and no forces or artificial fixed points are used in our formulation.

Figure 3.10 illustrates the idea of one placement iteration. In Figure 3.10(a), an initial placement is generated and cells are congested in the middle of the placement

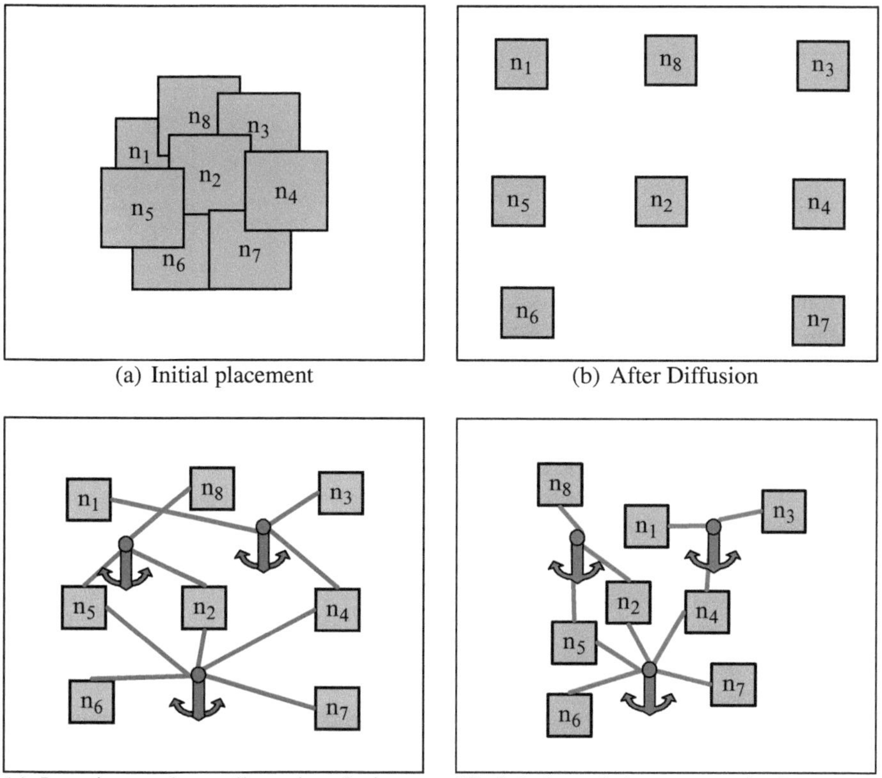

(a) Initial placement

(b) After Diffusion

(c) Inserting anchor cells and replacing a few multi-pin nets with virtual two-pin nets

(d) Solution of the new quadratic system, cells will not collapse back to the initial status due to anchor cells

Fig. 3.10. Anchor cell-based placement illustration (regular nets are not showed for simplicity).

region. After the preplacement, cells are spread, as shown in Figure 3.10(b). But the wire length after spreading could be very bad. It should be noted that cells have not actually moved yet in this step.

In Figure 3.10(c), we insert a few anchor cells to convert a few nets into the star model, one for each high-pin net. All anchor cells are locked once inserted. The locked anchor cells are used as fixed pins and their positions are updated into the quadratic system. After solving the new quadratic formulation, cells are rearranged, but will not collapse back to the initial placement due to the tension from their anchors, as shown in Figure 3.10(d). We can proceed to next iteration, or we can go over the anchoring cells insertion and wire length optimization substep multiple times before proceeding to the next iteration, to further reduce the wire length.

Figure 3.11 plots the first placement iteration of a circuit. In Figure 3.11(a), the wire length of the initial placement is 0.48×10^6, and cells are congested in the middle of the placement. After a few iterations of diffusion, cells are spread as shown

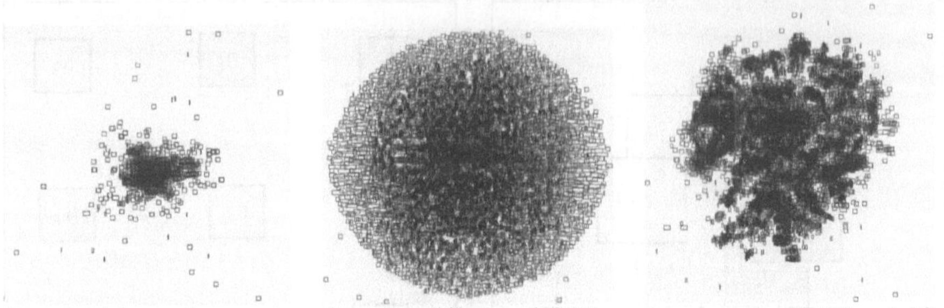

(a) Initial placement. HPWL: 0.48×10^6 (b) After diffusion. HPWL: 2.56×10^6 (c) Actual HPWL: 1.19×10^6

Fig. 3.11. One iteration of the diffusion guided placement.

in Figure 3.11(b). Although the diffusion explicitly controls the cell movement and improves the density distribution, it is not explicitly wire length aware. The total wire length increases to 2.56×10^6 in Figure 3.11(b). Once anchor cells are used. The new quadratic formulation leads to the placement in Figure 3.11(c), with the new HPWL 1.19×10^6, which improved significantly compared with that after the preplacement.

3.3.4 HPWL Transformation in a Quadratic System

A major weakness of a quadratic wire formulation is that the quadratic objective is an approximation of HPWL for a two pin nets. Transforming a multipin net into multiple two-pin nets may enlarge the gap between HPWL and the actual objective to optimize. To alleviate such a problem, existing techniques iteratively linearize the quadratic wire length objective [24]. Recently, the *Kraftwerk* proposes a method to linearize the quadratic objective into HPWL in the clique model-based transformation [22]. Here we propose a method to transform the quadratic objective into HPWL by using the star model-based transformation, which helps to reduce the gap between quadratic wire length and HPWL in the DPlace framework.

As the wire length minimization problem is independent in x and y directions, here we show the formulation in y direction only. Assuming net e is connected with n cells, and HPWL in direction y is L_e. We add a star cell s to decompose the net e into n two-pin connections. Let l_i denotes the distance between star s and cell i and let w_i denote the weight of each two-pin connection. We assign all cells into two sets based on if the cell n_i has a y coordinate large than that of star s. As a result, we have two sets, set $A = \{n_i : y_i > y_s\}$ and set $B = \{n_i : y_i < y_s\}$ for each star model transformation. We define the weight of each two pin net as follows.

$$w_i = \frac{L_{sA}}{S_{AB} \times |y_i - y_s|}, \forall n_i \in A$$

$$w_i = \frac{L_{sB}}{S_{AB} \times |y_i - y_s|}, \forall n_i \in B,$$

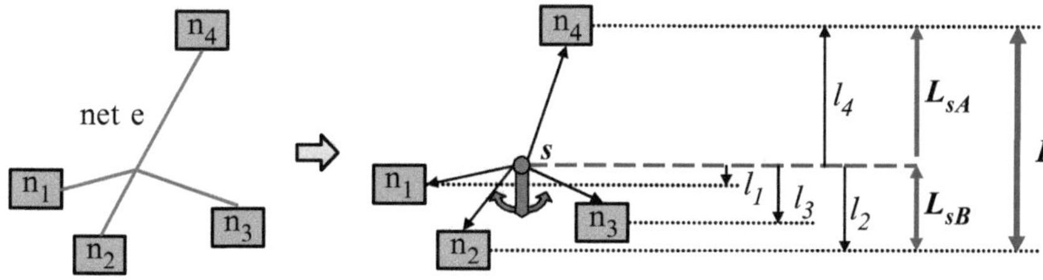

Fig. 3.12. Net weights computation. $A = \{n_4\}$, $B = \{n_1, n_2, n_3\}$ in this example.

where

$$\begin{aligned} S_{AB} &= 0.5 \sum_{n_i} |y_i - y_s| \\ L_{sA} &= \max\{y_i\} - y_s \\ L_{sB} &= y_s - \max\{y_i\} \\ L_e &= L_{sA} + L_{sB} \end{aligned} \tag{3.5}$$

The anchor cell s is placed at the gravity center of all cells on net e, and S_{AB} is defined as the half of the sum of all distances from cell i to the star. Star s splits the length L_e into two parts, L_{sA} and L_{sB}, as shown in Figure 3.12.

In the following, we show that the above net weighting strategy transforms the quadratic wire length objective into HPWL objective exactly.

$$\begin{aligned} \sum_{i=1}^{n} w_i(y_i - y_s)^2 &= \sum_{i \in A} \frac{(y_i - y_s)^2 \times L_{sA}}{|y_i - y_s| \times S_{AB}} + \sum_{i \in B} \frac{(y_i - y_s)^2 \times L_{sB}}{|y_i - y_s| \times S_{AB}} \\ &= \frac{L_{sA}}{S_{AB}} \sum_{i \in A} |y_i - y_s| + \frac{L_{sB}}{S_{AB}} \sum_{i \in B} |y_i - y_s| \\ &= L_{sA} + L_{sB} = L_e \end{aligned} \tag{3.6}$$

Figure 3.12 shows an example of 4-pin nets transformation.

3.3.5 Fixed Blockages

Fixed blockages are obstacles to cell spreading. Modern design may contain a large number of fixed blockages, which disrupt the cells from smooth spreading. Fixed blockages are density obstacles to prevent cells to pass over and cells are often placed on top of the fixed blockages in initial placement. If not properly handled, the wire length may grow dramatically while push cells passing over or out of blockages.

In DPlace, we use a contour-based density smoothing technique to alleviate the density obstacles. First, we identify large blockages, which are those fixed macros with width and height larger than a certain threshold, such as 1% size of the chip size. In the beginning of the global placement, we adjust the density on bins covered by

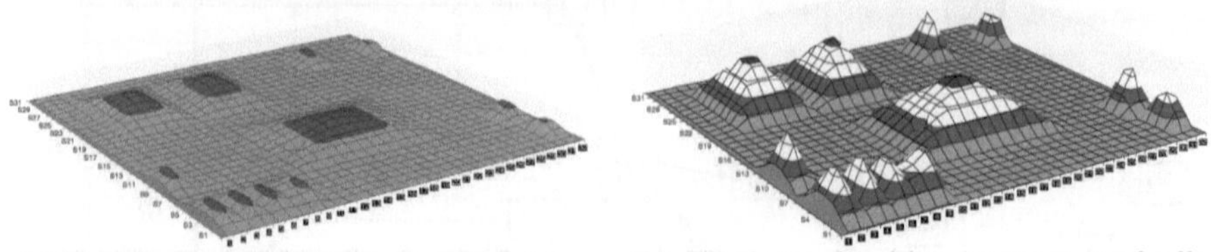

(a) Initial adjusted densities for blockages

(b) Blockage densities increase gradually during the placement

Fig. 3.13. Dynamic density on blockages.

blockages, and the adjusted density distribution is contour based. For a bin covered by a big blockage, the bin density is set to be proportional to the distance between the bin to the blockage boundary. Therefore the highest density is in the bin lying in the middle of the blockage.

In the earlier stages of the global placement, the adjusted fixed blockage density is set to a very small value to allow cells to flow over. As the cells spreading stabilizes, the adjusted density increases gradually, as shown in Figure 3.13(b). The density in the middle of the fixed blockage rises to push overlapping cells out of blockages smoothly. The diffusion-based preplacement pushes cells over blockage easily according to the adjusted density distribution.

3.3.6 Wire Length Improvement Heuristics

Beside the core techniques proposed above, there are more issues that will affect the quality of the placer. Pushing cells away from a region of congestion often contradicts with wire length optimization objective. To further improve the wire length, wire length improving heuristics are performed between each iteration. In certain extent, the wire length improvement heuristics largely determine the quality of the final HPWL. The inaccuracy of using the quadratic wire length as the objective is magnified in large-scale ISPD2005 benchmarks, which contains a large amount of fixed macros.

Therefore, the wire length improvement heuristics are crucial for the HPWL results in quadratic placement. In DPlace, the quadratic optimization step is fast and thus most of the CPU time is spent on wire length improving heuristics. In our experiments, the medium improvement heuristics used in FDP [12] was found effective in the earlier stages of the global placement. Similar technique can also be used in detailed placement for cell swapping. However, the medium improvement heuristic tends to create a lot of overlap in global placement. We use the iterative local refinement [11] to improve the density distribution and further reduce the wire length during the later stages of global placement.

3.4 Legalization and Detailed Placement

Legalization and detailed placement are nontrivial for the final wire length quality of the placer. Before legalization, we divide the placement region into regular bin structures and analyze the density overflow in each bin. We swap cells out of the overflowed bins and swap cells between bins if such a swap helps to further reducing the wire length. Once the bin density overflow is below a threshold, we run a Tetris [25] like legalization flow. We first legalize all movable macros such that no overlap exist between macros. Blockages/macros will split the placement region into row segments. We identify all row segments, sort cells and pack cells into the closet row segment with the minimum cost.

The detailed placement is based on the technique used in [26] and the standard cells are moved between different rows to improve the wire length. We use the global swapping technique to swap cells relatively larger distance to improve the wire length and we use a fixed window to slide through the placement region and test every cell pairs inside the window, and greedily swap two cells if such swapping helps reducing the total wire length.

3.5 Overall Algorithm

The overall algorithm of DPlace is summarized in Algorithm 1. The first iteration of the global placement stage is illustrated in Figure 3.10. In every global placement iteration, cells are diffused to reach a specified density distribution, and the anchor cell-based wire length optimization is performed m times to reduce the wire length. The larger m, the shorter the wire length, and the worse the density distribution. Therefore, m is normally less than 3. The legalization and detailed placement stages are divided in a fashion similar as most of existing placement tools. The placement is legalized before improving the wire length in the detailed placement.

3.6 Experiments

We test our placer on a Linux server with 3.4 GHz 64-bit Xeon processors. We give the wire length and runtime results on four set of benchmarks, the ISPD 2005 and ISPD 2006 placement contest benchmarks [1], and the PEKO-MS 2005/2006 benchmarks. We tested both the LASPack CG solver [27] and the Hybrid solver [28] as our quadratic system solver. Our experimental results are reported based on the Hybrid solver.

3.6.1 Advantages of our New Formulation

Table 3.1 shows the statistics of the new Hessian matrix $\mathbf{A}'$ used in our placer, vs. the Hessian matrix $\mathbf{A}$ in conventional formulation. Column $Size$ shows the dimension of the Hessian, and column $Non-0s$ shows the nonzero entries in the Hessian. Column

Algorithm 1 The DPlace

```
The global placement
    Build matrix A, and matrix A'
    Generate an initial quadratic placement with matrix A
    Repeat
        Do diffusion-based preplacement for k iterations
        Do m iterations
            Generate anchor cells and lock them at the gravity centers
            Compute HPWL transformations net weights, update A'x = b
            Solve x = A'^-1 b
        end
        if (In first a few iterations)
            Use medium improvement heuristic to repair wire length
        else if (Cells are roughly spread)
            Use iterative local refinement to repair wire length
    Until (reaches a desired density distribution)
    Further diffuse cells to remove remaining overlap
The legalization
    Legalize the macros
    legalize the standard cells
The detailed placement
    Further check the cell density and congestion
    Global cell swapping
    Greedy window-based cell swapping
```

Table 3.1. Statistics on new Hessian $\mathbf{A}'$ and the Hessian $\mathbf{A}$ for conventional formulation, and the quadratic solver runtime comparisons.

	Matrix A				Matrix A′				Solver
	Size	Non-0s	Precon(s)	Solve (s)	Size	Non-0s	Precon(s)	Solve(s)	speed-up
adaptec1	243K	196K	15.85	4.65	211K	430K	0.53	0.19	24.5×
adaptec2	355K	2,099K	25.61	7.38	254K	557K	0.90	0.30	24.6×
adaptec3	674K	3,713K	38.18	15.61	494K	1,131K	1.74	0.58	26.9×
adaptec4	508K	3,676K	38.42	15.51	451K	997K	1.97	0.49	31.7×
bigblue1	392K	2,287K	29.78	6.87	278K	603K	1.16	0.36	19.1×
bigblue2	729K	3,937K	47.79	22.78	535K	1,178K	2.29	0.82	27.8×
bigblue3	1,389K	7,290K	103.93	39.32	1,096K	2,714K	4.54	1.70	23.1×
bigblue4	2,831K	16,850K	221.47	75.70	2,169K	5,190K	10.66	3.91	19.4×
									24.6×

Precon. shows the CPU time to preconditioning each Hessian matrix. Same preconditioning quality targets are used for the comparison. Column *Solve* shows the CPU time to solve one iteration of the quadratic system. Comparing with the conventional Hessian **A**, the new Hessian $\mathbf{A}'$ is about 30% smaller on the dimension of the matrix. Furthermore, because $\mathbf{A}'$ is extremely sparse (Figure 3.9 and Table 3.1), the runtime

Table 3.2. Wire length and runtime results for ISPD 2005 benchmarks.

	HPWL ($\times 10^6$)	GP(s)	DP(s)	Total(s)
adaptec1	82.56	778	173	951
adaptec2	91.64	952	343	1295
adaptec3	229.63	2219	712	2931
adaptec4	201.42	1591	874	2465
bigblue1	100.14	1412	312	1724
bigblue2	173.51	2451	1114	3565
bigblue3	383.33	4814	1529	6343
bigblue4	926.53	15482	4870	20352

Table 3.3. Wire length and runtime results for ISPD 2006 benchmarks.

	HPWL ($\times 10^6$)	DHPWL ($\times 10^6$)	GP (s)	DP (s)	Total (s)
adaptec5	433.06	497.56	3276	1474	4750
newblue1	89.18	89.46	1227	578	1805
newblue2	215.12	217.19	1724	768	2492
newblue3	322.39	324.55	1929	1168	3097
newblue4	266.52	324.56	2141	1361	3502
newblue5	578.52	725.12	5233	1852	7085
newblue6	579.86	599.44	4712	2863	7575
newblue7	1089.15	1215.32	13625	4475	18100

to precondition and solve the new quadratic system are improved significantly. The quadratic solver achieved a 24× speed up on solving time.

3.6.2 ISPD Placement Contest Benchmarks

Table 3.2 gives the HPWL results of DPlace on ISPD 2005 contest benchmarks. We also show the runtime for the global placement and the detailed placement in Table 3.2.

The DPlace results on ISPD 2006 contest benchmarks are presented in Table 3.3. The placement objectives in ISPD 2006 placement contest include both the wire length and the density distribution. As HPWL stands for the half parameter wire length, we use DHPWL representing the density weighted HPWL, which is $HPWL(1 + Density\_Target\_penalty\_factor)$. The $Density\_Target\_penalty\_factor$ is the scaled density overflow used in the placement contest.

3.6.3 PEKO-MS Benchmarks

Tables 3.4 and 3.5 show the HPWL and runtime results of the PEKO-MS 2005 and 2006 benchmarks, which are transformed from ISPD 2005 and 2006 benchmarks in a way that the optimal wire length is known. Column $OPTWL$ stands for the known optimal wire length. $HPWL/OPT$ shows the ratio of the generated wire length against the optimal wire length. GP, DP, and $Total$ show the runtime of

Table 3.4. Wire length and runtime results for PEKO-MS-2005 benchmarks.

	OPTWL ($\times 10^6$)	HPWL ($\times 10^6$)	HPWL/OPT	GP(s)	DP(s)	Total (s)
adaptec1	20.06	29.53	1.47	269	240	509
adaptec2	24.97	40.56	1.62	349	257	606
adaptec3	40.95	72.59	1.77	521	485	1006
adaptec4	39.39	68.13	1.73	689	389	1078
bigblue1	20.86	31.62	1.52	325	247	572
bigblue2	42.26	67.89	1.61	835	312	1147
bigblue3	94.4	235.45	2.49	943	516	1459
bigblue4	171.48	377.94	2.20	2129	2429	4558
Average			1.80			

Table 3.5. Wire length and runtime results for PEKO-MS-2006 benchmarks.

	OPTWL ($\times 10^6$)	HPWL ($\times 10^6$)	DHPWL ($\times 10^6$)	HPWL/OPT	GP (s)	DP (s)	Total (s)
adaptec5	81.89	207.1	972.96	2.53	1051	599	1650
newblue1	20.5	49.58	107.13	2.42	627	321	948
newblue2	328.69	715.12	1603.56	2.18	1407	313	1720
newblue3	73.51	160.69	388.48	2.19	774	287	1061
newblue4	49.14	109.03	417.07	2.22	1222	295	1517
newblue5	102.08	302.33	1392.88	2.96	1183	811	1994
newblue6	90.66	207.09	556.05	2.28	1482	1140	2622
newblue7	206.18	763.08	2754.53	3.70	4242	2752	6994
Average				2.56			

global placement, the detailed placement and the total runtime, respectively. Circuits in the PEKO benchmark have no global nets and all local nets. As a results, cells have been roughly spread in the first iteration of the global placement. Therefore, the runtime on global placement for PEKO-MS 2005/2006 benchmarks is relatively small compared with that in ISPD 2005/2006. The PEKO-MS benchmarks show that there exists still obvious gap between the DPlace results and the known optimal wire length, both in global and detailed placement stage. Due to high percentage of local nets, the deficiency of the detailed placement is magnified.

3.7 Conclusions

In this chapter, we present a new quadratic placement tool, DPlace. DPlace uses the diffusion-based spreading technique to generate a golden placement for improved density distribution, and uses the anchor cells-based formulation to repair the wire length. Different from existing force directed approaches, we do not add forces or extra fixed points in the DPlace formulation. An anchor cell is the part of the internal net model, and the functions of anchor cells include both the net model transformation and cell movement control.

Furthermore, the Hessian matrix of the anchor cells-based quadratic formulation is extremely sparse. As a result, the runtime to solve such a linear system is improved by 24 times in our experiments. In DPlace framework, since it is possible to affix explicit cell movement control in the preplacement stage, our new formulation has the advantages for ECO and timing driven placement, in which precise cell movement control is important.

References

1. G.-J. Nam, C. J. Alpert, P. Villarrubia, B. Winter, and M. Yildiz, "The ispd2005 placement contest and benchmark suite," in *Proc. Int. Symp. on Physical Design*, (New York, NY, USA), pp. 216–220, ACM, 2005
2. G.-J. Nam, "Ispd 2006 placement contest: Benchmark suite and results," in *Proc. Int. Symp. on Physical Design*, (New York, NY, USA), pp. 167–167, ACM, 2006
3. TimberWolf Systems, Inc., "Timberwolf placement & global routing software package," in http://www2.twolf.com/benchmark.html
4. A.E. Caldwell, A.B. Kahng, and I.L.Markov, "Can recursive bisection alone produce routable, placements?," in *Proc. Design Automation Conf.*, pp. 477–482, 2000
5. M. Wang, X. Yang, and M. Sarrafzadeh, "Dragon2000: Standard-cell placement tool for large industry circuits," in *Proc. Int. Conf. on Computer Aided Design*, pp. 260–263, 2000
6. M.C. Yildiz and P.H. Madden, "Improved cut sequences for partitioning based placement," in *Proc. Design Automation Conf.*, (New York, NY, USA), pp. 776–779, ACM, 2001
7. J. Kleinhans, G. Sigl, F.M. Johannes, and K. Antreich, "GORDIAN: VLSI placement by quadratic programming and slicing optimization," *IEEE Trans. on Computer-Aided Design of Integrated Circuits and Systems*, vol. CAD-10, pp. 356–365, March 1991
8. H. Eisenmann and F. M. Johannes, "Generic global placement and floorplanning," in *Proc. Design Automation Conf.*, pp. 269–274, 1998
9. B. Hu and M. Marek-Sadowska, "Far: fixed-points addition & relaxation based placement," in *Proc. Int. Symp. on Physical Design*, (New York, NY, USA), pp. 161–166, ACM, 2002
10. A.B. Kahng and Q. Wang, "An analytic placer for mixed-size placement and timing-driven placement," in *Proc. Int. Conf. on Computer Aided Design*, pp. 565–572, November 2004
11. N. Viswanathan and C.C.N. Chu, "Fastplace: Efficient analytical placement using cell shifting, iterative local refinement and a hybrid net model," in *Proc. Int. Symp. on Physical Design*, pp. 26–33, 2004
12. K. Vorwerk, A. Kennings, and A. Vannelli, "Engineering details of a stable force-directed placer," in *Proc. Int. Conf. on Computer Aided Design*, 2004
13. T. Chan, J. Cong, and K. Sze, "Multilevel generalized force-directed method for circuit placement," in *Proc. Int. Symp. on Physical Design*, 2005
14. B. Yao, H. Chen, C.-K. Cheng, N.-C. Chou, L.-T. Liu, and P. Suaris, "Unified quadratic programming approach for mixed mode placement," in *Proc. Int. Symp. on Physical Design*, 2005
15. ISPD_2005_Placement_Contest, "http://www.sigda.org/ispd2005/ispd05/slides/10-1-placement-contest-ispd05.ppt," 2005
16. A.B. Kahng, S. Reda, and Q. Wang, "Aplace: A general analytic placement framework," in *Proc. Int. Symp. on Physical Design*, pp. 233–235, April 2005
17. T.-C. Chen, Z.-W. Jiang, T.-C. Hsu, H.-C. Chen, and Y.-W. Chang, "A high quality analytical placer considering preplaced blocks and density constraint," in *Proc. Int. Conf. on Computer Aided Design*, 2006

18. W.C. Naylor, R. Donelly, and L. Sha", "Non-linear optimization system and method for wire length and dealy optimization for an automatic electric circuit placer," US patent 6,301,693, 2001
19. T. Luo and D.Z. Pan, "Large scale placement with explicit cell movement control," in *Technical Report UT-CERC-06-01*, April 2006
20. F. Mo, A. Tabbara, and R. K. Brayton, "A force-directed macro-cell placer," in *Proc. Int. Conf. on Computer Aided Design*, p. 4, EECS, UC Berkeley, November 2000 A demo can be found at: http://www-cad.eecs.berkeley.edu/ fanmo/PlacementAlgorithm/index.html
21. ISPD_2002_Benchmark, "http://vlsicad.eecs.umich.edu/bk/ispd02bench/,"
22. P. Spindler and F. M. Johannes, "Fast and robust quadratic placement combined with an exact linear net model," in *Proc. Int. Conf. on Computer Aided Design*, 2006
23. H. Ren, D. Z. Pan, C. J. Alpert, and P. Villarrubia, "Diffusion-based placement migration," in *Proc. Design Automation Conf.*, June, 2005
24. G. Sigl, K. Doll, and F. M. Johannes, "Analytical placement: A linear or a quadratic objective function?," in *DAC '91: Proc. 28th Conf. on ACM/IEEE Design Automation*, (New York, NY, USA), pp. 427–432, ACM, 1991
25. D. Hill, "Method and system for high speed detailed placement of cells within an integrated circuit design," US patent 6,370,673, 2002
26. M. Pan, N. Viswanathan, and C.C.N. Chu, "An efficient and effective. detailed placement algorithm," in *Proc. Int. Conf. on Computer Aided Design*, 2005
27. LASpack, "http://www.mgnet.org/mgnet/codes/laspack/html/laspack.html," 1995
28. H. Qian and S.S. Sapatnekar, "A hybrid linear equation solver and its application in quadratic placement," in *Proc. Int. Conf. on Computer Aided Design*, 2005

4

Kraftwerk: A Fast and Robust Quadratic Placer Using an Exact Linear Net Model

Peter Spindler and Frank M. Johannes
Institute for Electronic Design Automation, Technische Universitaet Muenchen, Munich, Germany {peter.spindler, frank.johannes}@tum.de

Summary. This chapter describes the quadratic placer called "Kraftwerk." Kraftwerk is based on distributing the modules on the chip by using an additional force. The additional force is separated in this placer into two forces: hold force and move force. Both of these forces are determined without any heuristics. This novel systematic force modeling yields the robustness of our iterative placement algorithm by provably converging to an overlap-free placement.

In addition to Kraftwerk, an exact linear net model is proposed, which can be used by any quadratic placer. This new net model accurately expresses the half-perimeter wire length (HPWL) in the quadratic cost function of quadratic placement. The HPWL in general is a linear metric for the net length and represents a common and efficient estimation for the routed wire length.

The implementation details of Kraftwerk are presented at length in this chapter. Among others, a deterministic quality control is described to handle the important trade-off between CPU time and placement quality. The control of the module density is shown in order to distribute the modules on the chip according to a given module target density.

The experimental results on various modern benchmarks demonstrate that Kraftwerk offers both high-quality placements and excellent computational efficiency. Based on the ISPD 2005 contest benchmarks and compared to APlace, which produces the best net lengths in these benchmarks, Kraftwerk is 35× faster and has just 4% higher net lengths on average. Using the ISPD 2006 contest benchmarks, which evaluate a placer by its CPU time, net length, and compliance with a given module density, Kraftwerk provides excellent results. mPL and APlace are 9% and 22% worse, respectively. In the PEKO-MS ISPD 2005 benchmarks, Kraftwerk is supposed to be one of the fastest placers and produces placements with the best net lengths. mPL and APlace have 20% and 27% higher net length, respectively.

4.1 Introduction

As Moore's law is still valid [1], i.e., circuit sizes are doubled every 18 months, new fast and efficient placement algorithms with accurate net models will be needed for

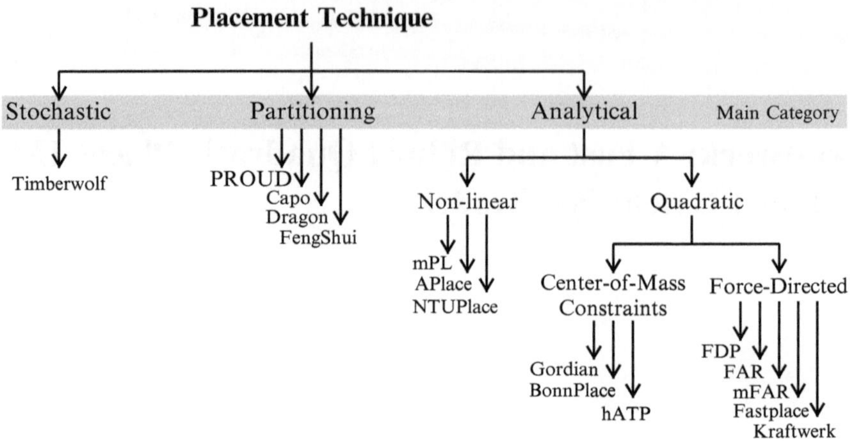

Fig. 4.1. Placement techniques and state-of-the-art placers.

the layout synthesis of next-generation VLSI circuits with tens of millions of standard cells.

Figure 4.1 displays that state-of-the-art placers can be classified in three main categories according to their placement technique.

(1) *Stochastic Approaches.*
Placers based on stochastic approaches often utilize Simulated-Annealing. This optimization method provably finds the global optimum but suffers from long run times. The best-known representative of stochastic placers is Timberwolf [2].

(2) *Partitioning Approaches.*
Another placement approach is to recursively partition the circuit and the placement area. PROUD [3] partitions the circuit based on the locations of the modules as determined by quadratic placement. The min-cut placers Capo [4], Dragon [5], and FengShui [6] partition the circuit based on a certain cost function, e.g., number of wires crossing a boundary of adjacent partitions.

(3) *Analytical Approaches.*
The core of all analytical placers is an objective function which is minimized by methods of mathematical analysis. Depending on the kind of objective function, analytical placers can be subdivided into two categories:

(a) *Nonlinear-Optimization-Based Placers.* The objective function is nonlinear, e.g., a log–sum–exponential function [7], which is minimized by nonlinear optimization techniques like conjugate-gradient optimization [8]. Examples of nonlinear-optimization-based placers are APlace [9], mPL [10], and NTUPlace [11].

(b) *Quadratic Placers.* The objective function is quadratic and can therefore be minimized efficiently by solving a system of linear equations. Quadratic placers are for instance Gordian [12], Kraftwerk [13], FAR [14], FastPlace [15], mFAR [16], BonnPlace [17], hATP [18], and FDP [19].

To cope with modern circuits having millions of modules, many placers combine different optimization techniques with a hierarchical approach, e.g., mFAR, APlace, mPL, NTUPlace, and hATP. The partitioning approaches PROUD, Capo, Dragon, and FengShui are per se hierarchical.

Quadratic placers are popular, because they allow good quality results at low CPU times. But they face two problems: First, a method is needed to model a realistic objective like a linear net length in the quadratic objective function. Second, a technique is necessary to reduce the module overlap, which usually exists in quadratic placement. Both problems are addressed in this chapter.

Depending on the technique to reduce the module overlap, quadratic placers can be divided into two categories (1) Constraint-based quadratic placers like Gordian, BonnPlace, and hATP, which achieve an overlap-free placement by center-of-mass constraints. To refine the center-of-mass constraints, these quadratic placers often partition the placement area recursively and assign modules to the placement partitions. (2) Force-directed quadratic placers like Kraftwerk, FAR, mFAR, FastPlace, and FDP, which utilize an additional force to distribute the modules on the chip and thus reduce the module overlap.

This chapter describes a force-directed quadratic placer. Different approaches appeared to implement the additional force needed for this category of quadratic placement. Kraftwerk (1998 version) [13] utilizes the module density to determine a constant additional force, which drives the modules from high to low-density regions. Please note that although this chapter has also the title "Kraftwerk," the placer described here differs a lot from the 1998 version [13]. FDP utilizes a similar approach as [13]. FAR calculates the additional force like [13] but models it by fixed points. mFAR uses two different fixed points to express the additional force. The "perturbing fixed points" reduce module overlap and are calculated heuristically by a local bin utilization. The "controlling fixed points" achieve the force equilibrium and are determined also by heuristics. FastPlace uses a similar technique for the additional force as mFAR.

This chapter presents a fast, robust, flat, iterative force-directed quadratic placer, which is unique because it does not resort to any heuristics and moreover the convergence to an overlap-free placement is guaranteed. In detail, our quadratic placer is characterized by the following enhancements to other force-directed quadratic placement approaches:

- We separate the additional force into two fundamental components: move force and hold force:
  - We use a generic demand-and-supply formulation of the placement problem and the potential formulation of [13], to calculate a nonheuristic move force which is implemented by target points.

  – To decouple the iterations of quadratic placement, we use a constant and nonheuristic hold force.
- Based on this new systematical force modeling, module positions can be computed efficiently. Moreover we prove that our quadratic placer converges to an overlap-free placement.
- As a result of the force separation, our placement algorithm can be easily restarted at any iteration without any initialization. This efficiently supports the engineering change order (ECO), where the circuit is slightly modified and needs to be placed again.
- To control the important trade-off between CPU time and quality of placement, we implement a deterministic quality control with just one single parameter.
- In order not to narrow the design space, we do not utilize a hierarchical approach, but place all modules simultaneously in each iteration, i.e., our placer is flat.

In addition to the heuristic-free force-directed quadratic placer, we describe a new linear net model. Quadratic placers usually formulate nets by the clique net model or by the equivalent star net model. Moreover the net weights are used to adapt the quadratic objective function to a linear objective. With the number of pins denoted by P, a common net weight for the clique net model is $1/P$ in order to adapt its quadratic cost function to the star net model [20], [15]. For additional linearization, Vygen et al. [21] use a net weight of $1/(P-1)$ and Kleinhans et al. [12] set the net weight to $2/P$. These approximation techniques express the linear net length in a quadratic cost function in a heuristic manner [22].

In this chapter, we present the new linear BoundingBox net model, which can be utilized universally in any quadratic placers. The following properties distinguish our new linear net model from previous net models.

- Exact and deterministic representation of the HPWL in a quadratic objective function. The HPWL is defined per net by the half-perimeter of the bounding box enclosing its pins. The HPWL is a linear metric for the net length and an efficient estimation for the routed wire length.
- Efficient removal of module overlap.
- Lower memory usage and runtime.

The rest of the chapter is organized as follows: Section 4.2 describes the new linear BoundingBox net model. The force-directed quadratic placer is presented in Section 4.3 then. Section 4.4 deals with the implementation details of our placer. Experimental results in various benchmarks are provided in Section 4.5, followed by the conclusion in Section 4.6.

4.2 Net Model

Quadratic placers in general are based on two-pin connections and a net model is necessary to represent the length of one net by just two pin connections. The sum of the weighted and squared Euclidean distance between the two-pin connections of one net n then creates the quadratic cost function Γ_n of one net:

$$\Gamma_n = \sum_{e=(i,j)\in E_n} \frac{w_{e,x}}{2}(x_i - x_j)^2 + \frac{w_{e,y}}{2}(y_i - y_j)^2 \tag{4.1}$$

The set E_n is the set of two-pin connections representing net n. The sum of the quadratic cost function of all nets $n = 1, 2, 3, ..., N$ form the quadratic cost function of the circuit: $\Gamma = \sum_{n=1}^{N} \Gamma_n$.

In this section we will present our new BoundingBox net model in comparison to the traditional clique net model. This new net model expresses the HPWL of one net, which is a linear net metric, exactly in the quadratic cost function of one net Γ_n. We will denote the number of pins of one net by P and the x-coordinate of pin $i = 1, 2, 3, ...P$ by x_i. Since (4.1) can be separated in x and y-direction: $\Gamma_n = \Gamma_{n,x} + \Gamma_{n,y}$, we will describe the net models only in x-direction. The the y-direction can be obtained similarly.

4.2.1 Clique Net Model

In the classical clique net model all possible two-pin connections of the net are used:

$$\Gamma_{n,x} = \Gamma_{C,x} = \frac{w_C}{2} \cdot \sum_{i=1}^{P} \sum_{j=i+1}^{P} (x_i - x_j)^2 \tag{4.2}$$

The net weight w_C of the clique model is specified by

$$w_C = \frac{1}{P} \cdot \frac{2}{P} \cdot \lambda \tag{4.3}$$

The factor $1/P$ adapts the clique model to the star model [20], [15]. The factor $2/P$ is to adjust number of connections of the clique to the number of connections in the corresponding spanning tree [12]. The additional net weight λ can be used to linearize the quadratic clique length $\Gamma_{C,x}$ [22].

The number of two-pin connections $\mathrm{NC_C}$ in the clique model is determined by

$$\mathrm{NC_C} = 0.5 \cdot P(P-1) \tag{4.4}$$

The squared clique length as expressed in the quadratic cost function (4.2) is one metric for the net length. Since the placement process should consider the routing process, the net length should reflect the routed wire length. Since the nets are routed with horizontal and vertical wires, the minimal routed wire length is the length of the rectilinear Steiner minimal tree (RSMT) [23]. But the RSMT problem is NP-hard and a very efficient and lower bound for the routed wire length is the HPWL. In detail, for a set of circuits the authors of [24] demonstrate that the HPWL differs from RSMT length by just 8% but is determined 1.4×10^5 times faster. Moreover, for nets with two or three pins, and most of the nets of a circuit are of this kind, the RSMT length is equal to the HPWL [23].

The HPWL is defined per net by the half-perimeter of the bounding box enclosing its pins. If this box has width w and height h

$$w = \max(x_i) - \min(x_i) \qquad h = \max(y_i) - \min(y_i) \tag{4.5}$$

then the HPWL is calculated by

$$\Gamma_{HPWL} = \Gamma_{HPWL,x} + \Gamma_{HPWL,y} = w + h \tag{4.6}$$

Since the HPWL is an efficient estimation for the routed wire length, the net length and thus the quadratic cost function of one net should reflect the HPWL. Figures 4.2 (a)–(d) illustrate the approximation error

$$\epsilon_{C,x} = \frac{\Gamma_{C,x}}{\Gamma_{HPWL,x}} - 1$$

between the cost function $\Gamma_{C,x}$ of the clique net model and the HPWL metric of randomly generated nets. To linearize the quadratic clique length (4.2), the additional net weight λ was set to

$$\lambda = \frac{10}{\Gamma_{HPWL,x} + 10}$$

Figures 4.2 (a)–(d) demonstrate the following properties of the approximation error $\epsilon_{C,x}$ of the clique net model:

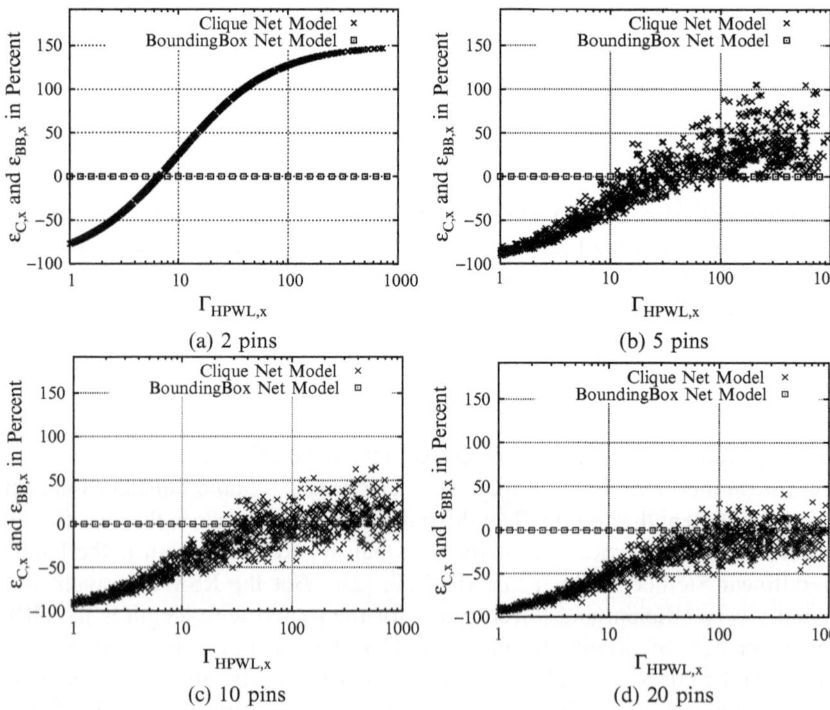

(a) 2 pins (b) 5 pins (c) 10 pins (d) 20 pins

Fig. 4.2. Approximation error $\epsilon_{C,x}$ and $\epsilon_{BB,x}$ of the clique net model and the BoundingBox net model if their quadratic cost functions $\Gamma_{C,x}$ and $\Gamma_{BB,x}$ are referred to the HPWL metric $\Gamma_{HPWL,x}$.

- $\epsilon_{C,x}$ is not constant over Γ_{HPWL}, x
- $\epsilon_{C,x}$ is up to 150%
- $\epsilon_{C,x}$ is spreading at nets with more than two pins
- $\epsilon_{C,x}$ is decreasing in proportion to the number of pins in a net

Even if the approximation error depends on the net weight λ, the above statements are valid in general. Therefore the clique net model approximates the HPWL metric and thus the routed wire length very inaccurately. Hence a new linear net model called BoundingBox net model is presented in the following that reproduces the HPWL without error.

4.2.2 BoundingBox Net Model

In the BoundingBox net model, not all possible two-pin connections of the net are used, but only a few characteristic ones, as illustrated in Figure 4.3(a): Pin a with lowest x-coordinate is connected with pin b with highest x-coordinate. This creates connection 1 with length $l_{x,1} = w$. The remaining $P - 2$ inner pins of the net are connected with both outer pins a and b. This creates connections j and $j + 1$ with $j = 2, 4, 6, ..., 2(P - 2)$ and $l_{x,j} + l_{x,j+1} = w$. Considering that the pins a and b are the bounds of the net's box, the BoundingBox net model is characterized that all its connections are joined with the bounds of this box. The number NC_{BB} of two-pin connections in the BoundingBox net model is

$$NC_{BB} = 1 + 2(P - 2) \tag{4.7}$$

Each connection $i = 1, 2, 3, ..., NC_{BB}$ has the weight $w_{x,i}$. The length of every connection is squared, weighted, and added to the quadratic cost function $\Gamma_{BB,x}$ of this net model:

$$\Gamma_{n,x} = \Gamma_{BB,x} = \frac{1}{2} \sum_{i=1}^{NC_{BB}} w_i \cdot l_{x,i}^2 \tag{4.8}$$

With each weight calculated by

$$w_{x,i} = \frac{2}{P - 1} \frac{1}{l_{x,i}} \tag{4.9}$$

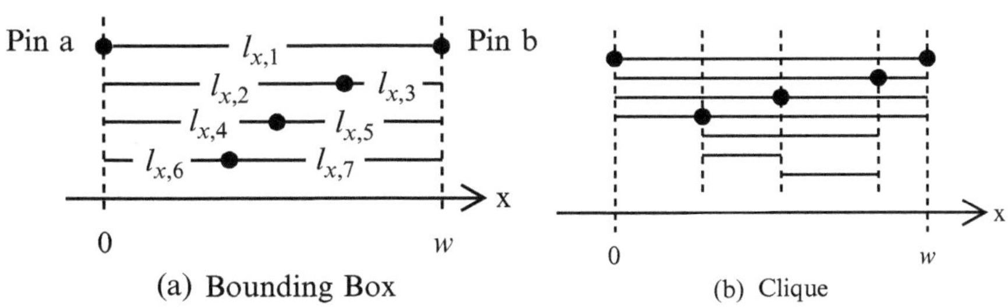

Fig. 4.3. BoundingBox and clique net model of a five pin net in x-direction.

the quadratic cost function $\Gamma_{\text{BB},x}$ of the BoundingBox net model is equal to the width w of the net's bounding box in x-direction.
Proof.

$$\begin{aligned}\Gamma_{BB,x} &= \frac{1}{2}\Big[w_{x,1}\cdot l_{x,1}^2 + \sum_{j=1}^{P-2}\Big(w_{x,2j}\cdot l_{x,2j}^2 + w_{x,2j+1}\cdot l_{x,2j+1}^2\Big)\Big] \\ &= \frac{1}{P-1}\Big[l_{x,1} + \sum_{j=1}^{P-2}\big(l_{x,2j} + l_{x,2j+1}\big)\Big] \\ &= \frac{1}{P-1}\big[w + (P-2)\cdot w\big] = w \end{aligned} \tag{4.10}$$

□

With $w = \Gamma_{HPWL,x}$ this yields an approximation error $\epsilon_{\text{BB},x} = \frac{\Gamma_{\text{BB},x}}{\Gamma_{HPWL,x}} - 1$ of zero in the BoundingBox net model as shown in Figs. 4.2 (a)–(d).

Please note that in each iteration of the quadratic placement algorithm as described in Sect. 4.2.3 (see also Figure 4.4), the weights $w_{x,i}$ as well as the bounds are determined for each net.

By creating the y-part of the BoundingBox net model similarly to the above described x-part, the cost function $\Gamma_{\text{BB}} = \Gamma_{\text{BB},x} + \Gamma_{\text{BB},y}$ of the BoundingBox net model is equal to the HPWL $\Gamma_{HPWL} = w + h$ of the net.

4.2.3 Advantages of the BoundingBox Net Model

- As the BoundingBox net model is based on two-pin connections, it can be used in any quadratic placer. Thus, all quadratic placers can now represent the linear net length measured by the HPWL exactly in the quadratic cost function. Furthermore, the HPWL is a common and efficient estimation of the routed wire length. An important disadvantage of quadratic placers compared to nonlinear-optimization-based placers like APlace [9] and mPL [10] has been eliminated in this way, while the CPU time advantage of quadratic placers is maintained.
- Next Sect. 4.3 describes that the quadratic cost function Γ is represented in a matrix–vector notation (4.12). There, the number of entries in the matrix is proportional to the number NC of two-pin connections. (4.4) describes that NC depends quadratically on the number P of pins in the clique model. Equation (4.7) shows that NC depends only linearly on P in the BoundingBox net model. Thus, NC and therefore the number of matrix entries is much smaller in the new net model. Therefore the memory usage of the matrix is much smaller and the matrix–vector multiplication is much faster. The last advantage of the BoundingBox net model yields in addition a lower CPU time to determine the minimal net length (4.13) as the CPU time of the conjugate-gradient solver depends on the CPU time to execute the matrix–vector multiplication.
- Next Sect. 4.3 will explain that there is usually a lot of overlap between the modules and quadratic placers reduce this overlap step-by-step by an additional force.

With two pins fixed and no additional forces, all remaining inner pins are located at the same position in the clique model. This is because the inner connections existing in this net model (see Figure 4.3(b)) pull the inner pins together. Hence the modules connected to the inner pins overlap a lot and the module overlap is hard to reduce in the clique model. The problematic inner connections do not exist in the BoundingBox net model and therefore this net model does not tend to glue the inner pins together, but supports the reduction of the module overlap much better than the clique model.

4.3 Quadratic Placement Methodology

A placement algorithm in general minimizes the net length under the constraint that the modules are placed overlap-free on the chip. Section 4.2 explains how the net length is expressed in the quadratic cost function Γ of quadratic placement. This section describes our fast and robust quadratic placer. First, we will discuss the representation and the characteristics of the quadratic cost function in general. Then we will explain how the module overlap is removed.

The quadratic cost function Γ expresses the net length and is the sum of the weighted and squared Euclidean distances between the set E of two-pin connections:

$$\Gamma = \sum_{e=(i,j)\in E} \frac{w_{e,x}}{2}(x_i - x_j)^2 + \frac{w_{e,y}}{2}(y_i - y_j)^2 = \Gamma_x + \Gamma_y \qquad (4.11)$$

The set E consists of the sets E_n (with $n = 1, 2, 3, ..., N$) of the two-pin connections of all N nets: $E = \{E_1 \cup E_2 \cup E_3 \cup \cdots \cup E_N\}$. As (4.11) can be separated in x and y-direction, we will describe in the following just the x-direction.

For sake of simplicity and without loss of generality, we assume that there is no offset between the pin position and the module position, i.e., the pin positions can be expressed directly by the module positions. Splitting all modules in M movable and F fixed modules, the x-positions of the movable modules can be collected in vector $\mathbf{x} = (x_1, x_2, x_3, ..., x_M)^{\mathrm{T}}$ and the quadratic cost function Γ_x can be written in matrix–vector notation [25]:

$$\Gamma_x = \frac{1}{2}\mathbf{x}^{\mathrm{T}}\mathbf{C}_x\mathbf{x} + \mathbf{x}^{\mathrm{T}}\mathbf{d}_x + \text{const} \qquad (4.12)$$

Here the matrix $\mathbf{C}_x$ is of dimension $M \times M$ and has the entry c_{ij} in row i and column j. The vector $\mathbf{d}_x$ is of dimension M and has entry $d_{x,i}$ in row i. To express the length $0.5\, w_{e,x}(x_i - x_j)^2$ between two movable modules i and j in the matrix–vector notation (4.12), the matrix entries c_{ii} and c_{jj} are increased by $w_{e,x}$ and the diagonal matrix entries c_{ij} and c_{ji} are decreased by $w_{e,x}$. If one module – let us say j – is fixed, the matrix entry c_{ii} is increased by $w_{e,x}$ and the vector entry $d_{x,i}$ is decreased by $w_{e,x} \cdot x_j$. The length between two fixed modules just contributes to the constant part of (4.12).

Since the net length is expressed by the cost function Γ, the module positions for minimal net length are found by minimizing Γ. For x-dimension, this is done by differentiating Γ with respect to x and solving the resulting system of linear equations with respect to $\mathbf{x}$. Denoting the vector differential operator

$$\left(\frac{\partial}{\partial x_1}, \frac{\partial}{\partial x_2}, \ldots, \frac{\partial}{\partial x_N}\right)^{\mathrm{T}}$$

by the nabla operator ∇_x, the minimization of the net length results in solving the following system of linear equations with respect to $\mathbf{x}$:

$$\nabla_x \Gamma = \nabla_x \Gamma_x = \mathbf{C}_x \mathbf{x} + \mathbf{d}_x = \mathbf{0} \tag{4.13}$$

Solving (4.13) can be done efficiently as the matrix $\mathbf{C_x}$ is highly sparse. A common and fast iterative method to solve highly sparse linear systems of equations is the conjugate-gradient technique.

Quadratic placement in general can be compared with a system of elastic springs: the quadratic cost function Γ_e of one two-pin connection $e = (i, j)$ is equal to the energy E_e of an elastic spring connecting the two pins i and j:

$$\Gamma_e = \frac{w_{e,x}}{2}(x_i - x_j)^2 + \frac{w_{e,y}}{2}(y_i - y_j)^2 = E_e = \frac{w}{2}l^2$$

Here we used the following substitutions: spring constant $w = w_{e,x} = w_{e,y}$, squared Euclidean spring elongation $l^2 = (x_i - x_j)^2 + (y_i - y_j)^2$. The quadratic cost function Γ is the sum of the quadratic cost functions Γ_e of all two-pin connections: $\Gamma = \sum_{e \in E} \Gamma_e$. Thus Γ represents the sum of all spring energies E_e, i.e., Γ reflects the total energy of the spring system. In general, the derivative of an energy with respect to x (or y) is the force in x- (or y-) direction. Therefore the derivative described by the nabla operator ∇_x of the cost function Γ, which represents the net length, is the net force in x-direction:

$$\nabla_x \Gamma = \mathbf{F}_x^{\text{net}} = \mathbf{C}_x \mathbf{x} + \mathbf{d}_x = \mathbf{0} \tag{4.14}$$

This net force is set to zero to find the minimum energy of the elastic spring system, which is equal to the minimum net length.

4.3.1 Additional Forces

With just net forces acting on the modules, the modules attract each other resulting in a lot of module overlap. Therefore the force-directed quadratic placement algorithm applies an additional force and reduces the module overlap in an iterative process.

In this chapter, we represent the module positions from last iteration in vector $\mathbf{x}'$, the module positions calculated in the current iteration in vector $\mathbf{x}$, and the change in module position between two iteration in vector $\Delta\mathbf{x}$:

$$\Delta\mathbf{x} = \mathbf{x} - \mathbf{x}' \tag{4.15}$$

The additional force to remove the module overlap is separated in our quadratic placer in two fundamental components: First, a hold force holding the modules in the current iteration and thus decoupling the iterations of the placement algorithm. Second, a move force moving the modules in the current iteration to reduce the module overlap.

Move Force

The placement problem in general can be formulated as a generic demand-and-supply system with the demand $D^{\mathrm{dem}}(x, y)$ and the supply $D^{\mathrm{sup}}(x, y)$ determining the distribution $D(x, y)$:

$$D(x, y) = D^{\mathrm{dem}}(x, y) - D^{\mathrm{sup}}(x, y) \tag{4.16}$$

Section 4.4.5 describes in detail how the demand is created from the distribution of the modules on the chip and how the supply is determined to control the module density. But the demand-and-supply system of (4.16) can be extended to optimize e.g., the temperature profile of the chip [26] or the routability of the chip.

The demand-and-supply system in general has to be balanced, e.g., the integral over the demand has to equal to the integral over the supply:

$$\int_{-\infty}^{\infty}\int_{-\infty}^{\infty} D^{\mathrm{dem}}(x, y)\,\mathrm{d}x\,\mathrm{d}y = \int_{-\infty}^{\infty}\int_{-\infty}^{\infty} D^{\mathrm{sup}}(x, y)\,\mathrm{d}x\,\mathrm{d}y \tag{4.17}$$

This is necessary to guarantee the convergence to a placement where the demand is adapted completely to the supply at each position (see Sect. 4.3.2). In other words, (4.17) has to be fulfilled to assure the convergence to an overlap-free placement.

The distribution $D(x, y)$ (4.16) of the demand-and-supply system can be viewed as a charge distribution [13], which creates an electrostatic potential Φ based on Poisson's equation:

$$\Delta\Phi = -D(x, y)\,. \tag{4.18}$$

The Poisson equation can be solved efficiently by a geometric multigrid solver [26,27].

In the electrostatic formulation (4.18), the potential Φ is high in regions where the distribution $D(x, y)$ is high, i.e., in high-density regions, and vice versa. Hence the gradient

$$\left(\frac{\partial\Phi}{\partial x}, \frac{\partial\Phi}{\partial y}\right)^{\mathrm{T}}$$

of the potential Φ can be used to move the modules away from high-density regions toward low-density regions and thereby reduce the overlap between the modules.

Therefore each module i gets a target point $\mathring{x}_i$, which is calculated by the module position x_i' and the gradient of the potential Φ:

$$\mathring{x}_i = x_i' - \frac{\partial}{\partial x}\Phi\Big|_{(x_i', y_i')} \tag{4.19}$$

Moreover, each module $i = 1, 2, 3, \ldots, M$ is connected to its target point by a spring with the spring constant $\mathring{w}_i$. This spring connection creates the move force $F_{x,i}^{\text{move}} = \mathring{w}_i(x_i - \mathring{x}_i)$. The move forces of all M movable modules are collected in the move force vector $\mathbf{F}_x^{\text{move}}$:

$$\mathbf{F}_x^{\text{move}} = \mathring{\mathbf{C}}_x \left(\mathbf{x} - \mathring{\mathbf{x}}\right) \tag{4.20}$$

The matrix $\mathring{\mathbf{C}}_x$ is a diagonal matrix with the spring constants as entries: $\mathring{\mathbf{C}}_x = \text{diag}(\mathring{w}_i)$. Collecting the gradients of Φ in x-dimension for all modules in vector Φ_x

$$\Phi_x = \left(\frac{\partial}{\partial x}\Phi\bigg|_{(x_1',y_1')}, \frac{\partial}{\partial x}\Phi\bigg|_{(x_2',y_2')}, \ldots, \frac{\partial}{\partial x}\Phi\bigg|_{(x_N',y_N')} \right)^{\text{T}} \tag{4.21}$$

the target point vector $\mathring{\mathbf{x}}$ can be calculated by $\mathring{\mathbf{x}} = \mathbf{x}' - \Phi_x$.

Hold Force

If the hold force $\mathbf{F}_x^{\text{hold}}$ is defined by the negative net force

$$\mathbf{F}_x^{\text{hold}} = -\left(\mathbf{C}_x \mathbf{x}' + \mathbf{d}_x\right) \tag{4.22}$$

and the sum of hold force and net force is set to zero, then the modules are held on their current positions, i.e., $\mathbf{x} = \mathbf{x}'$.

Proof.

$$\mathbf{F}_x^{\text{net}} + \mathbf{F}_x^{\text{hold}} = \mathbf{C}_x \mathbf{x} + \mathbf{d}_x - \mathbf{C}_x \mathbf{x}' - \mathbf{d}_x = \mathbf{0} \quad \Leftrightarrow \quad \mathbf{x} = \mathbf{x}' \tag{4.23}$$

□

Since all three components $\mathbf{C}_x$, $\mathbf{x}'$, and $\mathbf{d}_x$ of the hold force do not depend on $\mathbf{x}$, the hold force itself is constant.

Total Force

The net force, move force, and hold force add up to the total force $\mathbf{F}_\mathbf{x}$. The total force is then set to zero to get a placement with minimal net length and some overlap reduction:

$$\mathbf{F}_\mathbf{x} = \mathbf{F}_x^{\text{net}} + \mathbf{F}_x^{\text{move}} + \mathbf{F}_x^{\text{hold}} = \mathbf{0} \tag{4.24}$$

Altogether our systematic nonheuristic force-directed quadratic placement approach differs significantly from other force-directed quadratic placement approaches [13–16, 19], all of which either do not separate the additional force or use various heuristics to obtain the additional force. Contrary to that:

- We separate the additional force in hold force and move force
- We calculate the move force (4.20) nonheuristically by an electrostatic potential (4.18) and model it by target points (4.19)
- We represent the hold force by a nonheuristic constant force (4.22)

This results in the simple formulation of the total force (4.24) in our placer:

$$\mathbf{F_x} = \left(\mathbf{C}_x + \mathring{\mathbf{C}}_x\right) \Delta\mathbf{x} + \mathring{\mathbf{C}}_x \Phi_x = \mathbf{0} \tag{4.25}$$

The new module positions $\mathbf{x} = \mathbf{x}' + \Delta\mathbf{x}$ in x-dimension are efficiently computed by solving (4.25) for $\Delta\mathbf{x}$. To obtain the module positions in y-dimension, all described steps must be executed for y-direction.

4.3.2 Proof of Convergence

Please note that, in general, the placement problem is NP-hard and all placement approaches model the problem by algorithms which can be executed with polynomial time complexity [28]. In our placement approach we apply the demand-and-supply formulation (4.16), Poisson's equation (4.18), and the consequential force formulation (4.13), (4.20), (4.22), and (4.25). Our placer is unique because it does not resort to any heuristics. Therefore we can prove that our algorithm converges to a placement, in which the demand $D^{\text{dem}}(x, y)$ is adapted completely to the supply $D^{\text{sup}}(x, y)$ in each position (x, y):

$$\text{Convergence to:} \qquad D^{\text{dem}}(x, y) = D^{\text{sup}}(x, y) \qquad \text{for all} \quad (x, y) \tag{4.26}$$

If the demand $D^{\text{dem}}(x, y)$ is created by the distribution of the modules and the supply $D^{\text{sup}}(x, y)$ is at most one (as described in Sect. 4.4.5), the placement characterized by (4.26) reflects that there is no module overlap. This means that our placement algorithm converges to an overlap-free placement.

Sketch of proof:

1. With no move force, the modules are held with the hold force at their current position. This is shown in Sect. 4.3.1.
2. The distribution $D(x, y)$ creates the potential Φ (4.18), the gradient of the potential Φ is used to calculate the target points (4.19), and the target points form the move force (4.20). Thus the modules are moved away from high-density regions toward low-density regions.
3. Assuming that the supply $D^{\text{sup}}(x, y)$ is fixed and the demand $D^{\text{dem}}(x, y)$ is formed by the modules and thus moved by the move force, we can show that the demand is better adapted to the supply in each iteration. This iterative adaptation is based on (4.18), (4.19), (4.20), (4.22), (4.25), and on the idea of charge conservation. As the demand reflects the distribution of the modules on the chip, the statement of the iterative adaptation means that the module overlap is reduced in each iteration.
4. If the demand is adapted completely to the supply, i.e., (4.26) is fulfilled and the placement is overlap-free, our placement algorithm will not move the modules any more, i.e., it has reached its stable state. This is because (4.26) yields a distribution $D(x, y)$ which is zero at each position. Hence the potential Φ is

constant and thus the gradient of the potential Φ is zero. So (4.25) is transformed to $\left(\mathbf{C}_x + \mathring{\mathbf{C}}_x\right) \Delta\mathbf{x} = \mathbf{0}$. Consequently $\Delta\mathbf{x}$ is zero, reflecting that the modules are not moved any more. ■

Two remarks must be added to the proof of convergence:

- An assumption has to be made: two modules i and j may not have the same position: $(x_i, y_i) \neq (x_j, y_j)$. If they have exactly the same position then they will get the same move force and will probably be moved to the same position in the next iteration and hence the overlap between these two modules will not be removed. Practically this assumption has no impact on convergence since the module positions are calculated numerically and therefore will not be exactly the same.
  Even if two modules i and j have exactly the same position, the modules connected to these two modules i and j will probably move them to different positions in the next iteration.
- No theoretical statement to the iteration count of global placement can be made as this highly depends on the kind of circuit to be placed. But experiments showed that a practical bound of iteration count is around 5 if a high-spring constant (e.g., 1,000) of the target points is chosen.

4.4 Implementation Details

Figure 4.4 shows the complete algorithm of our placer. During the initial placement, a start solution is computed by minimizing the quadratic cost function Γ over a few iterations: $I_{\text{init}} \approx 5$. At this stage, the module overlap is not taken into account.

After that the module overlap is reduced iteratively in the global placement. Although the global placement converges to an overlap-free placement as proven above, it is stopped at a certain stopping criterion, e.g., module overlap $\leq 20\%$.

The global placement is terminated in standard-cell placement because it cannot arrange the modules on the chip rows which is needed to obtain a legal placement. This task and removing of the remaining module overlap is done during the final placement by "FindNextBestPlace": The next best place is sought for each module according to a certain cost function. This search takes around 10% of the CPU time of the global placement and therefore is very fast. Using the net length in HPWL metric as the cost function for "FindNextBestPlace" increases the total HPWL by around 2% compared to the last iteration of the global placement.

4.4.1 Engineering Change Order

The separation of the additional force in hold force and move force results in decoupling one iteration from the previous one. Therefore our global placement algorithm can be easily restarted at any iteration without special initializations. Thus ECO is

Initial Placement:
Place all modules in the center of the chip
for $i < I_{init}$ **do**
For x-direction: (similarly for y-direction)
Create $\mathbf{C}_x$, $\mathbf{d}_x$ (See Note 1 below)
Solve (4.13) for $\mathbf{x}$
$i = i + 1$

Global Placement:
repeat
Calculate potential Φ by (4.18)
For x-direction: (similarly for y-direction)
Create $\mathbf{C}_x$ (See Note 1 below), $\mathring{\mathbf{C}}_x = diag(\mathring{w}_i)$, Φ_x
Solve (4.25) for $\Delta\mathbf{x}$
Update module position $\mathbf{x}$ by $\Delta\mathbf{x}$
Quality control
until Module overlap $\leq 20\%$

Final Placement:
FindNextBestPlace

Note 1: If the BoundingBox net model is used, determine bound pins for each net and calculate every connection weight.

Fig. 4.4. Complete placement algorithm.

Table 4.1. Placing the gate sized circuit bigblue1 of the ISPD 2005 contest benchmarks from scratch and applying the ECO feature at different iterations. Net lengths and CPU times of the ECO results are compared to the result if the circuit is placed from scratch.

Mode	HPWL	CPU
From scratch	97.97 m	505 s
ECO at Iter 5	+0.05 %	–32%
ECO at Iter 10	+0.02 %	–42%
ECO at Iter 15	+0.04 %	–57%
ECO at Iter 20	+0.08 %	–72%
ECO at Iter 25	+0.07 %	–81%

efficiently supported. This means that after a small change in a circuit, e.g., after gate sizing, the circuit can be placed again without running the whole placement process from scratch, but starting immediately the placement algorithm from the last iteration.

Table 4.1 demonstrates the efficient support of ECO by our placement algorithm. Here we used the circuit bigblue1 of the ISPD 2005 contest benchmarks and changed it a bit by sizing about 10% of its modules. The first row in Table 4.1 shows the result in net length and in CPU time of this gate-sized circuit, in the case that our placement algorithm is run from scratch. The last rows in the table display the results if the ECO feature is used. Here we restarted the placement algorithm for the gate-sized circuit after loading the placement of the original circuit at different iterations: 5, 10, 15, 20, and 25. The results of Table 4.1 reveals that ECO can be applied at various iterations without harming the net length (increase below 0.1%) but with great improvement in the CPU time (up to 80%) compared if the changed circuit is placed from scratch.

4.4.2 Quality Control

In order to control the important trade-off between the quality of placement and the CPU time, a quality control procedure is called at the end of each iteration in global placement (see Figure 4.4).

This trade-off presents a challenge in everyday placement usage. On the one hand the deadline for placement can be near and therefore the chip has to be placed in short CPU time. On the other hand the quality of placement can be very important with no limit of CPU time.

Equation (4.11) shows that the quadratic cost function Γ consists of two-pin connections and each two-pin connection has the weights $w_{e,x}$ and $w_{e,y}$. These connection weights are changed in each iteration of global placement in order to adapt the quadratic cost function Γ to a realistic objective, e.g., to the net length measured in HPWL or to fulfill timing requirements. Section 4.2.2 describes the Bounding-Box net model, which uses the connection weights in order to model the HPWL in the quadratic cost function Γ. The authors of [29] show a technique to modify the connection weights in order to model timing requirements in the quadratic cost function Γ.

Hence the more iterations are spent in global placement, the better is the modeling of the real objective and the higher is the quality. On the other hand, the more iterations the global placement needs, the higher is the CPU time, since every single iteration takes a fix CPU time.

If the average module movement μ is controlled to be μ_{T} in every iteration, then the iteration count I_{global} of global placement is indirectly proportional to μ_{T}. This is because each module has to move a certain length Λ in global placement in order to get an overlap-free placement and thus following holds true:

$$\Lambda \approx \mu_{\mathrm{T}} \cdot I_{\mathrm{global}} \Leftrightarrow I_{\mathrm{global}} \propto \frac{1}{\mu_{\mathrm{T}}}. \tag{4.27}$$

To control the module movement μ to be μ_{T}, the target points' spring constants $\mathring{w}_i$ are used, since a target point with a small spring constant attracts its module less than with a high-spring constant.

Altogether, a certain target movement μ_{T} is set for the quality control and the average module movement μ is compared to μ_{T} in each iteration: if $\mu < \mu_{\mathrm{T}}$, then every $w_{\mathrm{T},i}$ is increased and if $\mu > \mu_{\mathrm{T}}$ then every $w_{\mathrm{T},i}$ is decreased. Details about this scaling process and the spring constants of the target points in general are explained in Sect. 4.4.3

Figure 4.5 shows that the presented quality control can efficiently govern the important trade-off between quality of placement and CPU time by using its only

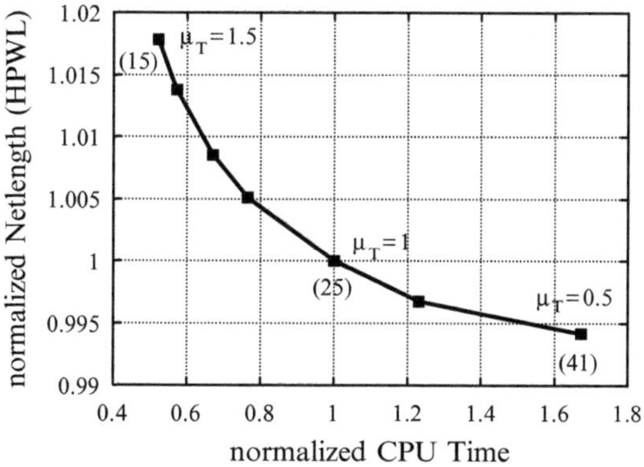

Fig. 4.5. Trade-off between quality of placement, measured in HPWL net length, and CPU time with quality control's parameter μ_T. The results are based on six circuits of the ISPD 2005 contest benchmarks. The values in the brackets express the average of iteration count I_global.

parameter μ_T. Here the quality of placement is measured in the HPWL of all nets. Compared to $\mu_\mathrm{T} = 20$, the CPU time can be decreased to 50% at $\mu_\mathrm{T} = 30$. At this point, the quality of placement is less than 2% worse than at the starting point. On the other side at $\mu_\mathrm{T} = 10$, the quality can be improved by around 0.5% at a CPU time increase of 70%. With a quality range of less than 2% and a CPU time range more than 100%, Figure 4.5 also demonstrates that our placement algorithm is very robust in quality of placement but flexible in CPU time.

4.4.3 Spring Constants of the Target Points

Please note that the following detailed description of the target points' spring constants does not affect the proof of convergence in Sect. 4.3.2 since this proof is independent of the target points' spring constant.

The spring constants of the target points $\mathring{w}_i$ are initialized with

$$\mathring{w}_i = \frac{1}{M} \qquad M\text{: Number of movable modules} \tag{4.28}$$

Then a function $\kappa(\mu)$ is used in every iteration to scale the spring constants $\mathring{w}'_i$ of the last iteration to the spring constants $\mathring{w}_i$ in the current iteration depending on the module movement μ, i.e., depending on the change of the modules positions during the last iteration:

$$\mathring{w}_i = \kappa(\mu) \cdot \mathring{w}'_i. \tag{4.29}$$

Figure 4.6 shows this scaling function $\kappa(\mu)$: At a smaller module movement μ than the target module movement μ_T, the scaling function $\kappa(\mu)$ is greater than one. With (4.29) the spring constants of the target points are increased. At a higher module movement μ than the target movement μ_T, the scaling function $\kappa(\mu)$ is smaller than one and the spring constants of the target points are decreased.

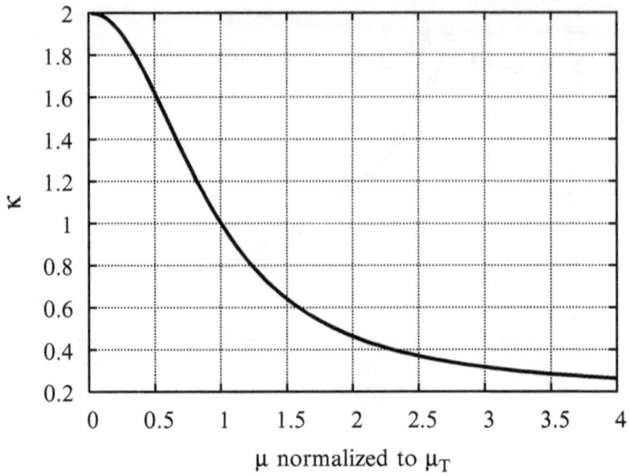

Fig. 4.6. Function $\kappa(\mu)$ to scale the target points' spring constants depending on the module movement μ.

In detail, the scaling function $\kappa(\mu)$ is defined by the following three equations (4.30)–(4.32). The scaling function is determined by a lower limit κ_l, by an upper limit κ_u, and by a sensitivity s at $\mu = \mu_\mathrm{T}$. Moreover, the offset o in (4.32) defines that the scaling function $\kappa(\mu)$ has to be one at $\mu = \mu_\mathrm{T}$, i.e., the spring constants $\mathring{w}_i$ do not change if the modules move with the target movement μ_T.

In Figure 4.6 showing the scaling function $\kappa(\mu)$ the lower limit is $\kappa_\mathrm{l} = 0.2$, the upper limit is $\kappa_\mathrm{u} = 2$, and the sensitivity is $s = 1$.

$$\kappa(\mu) = \frac{\kappa_\mathrm{u} + \kappa_\mathrm{l}}{2} + \frac{\kappa_\mathrm{u} - \kappa_\mathrm{l}}{2} \tanh\left(\hat{s} \cdot \ln \frac{\mu_\mathrm{T}}{\mu} - o\right). \tag{4.30}$$

$$\text{Sensitivity } s: \qquad -\frac{\partial}{\partial \mu}\kappa\bigg|_{\mu=\mu_\mathrm{T}} \stackrel{!}{=} s \quad \Leftrightarrow \quad \hat{s} = \frac{2}{\kappa_\mathrm{u} - \kappa_\mathrm{l}} s. \tag{4.31}$$

$$\text{Offset } o: \qquad \kappa(\mu = \mu_\mathrm{T}) \stackrel{!}{=} 1 \quad \Leftrightarrow \quad o = \operatorname{atanh}\left(\frac{2 - \kappa_\mathrm{u} - \kappa_l}{\kappa_\mathrm{u} - \kappa_\mathrm{l}}\right). \tag{4.32}$$

4.4.4 Convergence Plot

Figure 4.7 demonstrates the convergence of the placement algorithm by displaying the progression of different characteristic values over the iterations.

Progression of the Module Overlap Ω

At the start, the module overlap is high (almost 100%). Then it is decreasing continuously over the iterations, which demonstrates that step 3 of the convergence proof is justified by experiments. When the global placement process is stopped the module overlap is 20%.

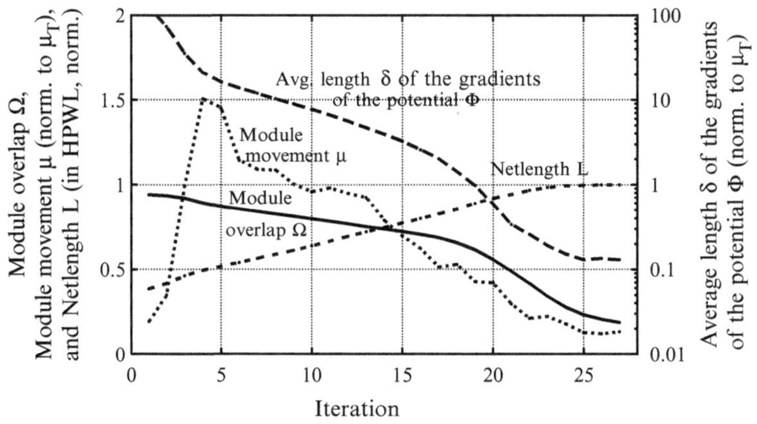

Fig. 4.7. Convergence plot displaying the progression of different characteristic values over the iterations: module overlap Ω, average Euclidean length δ of the gradients of the potential Φ, module movement μ and net length L. Used circuit: bigblue1 of the ISPD 2005 contest benchmarks.

Progression of the Average Length δ of the Gradients of the Potential Φ

The average Euclidean length δ of the gradients of the potential Φ correlates highly with the module overlap Ω: δ is very high at the start and it is then decreasing continuously over the iterations.

This is because the lower the module overlap Ω is, the more even is the distribution $D(x, y)$. As the potential Φ is determined by the distribution $D(x, y)$, a lower module overlap Ω yields also a more even potential Φ. Since the evenness of the potential Φ is represented in the average Euclidean length δ of the gradients of the potential Φ, a lower module overlap Ω results in a lower δ.

Progression of the Module Movement μ

At the start, the module movement μ is low. Then it is rapidly increasing up to 1.5 relative to the target module movement μ_{T}. After this peak, the module movement μ is controlled to be around μ_{T} using the spring constants of the target points and the scaling function $\kappa(\mu)$ as explained in Sect. 4.4.3.

But after around iteration 12, μ cannot be held at μ_T but is slowly decreasing. This is because the modules are moved by the target points. These target points are calculated by the module centers and the gradient of the potential Φ (4.19). Denoting the average Euclidean length of the potential Φ by δ (as done above), the modules can be moved at most the distance δ in one iteration. This means that δ is an upper bound for the average module movement μ. Since δ is around μ_{T} at iteration 12 and δ is decreasing at higher iterations, the module movement μ cannot be held at μ_{T} after iteration 12 but is slowly decreasing.

Progression of the Net Length L

The net length is continuously increasing over the iterations with an almost constant ascent up to iteration 22. After this, the module movement μ is very low and thus the change in the net length is almost zero. From start to end, the net length increases by around 60%.

Robustness

Figure 4.7 also demonstrates the robustness of our placement approach. With continuously decreasing module overlap μ, the average length δ of the gradients of the potential Φ is also continuously decreasing. Since δ is an upper bound for the module movement μ, the module movement μ has to decrease after a certain iteration (this is iteration 12 in Figure 4.7) and has to be zero if the module overlap is completely removed. This means that our placement algorithm is self-controlling for the module movement, which is one aspect of our robustness: whatever the spring constants $\mathring{w}_i$ of the target point is, i.e., however quality control is implemented, the algorithm converges to an overlap-free placement. In addition, at the last iterations of the convergence process the modules are moved continuously less until they are not moved anymore when the algorithm has converged.

4.4.5 Control of the Module Density

Section 4.3.1 describes that the distribution $D(x, y)$ is determined by the demand $D^{\mathrm{dem}}(x, y)$ and the supply $D^{\mathrm{sup}}(x, y)$: $D(x, y) = D^{\mathrm{dem}}(x, y) - D^{\mathrm{sup}}(x, y)$. There it is also explained that the distribution $D(x, y)$ creates an electrostatic potential Φ and the modules are moved based on the gradient of the potential Φ.

To remove the module overlap, the demand $D^{\mathrm{dem}}(x, y)$ has to reflect the distribution of the modules and the supply $D^{\mathrm{sup}}(x, y)$ has to represent the region, in which the modules should be placed and the density for the modules there.

Module Demand

To compute the module demand, a rectangle function $R(x, y; x_{\mathrm{ll}}, y_{\mathrm{ll}}, w, h)$ is needed, which is defined in the x–y-plane and has the parameters lower left corner $(x_{\mathrm{ll}}, y_{\mathrm{ll}})$, the width w, and the height h:

$$R(x, y; x_{\mathrm{ll}}, y_{\mathrm{ll}}, w, h) = \begin{cases} 1 & \text{if } 0 \le x - x_{\mathrm{ll}} \le w \wedge 0 \le y - y_{\mathrm{ll}} \le h \\ 0 & \text{otherwise} \end{cases} \tag{4.33}$$

The demand $D^{\mathrm{dem}}(x, y)$ for the M movable and F fixed modules is calculated using the rectangle function R and the information that a module m has the center position (x_m, y_m), width w_m, and height h_m:

$$D^{\text{dem}}(x,y) = \sum_{m=1}^{M+F} d_m \cdot R\left(x, y; x_m - \frac{w_m}{2}, y_m - \frac{h_m}{2}, w_m, h_m\right) \tag{4.34}$$

The individual module density d_m is usually one for each module. To avoid so-called "halos," i.e., free space around large modules, the individual module density d_m is scaled down according to the module area $A_m = w_m \cdot h_m$ for large modules.

$$d_m = \begin{cases} \sqrt{\frac{A_{\text{large}}}{A_m}}\,(1 - \text{td}) + \text{td} & A_m > A_{\text{large}} \quad \text{(i.e., large module)} \\ 1 & \text{otherwise} \end{cases} \tag{4.35}$$

td is the given target density for the modules. A_{large} is the area, beyond which a module is said to be large. This threshold area A_{large} can be defined for example by the row height h_{row}: $A_{\text{large}} = (5 \cdot h_{\text{row}})^2$

Module Supply

The sketch of proof in Sect. 4.3.2 states that our algorithm converges to a placement in which the (module) demand $D^{\text{dem}}(x,y)$ is adapted to the (module) supply $D^{\text{sup}}(x,y)$ in each position (x,y). Thus the module density, represented in the demand, is determined by the supply.

If the whole chip, which is defined by the lower left corner $(x_{\text{Chip}}, y_{\text{Chip}})$, the width w_{Chip} and the height h_{Chip}, is used for the module supply, then the module supply is

$$D^{\text{sup}}_{\text{intrinsic}}(x,y) = d_{\text{intrinsic}} \cdot R(x, y; x_{\text{Chip}}, y_{\text{Chip}}, w_{\text{Chip}}, h_{\text{Chip}}) \tag{4.36}$$

The demand-and-supply system has to balanced, i.e., (4.17) has to be fulfilled, which gives

$$\int_{-\infty}^{\infty}\int_{-\infty}^{\infty} D^{\text{sup}}_{\text{intrinsic}}(x,y)\,\text{d}x\,\text{d}y = \int_{-\infty}^{\infty}\int_{-\infty}^{\infty} D^{\text{dem}}(x,y)\,\text{d}x\,\text{d}y \tag{4.37}$$

$$d_{\text{intrinsic}} \cdot A_{\text{Chip}} = \sum_{m=1}^{M+F} d_m \cdot A_m \tag{4.38}$$

$$d_{\text{intrinsic}} = \frac{\sum_{m=1}^{M+F} d_m \cdot A_m}{A_{\text{Chip}}} \tag{4.39}$$

Therefore the modules are spread evenly over the whole chip with the intrinsic module density $d_{\text{intrinsic}}$ if the whole chip placement area $A_{\text{Chip}} = w_{\text{Chip}} \cdot h_{\text{Chip}}$ is used for the module supply.

Sometimes spreading the modules over the whole chip is not desired because the chip allows the modules to be packed with a higher target density td than $d_{\text{intrinsic}}$

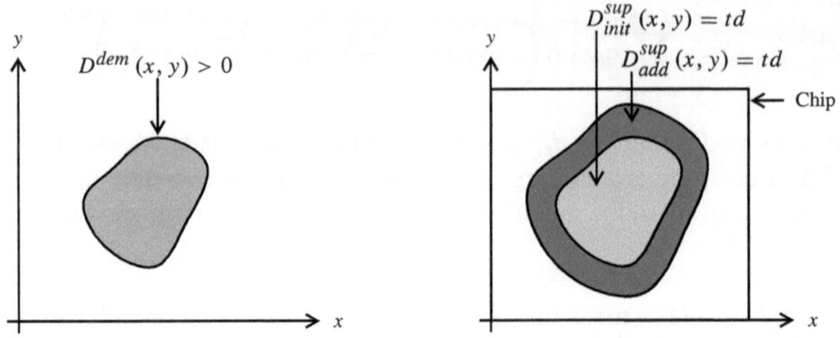

(a) Module demand $D^{\text{dem}}(x, y)$

(b) Initial module supply $D^{\text{sup}}_{\text{init}}(x, y)$ and additional module supply $D^{\text{sup}}_{\text{add}}(x, y)$

Fig. 4.8. Module supply $D^{\text{sup}}(x, y) = D^{\text{sup}}_{\text{init}}(x, y) + D^{\text{sup}}_{\text{add}}(x, y)$ to control the module density to be td.

in order to lower the net length. To control the module density to the given module target density td, the module supply $D^{\text{sup}}(x, y)$ has to be exactly td in the regions, where the modules should be placed, and zero elsewhere. The following two steps explain in detail how this module supply is created (see also Figure 4.8).

1. Create an initial module supply $D^{\text{sup}}_{\text{init}}(x, y)$ with the value td all-around where there is a module demand:

$$D^{\text{dem}}(x, y) > 0 \rightarrow D^{\text{sup}}_{\text{init}}(x, y) = \text{td} \tag{4.40}$$

2. Create an additional module supply $D^{\text{sup}}_{\text{add}}(x, y)$ with the value td around the initial module supply.

The module supply $D^{\text{sup}}(x, y)$ is then the initial and additional module supply:

$$D^{\text{sup}}(x, y) = D^{\text{sup}}_{\text{init}}(x, y) + D^{\text{sup}}_{\text{add}}(x, y) \tag{4.41}$$

The additional module supply is needed in order to get a balanced demand-and-supply system, i.e., to fulfill (4.17).

$$\int_{-\infty}^{\infty}\int_{-\infty}^{\infty} D^{\text{sup}}_{\text{add}}(x, y) + D^{\text{sup}}_{\text{init}}(x, y)\ \text{d}x\ \text{d}y \overset{!}{=} \int_{-\infty}^{\infty}\int_{-\infty}^{\infty} D^{\text{dem}}(x, y)\ \text{d}x\ \text{d}y \tag{4.42}$$

Figure 4.9 (c) shows the module density td vs. net length. The lowest net length is at the highest density td $=$ 1. With decreasing module density, the modules are spread more over the chip and hence the net length is increasing. At the intrinsic module density td $= \text{d}_{\text{intrinsic}} =$ 0.44, the modules are spread evenly over the chip and the net length has the highest value.

Figures 4.9 (a), (b), and (d) display the module congestion at different module densities td $=$ 0.44, 0.7, and 1, respectively. In each of the three module congestion

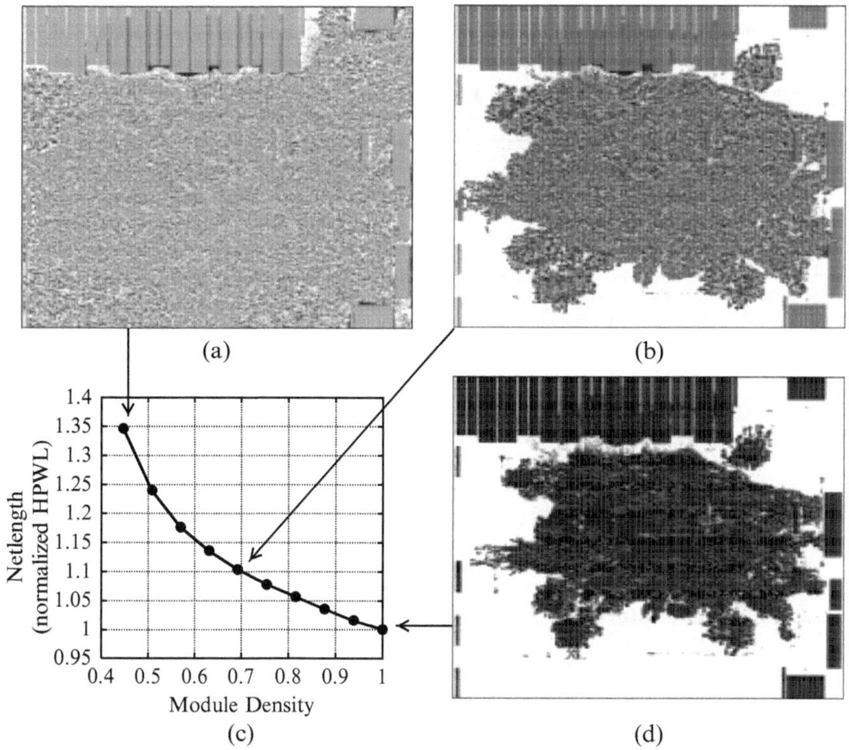

Fig. 4.9. Module density vs. net length (**c**). Module congestion of the final placements at different module target densities (**a**), (**b**), (**d**). *White* represents zero density, *black* represents a density of one. Used circuit: bigblue1 of the ISPD 2005 contest benchmarks.

plots, the region containing all the modules has a constant color which exactly represents the module density td. This demonstrates that the modules are spread with the given module density td. Therefore the above described two steps to create the module supply yields a precise control of the module density td.

4.5 Experimental Results

All presented results in this chapter were determined on an AMD Athlon Opteron 248 with 8 GB RAM and running at 2.2 GHz. Please note that 3.5 GB RAM were enough for the biggest circuit and that for a fair CPU time comparison, we used only one of the two available CPU cores. The net length of all circuits is measured by the HPWL representing a fast and accurate estimation for the routed wire length.

Please note that the results presented in this chapter are about 2% better in net length and 60% better in CPU time than the results presented in [30] and in [31]. This is because the recent version of our placer uses a greedy module swapping after

Table 4.2. The BoundingBox net model compared to the clique net model based on the eight circuits of the ISPD 2005 contest benchmarks.

	Clique Net Model		**BoundingBox Net Model**	
Circuit	HPWL [m]	CPU [s]	HPWL [m]	CPU [s]
adaptec1	90.11	374	82.67	256
adaptec2	101.55	460	93.03	344
adaptec3	251.88	641	229.36	687
adaptec4	218.66	829	200.85	714
bigblue1	108.79	679	97.97	505
bigblue2	171.21	1070	155.43	551
bigblue3	386.21	4078	344.94	2072
bigblue4	961.85	9682	859.18	4180
Average	1.000	1.000	0.907	0.696
Improvement to Clique			**9.93%**	**30.4%**

the final placement, which improves the net length, and it utilizes a cache optimized solver for the linear system of equations (4.25), which reduces the CPU time.

4.5.1 Clique and BoundingBox Net Model

Table 4.2 shows the comparison between the traditional clique net model and our new linear BoundingBox net model based on the eight circuits of the ISPD 2005 contest benchmarks. Since the number of connections is much smaller in our net model, the memory usage is about 75% lower and the CPU time is 30% lower using our net model.

In contrast to the clique net model, our new linear net model reflects the net length measured by the HPWL exactly in the quadratic cost function. Therefore net length is about 9% better in the BoundingBox net model than in the clique net model.

4.5.2 ISPD 2005 Contest Benchmarks

The results of our placer and other state-of-the-art placers on the ISPD 2005 contest benchmarks are given in Table 4.3. The results of the other placers are taken from [32]. The quality of placement for each placer is described in the column "Average" being the average ratio between the placer's HPWL of all five circuits compared to APlace's HPWL. Since the authors of [32] just mention that they use a 1.6 GHz machine, we scaled their CPU time results according to the ratio between the CPU frequencies: 1.6/2.2. Only the CPU times of APlace and Capo are given for the ISPD 2005 contest benchmarks in [32].

The leading APlace is on average 4% better than our placement approach but needs 35× more CPU time. Looking at single results reveals that our placer has the best result at bigblue3 circuit. Compared with Capo our placer is 12% better and 12× faster.

The presented results of our placer are based on the quality control parameter $\mu_T = 20$. As it is shown by the trade-off Figure 4.5, which is based on the same

Table 4.3. Results of our placer Kraftwerk compared to other state-of-the-art placers based on six contest-relevant circuits of the ISPD 2005 contest benchmarks [33].

Placer	Circuit						Average	CPU [h]
	adaptec2	adaptec4	bigblue1	bigblue2	bigblue3	bigblue4		
APlace	87.31	187.65	94.64	143.82	357.89	833.21	1.000	82.33
Kraftwerk	93.03	200.85	97.97	155.43	344.94	859.18	1.041	2.32
mFAR	91.53	190.84	97.70	168.70	379.95	876.26	1.064	n/a
Dragon	94.72	200.88	102.39	159.71	380.45	903.96	1.083	n/a
mPL	97.11	200.94	98.31	173.22	369.66	904.19	1.091	n/a
FastPlace	107.86	204.48	101.56	169.89	458.49	889.87	1.155	n/a
Capo	99.71	211.25	108.21	172.30	382.63	1098.76	1.166	27.49
NTUPlace	100.31	206.45	106.54	190.66	411.81	1154.15	1.206	n/a
FengShui	122.99	337.22	114.57	285.43	471.15	1040.05	1.494	n/a

circuits, the CPU time could be improved by 50% at a quality loss of 2%. On the other hand, the quality could be improved by 0.5% at a CPU time increase of 70%.

4.5.3 ISPD 2006 Contest Benchmarks

In the ISPD 2005 contest benchmarks, the modules were allowed to be packed at a density of 100%. This results in the lowest net length (see also Figure 4.9 (c)), but routing may not be feasible because of too high-wire density. Therefore each circuit of the ISPD 2006 contest benchmarks has an individual module density given, which has to be respected by the placement of the modules on the chip. Section 4.4.5 describes an efficient method to place the modules according to a given density by using the generic demand-and-supply system of our placer. In addition to the possibility that the modules can be packed with 100% density, the ISPD 2005 contest benchmarks also do not account for the CPU time needed to place the modules.

Therefore the ISPD 2006 contest benchmarks utilize three different quality factors: the net length measured in HPWL, the overflow of the module density beyond the given density, and the CPU time needed for placement.

The overflow factor is expressed in percent and reflects how good the given module density is respected by the placer. An overflow factor of 0% represents that the given module density is respected everywhere on the chip. The CPU factor is expressed in percent and is calculated by the logarithmic ratio of the placer's CPU time to the median[1] CPU time of all placers.

These three quality factors are combined to three different scoring functions to express the quality of a placer: HPWL, HPWL+Overflow, and HPWL+Overflow+CPU. All these three scoring functions are normalized to the best (minimal) achieved value.

[1] Please note that "median" is not equal to "average." Only for normal distributions they are the identical.

Table 4.4. Results of Kraftwerk in the ISPD 2006 contest benchmarks.

Circuit	HPWL (m)	Overflow factor (%)	CPU (s)	CPU factor (%)	Score: HPWL	Score: HPWL+ Overflow	Score: HPWL+ Overflow+ CPU
adaptec5	433.84	3.606	1618	−9.35	1.071	1.032	0.939
newblue1	65.92	0.415	603	−8.38	1.057	1.043	0.956
newblue2	203.91	1.286	508	−10.00*	1.033	1.082	0.975
newblue3	278.51	0.382	526	−10.00*	1.018	1.067	0.961
newblue4	304.24	1.709	1553	−8.63	1.068	1.033	0.945
newblue5	548.38	2.694	2622	−9.50	1.109	1.054	0.957
newblue6	528.59	1.702	2579	−9.89	1.048	1.036	0.936
newblue7	1126.58	3.155	4828	−9.06	1.053	1.051	0.958
Average		**1.869**		**−9.35**	**1.057**	**1.050**	**0.953**

*As required in these benchmarks, we limited the CPU factor to ±10%. The "raw" CPU-factors are −13.50% and −10.98%, respectively

Table 4.4 displays detailed results of our placer in the ISPD 2006 contest benchmarks. Please note that our values of the three scoring functions are normalized to the original best values as described in [31]. Therefore our placement approach achieves in the scoring function HPWL+Overflow+CPU a value below one in some circuits. On the average, our placer has an overflow factor of 1.87%, which demonstrates that the given module density is respected in a very good manner.

The original results of the ISPD 2006 contest benchmarks are based on using an AMD Athlon Opteron 252 running at 2.6 GHz. Our results as presented in Table 4.4 are based on an Opteron 248 running at 2.2 GHz. The ratio in the SPEC CPU 2000 benchmarks [34] of those two machines is 0.84 on average. Therefore we scaled our CPU times with the factor 0.84 to determine our CPU factors. The average CPU factor of −9.4% reflects that our placer is in the median more than 4× faster than the other placers.

The average values of other state-of-the-art placers in the three scoring functions are shown in Table 4.5. For comparison between the placers, the scoring function HPWL can be ignored because it does not take into account if the given module density is respected. The scoring function HPWL+Overflow is more of theoretical issue because it does not consider the CPU time needed for placement. Therefore the most realistic scoring function to compare the quality of a placer is HPWL+Overflow+CPU, because it reflects all important facts of a placer in every days usage.

Based on the most realistic scoring function HPWL+Overflow+CPU, our placer is the best. The next best placer mPL has a 9% higher value in this scoring function. Using the more theoretical scoring function HPWL+Overflow, our placer is the third best. mPL and NTUPlace are 2.8% and 2% better, mFAR and APlace are both 5.4% worse.

Therefore Tables 4.3 and 4.5 demonstrate the same characteristic of our placer: Under practical issues, i.e., considering net length, module density, and CPU time,

Table 4.5. Results of our placer Kraftwerk and other state-of-art placers in the ISPD 2006 contest benchmarks. Please note that our results as published in the original contest results are slightly different because they are based on an older version of our placer.

Placer	HPWL	HPWL+ Overflow	HPWL+ Overflow+ CPU
Kraftwerk	1.057	1.050	0.953
mPL	1.035	1.020	1.040
NTUPlace	1.016	1.029	1.049
mFAR	1.108	1.107	1.108
APlace	1.097	1.107	1.165
Dragon	1.331	1.300	1.232
Fastplace	1.177	1.392	1.329
DPlace	1.343	1.414	1.364
Capo	1.375	1.344	1.385

our placer is the best. Ignoring the CPU time and hence accounting a more theoretical quality, our placer is almost the best.

4.5.4 PEKO-MS ISPD 2005 Benchmarks

The authors of [35] present a set of netlist transformations called "monotone chains", which can be utilized to create circuits with known optimal or provably near-optimal placements.

Based on precomputed placements of the ISPD 2005 contest benchmarks and the monotone chains of [35], the PEKO mixed size ISPD 2005 benchmarks are created, which have provably near-optimal placements. These benchmarks have the attribute "mixed size" (MS), because they consists of millions of small movable modules as well as of some big movable modules. Furthermore the white space of the PEKO-MS ISPD 2005 benchmarks can be parametrized, i.e., the module density can be adjusted.

In our results of the PEKO-MS ISPD 2005 benchmarks, we set white space to the maximal value, i.e., the module density is minimal and the modules are spread over the whole chip. We use the maximal value of white space because the placements on which the PEKO-MS ISPD 2005 benchmarks are based have the same property, i.e., their modules are also spread over the whole chip.

Table 4.6 presents detailed results of our placer in the PEKO-MS ISPD 2005 benchmarks. The last column "Optimal" in this table shows the ratio of the net length between our final placements and the probably near-optimal final placements. On average we are 25.2% away from the provably near-optimal net length.

Table 4.7 displays that our placer produces placements with the best published net length in the PEKO-MS ISPD 2005 benchmarks. The other placers mPL, APlace and Capo have a 20%, 27%, and 57% higher net lengths, respectively. Unfortunately, the CPU times of the other placers are yet unknown. But we believe that our placer

Table 4.6. Results of Kraftwerk in the original and the PEKO-MS version of the ISPD 2005 contest benchmarks.

	ISPD 2005		**PEKO-MS ISPD 2005**			
	Initial Spread	CPU (s)	Initial Spread	CPU (s)	HPWL (m)	Optimal
Adaptec1	0.319	256	0.978	175	23.79	1.186
Adaptec2	0.540	344	0.976	225	29.21	1.170
Adaptec3	0.615	687	0.970	410	49.51	1.209
Adaptec4	0.474	714	1.008	357	48.59	1.234
Bigblue1	0.154	505	1.024	231	25.40	1.218
Bigblue2	0.701	551	0.898	471	61.07	1.445
Bigblue3	0.577	2072	0.842	1311	119.49	1.266
Bigblue4	0.714	4180	0.843	4092	221.43	1.291
Average	**1.000**	**1.000**	**2.395**	**0.670**		**1.252**

Table 4.7. Results of our placer Kraftwerk and other state-of-the-art placers in the PEKO-MS ISPD 2005 benchmarks. Results other than of Kraftwerk are taken from [35].

Placer	Average HPWL	Kraftwerk's Improvement (%)
Kraftwerk	1.252	
mPL	1.510	20.61
APlace	1.590	27.00
Capo	1.960	56.55

maintained to be one of the fastest as our CPU times placing the PEKO-MS ISPD 2005 benchmarks is 33% lower than placing the ISPD 2005 contest benchmarks and we have one of the lowest CPU times in the ISPD 2005 contest benchmarks.

This outstanding result of our placer in the net length may be surprising since the results in the ISPD 2005/06 contest benchmarks (see Tables 4.3 and 4.5) reveal that our placer usually is one of the fastest but does not have quite the best net length. Therefore we will discuss in the following some insights of our placer and the differences of the original and the PEKO-MS ISPD 2005 benchmarks.

Figure 4.4 shows that our placement algorithm computes an initial placement by just minimizing the net length. This usually results in a lot of module overlap because the modules are concentrated highly somewhere on the chip. Therefore the module are spread over the chip in global placement by utilizing the additional move and hold forces. If the module overlap is low enough, the spreading is stopped and the final placement is done by "FindNextBestPlace."

Hence the more the modules are spread already in the initial placement, the less the modules have to be moved during global placement and the better is the final placement in terms of net length. Therefore one key issue of our placement algorithm is the modules' spread in the initial placement.

Figure 4.10 (a) demonstrates that in the initial placement of circuit bigblue1 of the ISPD 2005 contest benchmarks, the modules are concentrated highly at the center

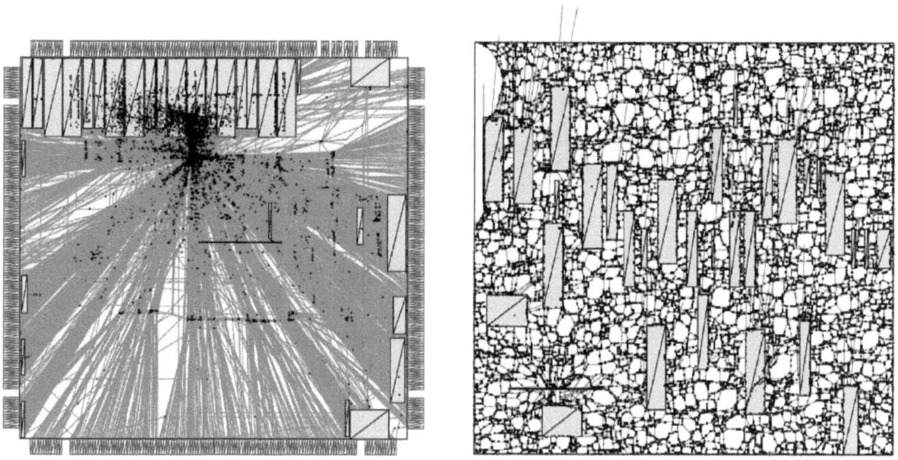

(a) Initial placement in the original version. Spread = 0.154

(b) Initial placement in the PEKO-MS version. Spread = 1.024

Fig. 4.10. Initial placements of the circuit bigblue1 at the original and the PEKO-MS version of the ISPD 2005 benchmarks. Modules are black, nets are gray. Spread is calculated by (4.43).

of the chip, i.e., they are spread not very well on the chip. Considering the same circuit of the PEKO-MS ISPD 2005 benchmarks, the modules are evenly spread over the whole chip in the initial placement, as displayed in Figure 4.10 (b).

To measure the spread of the modules, we use the following formula:

$$\text{Spread} = \sqrt{\frac{\text{var}(X) + \text{var}(Y)}{\frac{1}{12}\left(w_{\text{Chip}}^2 + h_{\text{Chip}}^2\right)}}. \tag{4.43}$$

The terms var(X) and var(Y) are the variances of the module positions in x- and y-direction. w_{Chip} and h_{Chip} represent the width and the height of the chip. Thus Spread is one if the modules are spread evenly over the whole chip area and Spread is smaller than one if the modules are concentrated somewhere on the chip.

Table 4.6 displays in the column "Initial Spread" the modules' spread in the initial placements of the circuits of the original and the PEKO-MS version of the ISPD 2005 contest benchmarks. On average, the modules are spread 2.4× more in PEKO-MS ISPD 2005 benchmarks. Thus the global placement has to move the modules very little and the final placements are almost the same as the initial placements, whereas the initial placements are based on just minimizing the net length. Therefore the net lengths of our final placements in the PEKO-MS ISPD 2005 benchmarks are amazingly good and better than in the ISPD 2005 contest benchmarks (compared to the net lengths of other placers).

But why are the initial placements in the PEKO-MS ISPD 2005 benchmarks that good, i.e., why are the modules spread so highly there? We believe that this depends mainly on the "monotone chains" presented in [35]. These sets of netlist

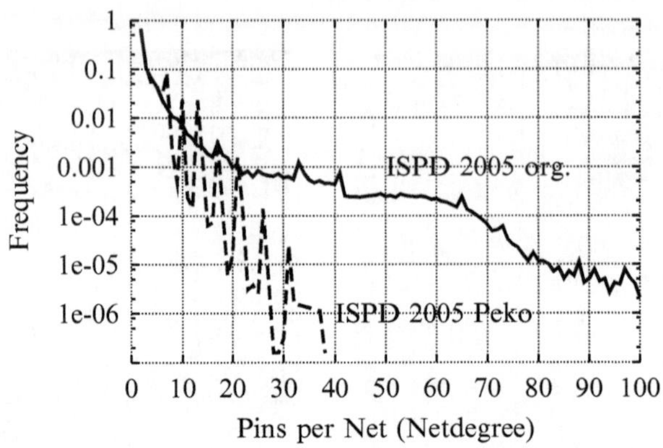

Fig. 4.11. Net statistic of the original and the PEKO-MS version of the ISPD 2005 benchmarks.

transformations are used beside other methods to convert the original ISPD 2005 contest benchmarks to the PEKO-MS versions. Figure 4.11 shows the net statistic, i.e., the frequency of nets with a certain net degree, of the original and the PEKO-MS ISPD 2005 benchmarks. The comparison between both net statistics reveals that the PEKO-MS version differs much in the frequency of nets with more than five pins. Moreover the PEKO-MS version has no nets with 40 or more pins. Therefore the "monotone chains" make big changes in the netlist and the circuits based on these netlist transformations have the property that just minimizing the net length (as done in our initial placement) yields an almost even module spreading. This property of the PEKO-MS ISPD 2005 benchmarks is much in favor of our placement algorithm.

4.5.5 PEKO-MS ISPD 2006 Benchmarks

Similar to the PEKO-MS ISPD 2005 benchmarks, the PEKO-MS ISPD 2006 benchmarks are created based on the ISPD 2006 contest benchmarks and the netlist transformation described in [35].

One difference between the PEKO-MS ISPD 2006 benchmarks and the 2005 version is that there are nets with up to 1,000 pins in the 2006 version. However both PEKO benchmarks are characterized that the net frequency is not continuously decreasing with the number of pins, as in the original ISPD 2005 and 2006 benchmarks, but there are big peaks. This is displayed in Figures 4.11 and 4.12.

The detailed results of our placer for the PEKO-MS ISPD 2006 benchmarks are represented in Table 4.8. The comparison in "Initial Spread" and "CPU" is the same as in the 2005 version: around 2.3× higher spread and 18% lower CPU time in the PEKO-MS ISPD 2006 benchmarks than in the original versions. The optimality gap of Kraftwerk in the net length using PEKO-MS ISPD 2006 benchmarks is 26.7% on average, as displayed in "Optimal." This is comparable to that in the 2005 version.

The ISPD 2006 contest benchmarks reflect in the "overflow" factor the quality, how good a placer respects a given module density. Therefore we also determined the

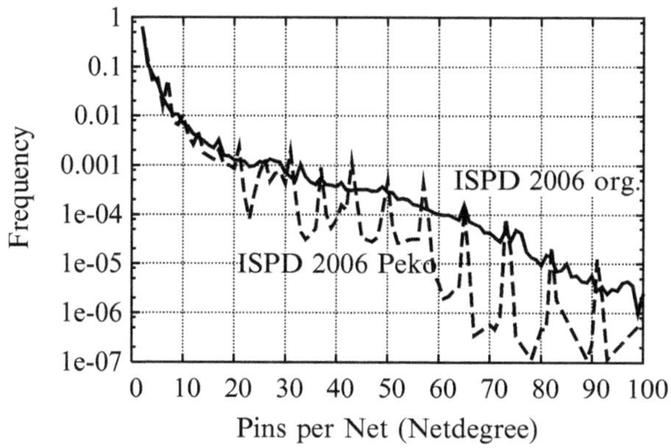

Fig. 4.12. Net statistic of the original and the PEKO-MS version of the ISPD 2006 benchmarks.

Table 4.8. Results of Kraftwerk in the original and the PEKO-MS version of the ISPD 2005 contest benchmarks.

	ISPD 2006		**PEKO-MS ISPD 2006**					
							Overflow	
Circuit	Initial Spread	CPU (s)	Initial Spread	CPU (s)	HPWL (m)	Optimal	Kraftwerk (%)	Optimal (%)
Adaptec5	0.583	1618	1.032	995	95.89	1.171	36.37	9.99
Newblue1	0.131	603	0.877	602	30.51	1.488	11.60	1.73
Newblue2	0.505	508	0.825	518	42.37	1.289	50.87	10.29
Newblue3	0.677	526	0.822	513	81.20	1.105	53.27	9.55
Newblue4	0.516	1553	0.952	850	60.58	1.233	36.26	9.26
Newblue5	0.516	2622	0.963	2738	140.04	1.372	66.94	9.58
Newblue6	0.571	2579	0.979	1329	111.12	1.226	51.88	8.36%
Newblue7	0.833	4828	0.989	4168	258.13	1.252	46.07	7.07
Average	**1.000**	**1.000**	**2.239**	**0.823**		**1.267**	**44.16**	**8.23**

overflow for the PEKO-MS ISPD 2006 benchmarks. The high value of 44.2% is not typical for Kraftwerk, as our placer has on average an overflow of 1.87% in the original ISPD 2006 contest benchmarks (see Table 4.4). We computed the overflow of the optimal placements of the PEKO-MS ISPD 2006 benchmarks and detected considerably high-overflow values. Thus this strange behavior of overflow is not due to the placer but more due to characteristics of the PEKO-MS benchmarks. Especially all standard cells have the same height and width in the PEKO-MS benchmarks and the row grid, on which the modules can be placed, is the same as the width of the standard cells. Hence the coarse row grid makes it hard to respect the module density everywhere on the chip and accordingly the "overflow" factor is bigger than in the original benchmarks.

In summary, the results in initial spread, CPU time, and overflow are totally different between both PEKO-MS benchmarks (2005 and 2006) and the original benchmarks. The same holds true for the characteristics in net statistics, module dimensions, and row grid. Thus the PEKO-MS benchmarks are somewhat artificial

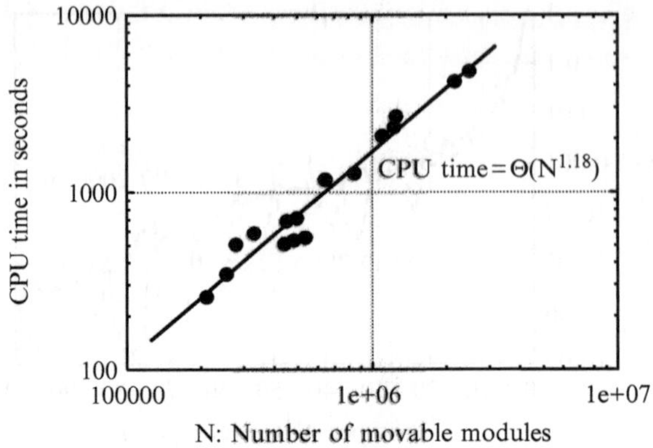

Fig. 4.13. Average computational complexity Θ. Results are based on 16 circuits of the ISPD 2005 and 2006 contest benchmarks.

and do not reflect the behavior of real benchmarks. Therefore the benefit of the netlist transformations, which the PEKO-MS benchmarks are based on and which are described in [35], is not clear.

4.5.6 Computational Complexity

To estimate the CPU time to place the next generation circuits by our placer, we determined the average computational complexity of our placer. This was done by placing the circuits of the ISPD 2005 contest benchmarks and ISPD 2006 contest benchmarks. The module density was set to 100% in both benchmarks, because it has an effect on the CPU time: the higher the module density, the less the modules have to be spread over the chip by global placement, and thus the lower is the CPU time.

Figure 4.13 shows the CPU time vs. the number N of movable modules in a double logarithmic scale. The regression line through the results of the 16 circuits of the ISPD2005/06 contest benchmarks gives an average computational complexity of $\Theta(N^{1.18})$. This represents that our placer has an almost linear complexity. Therefore the CPU times of future circuits will increase nearly linearly with the number of modules.

4.6 Conclusion

This chapter presented the force-directed quadratic placer Kraftwerk, which is based on two fundamentals. One fundamental is a novel nonheuristic force modeling, which gives robustness by the guaranteed convergence to an overlap-free placement. In addition, the introduced force separation yields an efficient support of the ECO. The other fundamental of Kraftwerk is the usage of an exact linear net model, which can be utilized by all quadratic placers and which expresses the net length measured in HPWL precisely in the quadratic cost function. Here the HPWL is a linear net metric and an efficient estimation of the routed wire length.

Beside the fundamentals of Kraftwerk, a detailed insight in its implementation was given in this chapter. So the quality control was presented and the control of the module density was shown. The control of the module density represents the basic placement problem. As Kraftwerk is based on a general demand-and-supply system, our placement approach can be easily extended to solve not only the basic placement but complexer placement problems like the optimization of the temperature profile of a chip or the routability of a chip.

The overall high quality and the low CPU time of Kraftwerk were demonstrated in this chapter by using various realistic and state-of-the-art benchmarks. To explain Kraftwerk's outstanding good net length in the PEKO-MS ISPD 2005 benchmarks, the characteristics of these benchmarks were discussed.

Having an almost linear computational complexity, Kraftwerk will remain one of the fastest placers in the future.

References

1. International technology roadmap for semiconductors. `http://public.itrs.net`
2. W.-J. Sun and C. Sechen. Efficient and effective placement for very large circuits. In *IEEE/ACM International Conference on Computer-Aided Design (ICCAD)*, pages 170–177, 1993
3. R.-S. Tsay, E.S. Kuh, and C.-P. Hsu. PROUD: A sea-of-gates placement algorithm. *ieeedesigntest*, pages 44–56, December 1988
4. J.A. Roy, D.A. Papa, S.N. Adya, H.H. Chan, A.N. Ng, J.F. Lu, and I.L. Markov. Capo: Robust and scalable open-source min-cut floorplacer. In *ACM/SIGDA International Symposium on Physical Design (ISPD)*, pages 224–226, 2005
5. T. Taghavi, X. Yang, and B.-K. Choi. Dragon2005: Large-scale mixed-size placement tool. In *ACM/SIGDA International Symposium on Physical Design (ISPD)*, pages 245–247, 2005
6. A.R. Agnihorti, S. Ono, C. Li, M.C. Yildiz, A. Khathate, C.-K. Koh, and P.H. Madden. Mixed block placement via fractional cut recursive bisection. *IEEE Transactions on Computer-Aided Design of Circuits and Systems*, 24(5):748–761, May 2005
7. W. Naylor, R. Donelly, and L. Sha. Non-linear optimization system and method for wire length and delay optimization for an automatic electric circuit placer. *U.S. Patent 6301693*, October 2001
8. K.G. Murty and F.-T. Yu. Linear complementary, linear and nonlinear programming. `http://ioe.engin.umich.edu/people/fac/books/murty/linear_complementarity_webbook/`
9. A.B. Kahng and Q. Wang. Implementation and extensibility of an analytic placer. *IEEE Transactions on Computer-Aided Design of Circuits and Systems*, 24(05):734–747, May 2005
10. T. Chan, J. Cong, and K. Sze. Multilevel generalized force-directed method for circuit placement. In *ACM/SIGDA International Symposium on Physical Design (ISPD)*, pages 185–192, 2005

11. T.-C. Chen, Z.-W. Jiang, T.-C. Hsu, H.-C. Chen, and Y.-W. Chang. A high-quality mixed-size analytical placer considering preplaced blocks and density constraints. In *IEEE/ACM International Conference on Computer-Aided Design (ICCAD)*, pages 187–192, 2006
12. J.M. Kleinhans, G. Sigl, F.M. Johannes, and K.J. Antreich. GORDIAN: VLSI placement by quadratic programming and slicing optimization. *IEEE Transactions on Computer-Aided Design of Circuits and Systems*, CAD-10(3):356–365, March 1991
13. H. Eisenmann and F.M. Johannes. Generic global placement and floorplanning. In *ACM/IEEE Design Automation Conference (DAC)*, pages 269–274, June 1998
14. B. Hu and M. Marek-Sadowska. FAR: Fixed-points addition & relaxation based placement. In *ACM/SIGDA International Symposium on Physical Design (ISPD)*, pages 161–166, 2002
15. N. Viswanathan and C. C.-N. Chu. Fastplace: Efficient analytical placement using cell shifting, iterative local refinement and a hybrid net model. *IEEE Transactions on Computer-Aided Design of Circuits and Systems*, 24(5):722–733, May 2005
16. B. Hu and M. Marek-Sadowska. Multilevel fixed-point-addition-based vlsi placement. *IEEE Transactions on Computer-Aided Design of Circuits and Systems*, 24(8):1188–1203, August 2005
17. U. Brenner and M. Struzyna. Faster and better global placement by a new transportation algorithm. In *ACM/IEEE Design Automation Conference (DAC)*, pages 591–596, June 2005
18. G.-J. Nam, S. Reda, C.J. Alpert, P.G. Villarrubia, and A.B. Kahng. A fast hierarchical quadratic placement algorithm. *IEEE Transactions on Computer-Aided Design of Circuits and Systems*, 25(4):678–691, April 2006
19. A. Kennings and K.P. Vorwerk. Force-directed methods for generic placement. *IEEE Transactions on Computer-Aided Design of Circuits and Systems*, 25(10):2076–2087, October 2006
20. M.C. Van Lier and R.H.J.M. Otten. Planarization by transformation. *IEEE Transactions on Circuits and Systems CAS*, 20(2):169–171, March 1973
21. J. Vygen. Algorithms for large-scale flat placement. In *ACM/IEEE Design Automation Conference (DAC)*, pages 746–751, 1997
22. G. Sigl, K. Doll, and F.M. Johannes. Analytical placement: A linear or a quadratic objective function? In *ACM/IEEE Design Automation Conference (DAC)*, pages 427–432, San Francisco, 1991
23. M. Hanan. On Steiner's problem with rectiliner distance. *SIAM Journal of Applied Mathemetics*, 14(2):255–265, 1966
24. C. Chu. FLUTE: Fast lookup table based wirelength estimation technique. In *IEEE/ACM International Conference on Computer-Aided Design (ICCAD)*, pages 696–701, 2004
25. K.M. Hall. An r-dimensional quadratic placement algorithm. *Management Science*, 17(3):219–229, November 1970
26. B. Obermeier and F.M. Johannes. Temperature-aware global placement. In *Asia and South Pacific Design Automation Conference*, volume 1, pages 143–148, Yokohama, Japan, January 2004
27. M. Kowarschik and C. Weiß. DiMEPACK – A Cache-optimized multigrid library. In H.R. Arabnia, editor, *Proceedings of the International Conference on Parallel and Distributed Processing Techniques and Applications (PDPTA*, pages 425–430. CSREA Press, June 2001
28. W.E. Donath. Complexity theory and design automation. In *ACM/IEEE Design Automation Conference (DAC)*, volume 19, pages 412–419, 1980

29. B. Obermeier and F. M. Johannes. Quadratic placement using an improved timing model. In *ACM/IEEE Design Automation Conference (DAC)*, pages 705–710, San Diego, June 2004
30. P. Spindler and F.M. Johannes. Fast and robust quadratic placement based on an accurate linear net model. In *IEEE/ACM International Conference on Computer-Aided Design (ICCAD)*, 2006
31. International symposium on physical design. `http://www.ispd.cc`
32. A.B. Kahng, S. Reda, and Q. Wang. Architecture and details of a high quality, large-scale analytical placer. In *IEEE/ACM International Conference on Computer-Aided Design (ICCAD)*, pages 890–897, 2005
33. G.-J. Nam, C.J. Alpert, P. Villarrubia, B. Winter, and M. Yildiz. The ISPD2005 placement contest and benchmark suite. In *ACM/SIGDA International Symposium on Physical Design (ISPD)*, pages 216–219, May 2005
34. Standard Performance Evaluation Corporation. SPEC CPU 2000. `http://www.spec.org/cpu2000`
35. J. Cong, M. Romesis, J.R. Shinnerl, K. Sze, and M. Xie. Locality and utilization in placement suboptimality. Technical report, UCLA Computer Science Department, 2006

Part III

Top-Down Partitioning-Based Techniques

5

Capo: Congestion-Driven Placement for Standard-cell and RTL Netlists with Incremental Capability

Jarrod A. Roy, David A. Papa and Igor L. Markov
The University of Michigan, Department of EECS, 2260 Hayward Ave., Ann Arbor, MI 48109-2121
{royj, iamyou, imarkov}@eecs.umich.edu

Summary. In this chapter, we describe the robust and scalable academic placement tool Capo. Capo uses the min-cut placement paradigm and performs (a) scalable multiway partitioning, (b) routable standard-cell placement, (c) integrated mixed-size placement, (d) wire length-driven fixed-outline floorplanning as well as (e) incremental placement.

5.1 Introduction

The success of min-cut techniques in fixed-die placement is based on the speed and strength of multilevel hypergraph partitioners, the convenient top-down framework that efficiently captures available on-chip resources, and the fact that modern VLSI circuits admit a large number of good placements, which include slicing placements. The recent trend for large amounts of whitespace, clearly visible in the ISPD05 and ISPD06 contest benchmarks, particularly increases the flexibility in the placement problem.

The earliest work describing the Capo placer was a paper from ISPD 1999 describing the end-case placers and optimal partitioners as well as terminal propagation with inessential nets used in Capo [13]. The Capo placer, first released at DAC 2000 [11], sought to produce routable placements with a pure min-cut algorithm. To this end, Capo 8.0 was successful for most industrial benchmarks evaluated, even though it did not build or use congestion maps. For example, it produced a routable placement of an industrial design with 200K cells in 1.5 h on a single-processor workstation. Capo's routability was evaluated with a full-fledged router and demonstrated that early estimators of routability may produce misleading results [11].

Capo's overall performance was on par with commercial tools, however an ISPD 2002 paper [40] proposed a new set of benchmarks on which Capo was less successful compared to a newer tool, Dragon. Dragon found routable placements in most cases by building congestion maps and biasing the placement process accordingly. This suggested that congestion-driven placement was far from solved and several papers in 2003–2005 and later reported even better results [1,5,23,27].

Earlier versions of Capo distributed whitespace approximately uniformly, according to the hierarchical whitespace distribution formula from [15]. However more recent work [4] introduces tunable whitespace distribution for improved wire length, while preserving a minimum amount of local whitespace in most regions to ensure routability. Whitespace allocation and detail placement have been further improved by analyzing the performance of Capo on *feature benchmarks* [32] designed to stress different aspects of placers.

Unlike Dragon and FengShui [5], Capo does not explicitly use multiway partitioning. The addition of *placement feedback* [24] counteracts this potential limitation. Additionally, cutline shifting in recursive bisection adds flexibility in partition shapes and sizes, as well as whitespace allocation; this is not readily available in direct min-cut multiway partitioning.

The most recent work on Capo has been on improving Capo's performance on routing benchmarks and difficult instances of floorplanning and mixed-size placement, and transforming Capo into an incremental placement tool. As of Fall 2006, Capo produces the best published routed wire length on several suites of routing benchmarks by directly optimizing Steiner wire length (StWL) and cutline shifting based on congestion [35]. Capo also performs efficiently with good solution quality on difficult instances of floorplacement which are not legally placeable by several other academic techniques [31]. Incremental placement in Capo consists of simulating the decisions a min-cut placer may have made to produce a given initial placement [36]. For each decision that is made, Capo chooses to accept or reject the decision. Accepting a particular decision means continuing the simulation of decisions whereas rejecting a decision results in replacement of a part of the design from scratch. Empirical results show that Capo's incremental placement moves objects minimally, produces solutions with good HPWL, and runs faster than other available legalization techniques [36].

Using the min-cut floorplacement algorithm from [34] and improvements introduced in [31, 35, 36], Capo 10 performs (a) scalable multiway partitioning, (b) routable standard-cell placement, (c) integrated mixed-size placement, (d) wire length-driven fixed-outline floorplanning, and (e) incremental placement. Capo was used by Synplicity in the Amplify ASIC product. In particular, Amplify ASIC RC targeted LSI Logic's RapidChip architecture. Most RapidChip designs produced were placed with Capo, and successful customers include companies such as HP, SGI, CISCO, Nortel Networks, Raytheon, Seagate, 3COM, Alcate, Hitachi, Fujitsu, IP Wireless, Cryptek, etc. Source code and executables of Capo 10 are available at `http://vlsicad.eecs.umich.edu/BK/PDtools/`.

5.2 Min-Cut Placement in Capo

5.2.1 Row-Based Placement

Internally, Capo's placement representation closely resembles the LEF/DEF and Bookshelf [14] file formats, which represent row information in standard-cell layout. Configurations of rows supply constraints for cell placement. Each row consists of nonoverlapping subrows aligned to the coordinate of the row. All subrows in a row share the same coordinate, height, site width, and site spacing. Placement instances in the Bookshelf format consist of several rows composed of one or more subrows.

Fixed objects may displace sites in the core region. Since fixed objects prevent standard cells from being placed in those sites, they are *obstacles*. To prevent the placer from using sites occupied by obstacles, one solution is to remove the sites beneath all fixed objects. Capo accomplishes this by fracturing the rows containing the occupied sites into subrows, excluding the sites beneath the obstacle [11, Sect. 4.2]. The result is a row-based placement structure containing only legal locations for placing standard cells.

5.2.2 Min-Cut Bisection

Top-down placement algorithms seek to decompose a given placement instance into smaller instances by subdividing the placement region, assigning modules to subregions and cutting the netlist hypergraph [11] (see Figure 5.1). Min-cut placers generally use either bisection or quadrisection to divide the placement area and netlist. Capo uses bisection as it allows for greater flexibility in cutline shifting to adapt to changing partition sizes [11, Sect. 3.2].

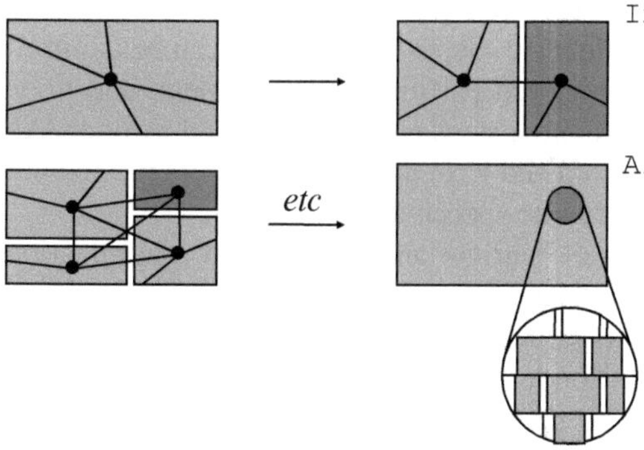


```
Variables:
   A queue of blocks
Initialization:
   A single block represents
   the original placement
   problem
Algorithm:
   while (queue not empty)
   dequeue a block
   if (small enough) consider
   endcase
   else {
   bipartition into smaller
   blocks
   enqueue each block
   }
```

Fig. 5.1. High-level outline of the top-down partitioning-based placement process [13]. © 2000 IEEE.

Each hypergraph partitioning instance is induced from a rectangular region, or bin, in the layout. In this context a *placement bin* represents (a) a placement region with allowed module locations (*sites*), (b) a collection of circuit modules to be placed in this region, (c) all signal nets incident to the modules in the region, and (d) fixed cells and pins outside the region that are adjacent to modules in the region (*terminals*). Top-down placement can be viewed as a sequence of passes where each pass examines all bins and divides some of them into smaller bins.

Capo implements three types of min-cut partitioners – optimal (branch-and-bound [13]), middle-range (Fiduccia–Mattheyses [12]) and large-scale (multilevel Fiduccia–Mattheyses partitioner MLPart [10]). Bins with seven or fewer cells use an optimal end-case placer. This variety of algorithms facilitates partitioning with small tolerance, allowing Capo to distribute the available whitespace uniformly [15] so as to facilitate routing. This provides a convenient baseline for further wire length improvement [4] by nonuniform distribution (this configuration is now used by default).

The efficiency of the partitioners and placers implemented in Capo as well as the min-cut placement framework are directly responsible for Capo's speed and scalability. To this end, large-scale partitioning is performed in $O(P \log P)$ time, where P is the number of pins in the hypergraph. The overall run-time spent on middle-range partitioning (FM) scales linearly, and so do cumulative run-times of all calls to optimal partitioning and placement. Further complexity analysis shows that Capo's asymptotic run-time scales as $O(P \log^2 P)$ on standard-cell designs.

5.3 Floorplacement

From an optimization point of view, floorplanning and placement are very similar problems – both seek nonoverlapping placements to minimize wire length. They are mostly distinguished by scale and the need to account for shapes in floorplanning, which calls for different optimization techniques. Netlist partitioning is often used in placement algorithms, where geometric shapes of partitions can be adjusted. This considerably blurs the separation between partitioning, placement, and floorplanning, raising the possibility that these three steps can be performed by one CAD tool. The authors of [34] develop such a tool and term the unified layout optimization *floorplacement* following Steve Teig's keynote speech at ISPD 2002.

Min-cut placers scale well in terms of runtime and wire length minimization, but cannot produce nonoverlapping placements of modules with a wide variety of sizes. On the other hand, annealing-based floorplanners can handle vastly different module shapes and sizes, but only for relatively few (100–200) modules at a time. Otherwise, either solutions will be poor or optimization will take too long to be practical. The loose integration of fixed-outline floorplanning and standard-cell placement proposed in [3] suffers from a similar drawback because its single top-level floorplanning step may have to operate on numerous modules. Bottom-up clustering can improve the scalability of annealing, but not sufficiently to make it competitive with other approaches. The work in [34] applies min-cut placement as much as possible and delays explicit floorplanning until it becomes necessary. In particular, since min-cut placement generates a slicing floorplan, it is viewed as an implicit floor planning step, reserving explicit floorplanning for "local" nonslicing block packing.

Placement starts with a single placement bin representing the entire layout region with all the placeable objects initialized at the center of the bin. Using min-cut partitioning, the bin is split into two bins of similar sizes, and during this process the cutline is adjusted according to actual partition sizes. Applying this technique recursively to bins (with terminal propagation) produces a series of gradually refined slicing floorplans of the entire layout region. In very small bins, all cells can be placed by a branch-and-bound end-case placer [13]. However, this scheme breaks down on modules that are larger than their bins. When such a module appears in a bin, recursive bisection cannot continue, or else will likely produce a placement with overlapping modules. Indeed, the work in [26] continues bisection and resolves resulting overlaps later. In this technique, one switches from recursive bisection to "local" floorplanning where the fixed outline is determined by the bin. This is done for two main reasons (a) to preserve wire length [9], congestion [8] and delay [21] estimates that may have been performed early during top-down placement, and (b) avoid legalizing a placement with overlapping macros.

While deferring to fixed-outline floorplanning is a natural step, successful fixed-outline floorplanners have appeared only recently [2]. Additionally, the floor planner may fail to pack all modules within the bin without overlaps. As with any constraint-satisfaction problem, this can be for two reasons: either (a) the instance is unsatisfiable, or (b) the solver is unable to find any of existing solutions. In this case, the technique undoes the previous partitioning step and merges the failed bin with its sibling bin, whether the sibling has been processed or not, then discards the two bins. The merged bin includes all modules contained in the two smaller bins, and its rectangular outline is the union of the two rectangular outlines. This bin is floorplanned, and in the case of failure can be merged with its sibling again. The overall process is summarized in Figure 5.2 and an example is depicted in Figure 5.3.

It is typically easier to satisfy the outline of a merged bin because circuit modules become relatively smaller. However, Simulated Annealing takes longer on larger bins and is less successful in minimizing wire length. Therefore, it is important to floorplan at just the right time, and the algorithm determines this point by backtracking. Backtracking does incur some overhead in failed floorplan runs, but this overhead is tolerable because merged bins take considerably longer to floorplan. Furthermore, this overhead can be moderated somewhat by careful prediction.

For a given bin, a floorplanning instance is constructed as follows. All connections between modules in the bin and other modules are propagated to *fixed terminals* at the periphery of the bin. As the bin may contain numerous standard cells, the number of movable objects is reduced by conglomerating standard cells into soft placeable blocks. This is accomplished by a simple bottom-up connectivity-based clustering [25]. The existing large modules in the bin are usually kept out of this clustering. To further simplify floor planning, soft blocks consisting of standard cells are artificially downsized, as in [4]. The clustered netlist is then passed to the

```
Variables: queue of placement bins
Initialize queue with top-level placement bin
1  While (queue not empty)
2    Dequeue a bin
3    If (bin has large/many macros or is marked as merged)
4      Cluster std-cells into soft macros
5      Use fixed-outline floorplanner to pack
         all macros (soft+hard)
6      If fixed-outline floorplanning succeeds
7        Fix macros and remove sites underneath the macros
8      Else
9        Undo one partition decision. Merge bin with sibling
10       Mark new bin as merged and enqueue
11   Else if (bin small enough)
12     Process end case
13   Else
14     Bi-partition the bin into smaller bins
15     Enqueue each child bin
```

Fig. 5.2. Min-cut floorplacement [34]. *Bold-faced lines* 3–10 are different from traditional min-cut placement. © 2006 IEEE.

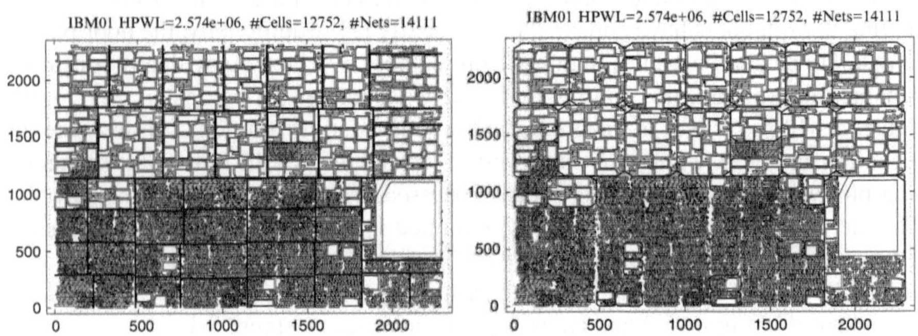

Fig. 5.3. Progress of mixed-size floorplacement on the IBM01 benchmark from IBM-MSwPins [34]. The picture on the left shows how the cutlines are chosen during the first six layers of min-cut bisection. On the right is the same placement but with the floorplanning instances highlighted by "rounded" rectangles. Floorplanning failures can be detected by observing nested rectangles. © 2006 IEEE.

fixed-outline floorplanner Parquet [2], which sizes soft blocks and optimizes block orientations. After suitable locations are found, the locations of all large modules are returned to the top-down placer and are considered fixed. The rows below those modules are fractured and their sites are removed, i.e., the modules are treated as fixed obstacles. At this point, min-cut placement resumes with a bin that has no large modules in it, but has somewhat nonuniform row structure. When min-cut placement is finished, large modules do not overlap by construction, but small cells sometimes overlap (typically below 0.01% by area). Those overlaps are quickly detected and removed with local changes.

Since the floorplacer includes a state-of-the-art floorplanner, it can natively handle pure block-based designs. Unlike most algorithms designed for mixed-size placement, it can pack blocks into a tight outline, optimize block orientations, and tune aspect ratios of soft blocks. When the number of blocks is very small, the algorithm applies floorplanning quickly. However, when given a larger design, it may start with partitioning and then call fixed-outline floorplanning for separate bins. As recursive bisection scales well and is more successful at minimizing wire length than annealing-based floorplanning, the proposed approach is scalable and effective at minimizing wire length.

5.3.1 Empirical Boundary Between Placement and Floorplanning

By identifying the characteristics of placement bins for which the algorithm calls floorplanning, one can tabulate the empirical boundary between placement and floorplanning. Formulating such ad hoc thresholds in terms of dimensions of the largest module in the bin, etc., allows one to avoid unnecessary backtracking and decrease the overhead of floorplanning calls that fail to satisfy the fixed outline constraint because they are issued too late. In practice, issuing floorplanning calls too early (i.e., on larger bins) increases final wire length and sometimes runtime. To improve wire length, the ad hoc tests for large modules in bins (that trigger floorplanning) are deliberately conservative.

These conditions shown in Table 5.1 were derived by closely monitoring the legality of floorplanning and min-cut placement solutions. When a partitioned bin yields an illegal placement solution it is clear that the bin should have been floorplanned and a condition should be derived. When a call to floorplanning fails to satisfy the fixed outline constraint the placer has to backtrack. To avoid paying this penalty, a condition should be derived to allow for floorplanning the parent bin and prevent the failure.

Table 5.1. Floorplanning conditions used in floorplacement [34]. Test 1 is the most fundamental, for if a bin meeting test 1 were not floorplanned, a failure would be guaranteed at the next level. Tests 2–6 detect bins dominated by large macros. Test 7 is a base case where only one module exists, but it is large.

Floorplanning conditions for floorplacement	
N, n:	The numbers of large modules and movable objects in a given bin.
$A(m)$:	The area of the m largest modules in a given bin, $m \leq n$.
C:	The capacity of a given bin.
Test 1.	At least one large module does not fit into a potential child bin.
Test 2.	$N \leq 30$ and $A(N) < 0.80 * A(n)$ and $A(n) > 0.6 * C$.
Test 3.	$N \leq 15$ and $A(N) < 0.95 * A(n)$ and $A(n) > 0.6 * C$.
Test 4.	$A(50) < 0.85 * C$.
Test 5.	$A(10) < 0.60 * C$.
Test 6.	$A(1) < 0.30 * C$ and $N = 1$.
Test 7.	$N = n = 1$.

These conditions are refined to prevent floorplanning failure by visual inspection of a plot of the resulting parent bin and formulating a condition describing its composition. An example of such a plot is shown in Figure 5.3. Floorplanned bins are outlined with rounded rectangles. Nested rectangles indicate a failed floorplan run, followed by backtracking and floorplanning of the larger parent bin. In our experience, these tests are strong enough to ensure that at most one level of backtracking is required to prevent overlaps between large modules.

5.4 Flexible Whitespace Allocation

The min-cut bisection-based placement framework offers much flexibility in whitespace allocation. This section describes uniform allocation of whitespace for min-cut bisection placement and two more sophisticated whitespace allocation techniques, minimum local whitespace and safe whitespace, that can be used for nonuniform whitespace allocation and satisfying whitespace constraints [37].

5.4.1 Uniform Whitespace

A natural scheme for managing whitespace in top-down placement, uniform whitespace allocation, was introduced and analyzed in [15]. Let a placement bin which is going to be partitioned have *site area* S, *cell area* C, *absolute whitespace* $W = \max\{S - C, 0\}$ and *relative whitespace* $w = W/S$. A bipartitioning divides the bin into two child bins with *site areas* S_0 and S_1 such that $S_0 + S_1 = S$ and *cell areas* C_0 and C_1 such that $C_0 + C_1 = C$. A partitioner is given cell area targets T_0 and T_1 as well as a tolerance τ for a particular bipartitioning instance. In many cases of bipartitioning, $T_0 = T_1 = C/2$, but this is not always true [6]. τ defines the maximum percentage by which C_0 and C_1 are allowed to differ from T_0 and T_1, respectively.

The work in [15] bases its whitespace allocation techniques on *whitespace deterioration*: The phenomenon that discreteness in partitioning and placement does not allow for exact uniform whitespace distribution. The whitespace deterioration for a bipartitioning is the largest α, such that each child bin has at least αw relative whitespace. Assuming nonzero relative whitespace in the placement bin, α should be restricted such that $0 \leq \alpha \leq 1$ [15]. The authors note that $\alpha = 1$ may be overly restrictive in practice because it induces zero tolerance on the partitioning instance but $\alpha = 0$ may not be restrictive enough as it allows for child bins with zero whitespace, which can improve wire length but impair routability [15].

For a given block, feasible ranges for partition capacities are uniquely determined by α. The partitioning tolerance τ for splitting a block with relative whitespace w is $(1 - \alpha)w/1 - w$ [15]. The challenge is to determine a proper value for α. First assume that a bin is to be partitioned horizontally n times more during the placement process. n can be calculated as $\lceil \log_2 R \rceil$ where R is the number of rows in the placement bin [15]. Assuming end-case bins have $\alpha = 0$ since they are not further partitioned, $\overline{w}$, the relative whitespace of an end-case bin, is determined to be $\overline{\tau}/\overline{\tau} + 1$ where $\overline{\tau}$ is the tolerance of partitioning in the end-case bin [15].

Assuming that α remains the same during all partitioning of the given bin gives a simple derivation of $\alpha = \sqrt[n]{\overline{w}/w}$ [15]. A more practical calculation assumes instead that τ remains the same over all partitionings. This leads to $\tau = \sqrt[n]{1 - \overline{w}/1 - w} - 1$ [15]. $\overline{w}$ can be eliminated from the equation for τ and a closed form for α based only w and n is derived to be [15]

$$\alpha = \frac{\sqrt[n+1]{1-w} - (1-w)}{w(\sqrt[n+1]{1-w})}.$$

If a bin has a user-defined "small" amount of whitespace or less, Capo attempts to divide the cell area approximately in half, within a given tolerance. The appropriate partitioning tolerance is chosen based on whitespace deterioration as calculated above. After a partitionment (i.e., a partitioning solution) is computed, the geometric cutline for the bin is positioned so that each side of the cutline has an equal percentage of whitespace. As tolerance is calculated assuming a fixed cutline, the cutline is shifted to make whitespace more uniform. Such whitespace allocation generally produces routable placements, at the cost of increased wire length.

5.4.2 Minimum Local Whitespace

If a placement bin has more than a user-defined minimum local whitespace (`minLocalWS`), partitioning will define a tentative cutline that divides the bin's placement area in half. Partitioning targets an equal division of cell area, but is given more freedom to deviate from its target. Tolerance is computed so that with whitespace deterioration, each descendant bin of the current bin will have at least `minLocalWS` [37].

The assumption that the whitespace deterioration, α, in end-case bins is 0 made in [15] and presented in Sect. 5.4 no longer applies, so the calculation of α must change. Since we want all child bins of the current bin to have `minLocalWS` relative whitespace, in particular end-case bins must have at least `minLocalWS` and thus we may set $\overline{w} =$ `minLocalWS`, instead of a function of τ. Using the assumption that α remain constant during partitioning, α can be calculated directly as $\alpha = \sqrt[n]{\overline{w}/w}$ [15]. With the more realistic assumption that τ remain constant, τ can be calculated as $\tau = \sqrt[n]{1 - \overline{w}/1 - w} - 1$ [15]. Knowing τ, α can be computed as [15]

$$\alpha = (\tau + 1) - \frac{\tau}{w}.$$

After a partitionment is calculated, the cutline is shifted to ensure that `minLocalWS` is preserved on both sides of the cutline. If the minimum local whitespace is chosen to be small, one can produce tightly packed placements which greatly improves wire length.

5.4.3 Safe Whitespace

The last whitespace allocation mode is designed for bins with "large" quantities of whitespace. In safe whitespace allocation, as with minimum local whitespace allocation, a tentative geometric cutline of the bin is chosen, and the target of partitioning

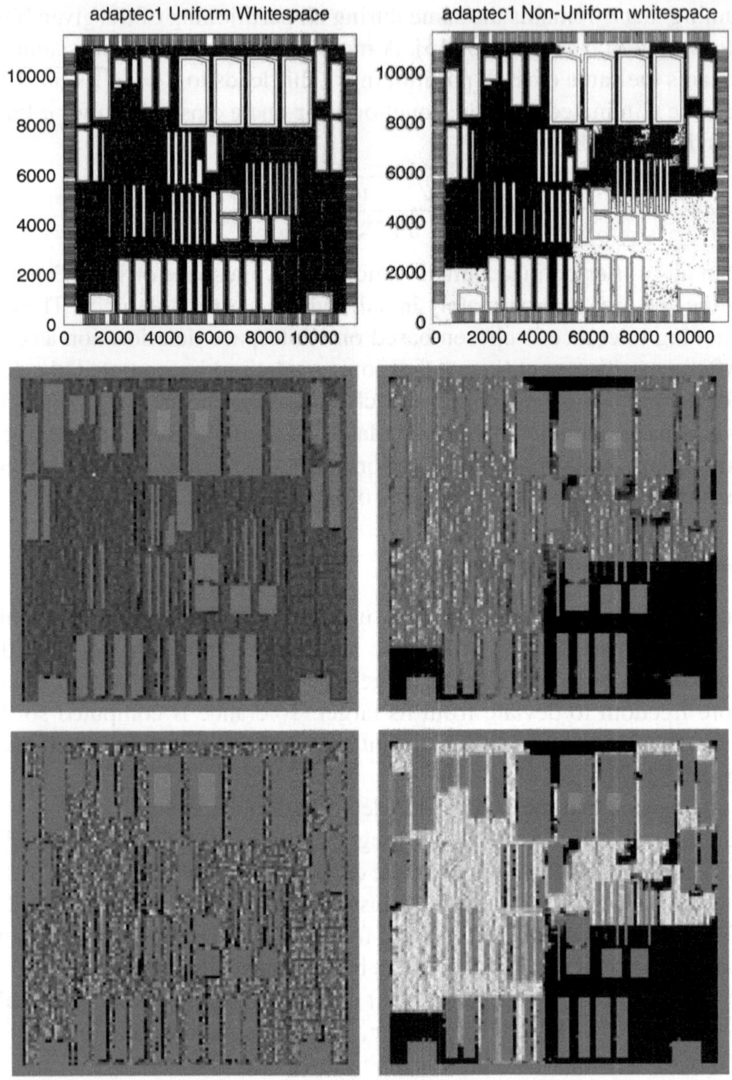

Fig. 5.4. The top row shows Capo 10 global placements of the contest benchmark adaptec1 with uniform whitespace allocation (*left*) and nonuniform whitespace allocation (*right*). Fixed obstacles are drawn with *double lines*. The *middle* and *bottom rows* depict the local utilization the placements. *Lighter* areas of the placement signify regions that violate the target placement density whereas *darker* areas have utilization below the target. Areas with no placeable area (such as those with fixed obstacles) are shaded as if they exactly meet the target density. The target placement density for the middle row is 90% and the bottom row is 60% (adaptec1 has 57.34% utilization). The HPWL for the uniform and nonuniform placements are 10.7e7 and 9.0e7, respectively. As the intensity maps show, when 60% utilization is the target, uniform whitespace allocation is much more appropriate than 12% minimum local whitespace. On the other hand, 12% minimum local whitespace has much better wire length is appropriate when the target is 90% utilization.

is an equal bisection of the cell area. The difference in safe whitespace allocation mode is that the partitioning tolerance is much higher. Essentially, any partitioning solution that leaves at least `safeWS` on either side of the cutline is considered legal. This allows for very tight packing and reduces wire length, but is not recommended for congestion-driven placement [37].

Figure 5.4 illustrates uniform and nonuniform whitespace allocation. The top row shows global placements with uniform (left) and nonuniform (right) whitespace allocation on the ISPD 2005 contest benchmark adaptec1 (57.34% utlization) [30]. In the nonuniform placement shown, the minimum local whitespace is 12% and safe whitespace is 14%. The middle and bottom rows show intensity maps of the local utilization of each placement. Lighter areas of the intensity maps signify violations of a given target placement density; darker areas have utilization below the target. Regions completely occupied by fixed obstacles are shaded as if they exactly meet the target density. The target densities for the middle and bottom rows are 90% and 60%, respectively. Note that uniform whitespace produces almost no violations when the target is 90% and relatively few when the target is 60%. The nonuniform placement has more violations as compared to the uniform placement especially when the target is 60%, but remains largely legal with the 90% target density.

5.5 Detail Placement

Capo uses several different techniques to further reduce HPWL after global placement such as the sliding window optimizer RowIroning and a greedy cell movement scheme described below. In addition, Capo 10 performs optimal whitespace allocation using min-cost network flows without changing relative cell ordering [7, 38].

5.5.1 RowIroning

In RowIroning, optimal placers based on branch-and-bound and dynamic programming techniques replace windows of cells and whitespace chosen from the placement area [12]. These placers pack cells, and whitespace is represented by fake cells. To model whitespace accurately, one fake cell per site is needed, but Capo evenly divides contiguous regions of whitespace into at most three fake cells to limit runtime. This window of local improvement moves over all cells in left-to-right and top-to-bottom order (or the opposite directions).

5.5.2 Optimal Branch-and-Bound Placement

In the top-down partitioning-based placement approach, the original placement problem (considered as a "bin") is partitioned into two subproblems (sub-bins) and then recursively into smaller and smaller subproblems (recall Figure 5.1). Eventually, wire length can be directly optimized for bins with few nodes. We now describe *optimal placers* that operate on arbitrary single-row end-case instances given by[1]:

[1] End-cases have only one row because Capo preferentially splits small multirow blocks between rows.

- A hypergraph with nodes (cells) having (x, y)-dimensions. All cell heights are assumed equal to the row height.
- Every hyperedge has a bounding box of fixed pin locations corresponding to the external terminals incident to that net.
- Each hyperedge-to-node connection has a *pin offset* relative to the cell origin.
- A placement region, i.e., a subrow of a certain length.[2]

Additionally assuming the uniform distribution of whitespace, we can consider placement solutions as permutations of hypergraph nodes. The end-case placement problem thus naturally lends itself to enumeration and branch-and-bound. Implementations based on enumeration do not appear competitive in this context and will not be covered further.

In our branch-and-bound placer, nodes are added to the placement one at a time, and the bounding boxes of incident edges are extended to include the new pin locations. The branch-and-bound approach relies on computing, from a given partial placement, a lower bound on the wire length of any completion of the placement. Extensions of the current partial solution are considered only as long as this lower bound is smaller than the cost of the best seen complete solution.

One difficulty in applying branch-and-bound to end-case placement is varying cell widths. We restrict cells in the small instance to be packed with a fixed-size space between neighbors, i.e., whitespace is distributed equally between them. Replacing a cell with a cell of different width will change the location of at least one neighbor, triggering bounding box recompilations for incident nets. To simplify maintenance, the nodes are packed from left to right and always added to or removed from the right end of the partially-specified permutation. Such a lexicographic ordering naturally leads to a stack-driven implementation, where the states of incident nets are "pushed" onto stacks when a node is appended on the right side of the ordering, and "popped" when the node is removed. Bounding entails "popping" nodes at the end of a partial solution before all lexicographically greater partial solutions have been visited. Pseudocode is provided in Figure 5.5.

5.5.3 Greedy Cell Movement

Capo makes use of a gridded greedy movement technique to improve both wire length and whitespace distribution. A grid is imposed on the placement region to analyze local placement density. For cells that are in regions with density violations, candidate legal new locations are found in areas of lower density violation. Candidate moves are ranked by how well they alleviate the violations and how they affect wire length. Moves are made until a threshold of improvement is reached. We have found this to be a fast and effective method of removing density violations without adversely affecting wire length.

[2] For unfortunately short subrows that cannot accommodate all cells without overlaps, our end-case placer first minimizes overlap, then wire length

Single Row Placement Branch-and-Bound Input and Data Structures		
Input	cellWidth[0..N]	width of each cell
	pinOffsets[cellId][netId]	pin-offsets for each cell-pin pair
	terminalBoxes[netId]	bounding boxes of net terminals
	RowBox	bounding box of the row
Data Struct	nodeQueue =[0....N-1]	inverse initial ordering
	nodeStack=< *empty* >	placement ordering
	counterArray=< *empty* >	loop counter array
	idx=$N-1$	index
	costSoFar= 0	cost of the current placement
	bestYetSeen = Infinite	cost of best placement yet found
	nextLoc = row's left edge	location to place next cell at

Single-Row Placement with Branch-and-Bound : Algorithm

```
 1 while(idx < numCells)
 2 {
 3   s.push(q.dequeue()) // add a cell at nextLoc (the right end)
 4   c[idx] = idx
 5   costSoFar = costSoFar + cost of placing cell s.top()
 6   nextLoc.x = nextLoc.x + cellWidth[s.top()]
 7
 8   if(costSoFar ≤ bestCostSeen) bound
 9    c[idx] = 0
10
11   if(c[idx] == 0) // the ordering is complete or has been bounded
12   {
13     if(idx == 0 and costSoFar < bestCostSeen)
14     {
15       bestCostSeen = costSoFar
16       save current placement
17     }
18     while(c[idx] == 0)
19     {
20       costSoFar = costSoFar - cost of placing cell s.top()
21       nextLoc.x = nextLoc.x - cellWidth[s.top()]
22       q.enqueue(s.pop()) // remove the right-most cell
23       idx++
24       c[idx]- -
25     }
26   }
27   idx- -
28 }
```

Fig. 5.5. Branch-and-Bound algorithm for single-row placement is produced from a lexicographic enumeration of placement orderings by adding code for *bounding* in lines 8 and 9 (in bold) [13]. © 2000 IEEE.

5.6 Placement for Routability

With uniform whitespace allocation, Capo typically produces routable placements, but some congested areas remain. Capo 10 implements a whitespace allocation scheme described in [35] to improve placement routability. This technique uses a congestion map to estimate routing congestion after each layer of min-cut placement. Based on the congestion estimates, whitespace is allocated preferentially to areas of high congestion through cutline shifting. Coupled with other techniques from ROOSTER [35], Capo 10 outperforms best published routed wirelengths and via counts as of Fall 2006.

5.6.1 Optimizing Steiner Wire length

Weighted terminal propagation as described in [17] is sufficiently general to account for objectives other than HPWL such as StWL [35]. StWL is known to correlate with final rWL more accurately than HPWL and the authors of [35] hypothesize that if StWL could be directly optimized during global placement, one may be able to enhance routability and reduce rWL.

When bipartitioning a bin, the pins for a particular net may all fall into one partition (leaving the net uncut) or be split amongst both partitions (cutting the net). We will refer to the two possible partitions as partition 1 and partition 2. When using weighted terminal propagation from [17], one must calculate three costs per net per partitioning instance: w_1, w_2, and w_{12}. These costs represent the cost of the pins of a net all being placed in partition 1, partition 2, or split between both, respectively.

The points required to calculate w_1 for a given net are the terminals on the net (pins not allowed to move) plus the center of partition 1. Similarly, the points required to calculate w_2 are the terminals plus the center of partition 2. Lastly, the points to calculate w_{12} are the terminals on the net plus the centers of both partitions. See Figure 5.6 for an example of calculating these three costs. Clearly the HPWL of the set of points necessary to calculate w_{12} is at least as large as that of w_1 and w_2 since it contains an additional point. By the same logic, StWL also satisfies this relationship since RSMT length can only increase with additional points. Since StWL is a valid

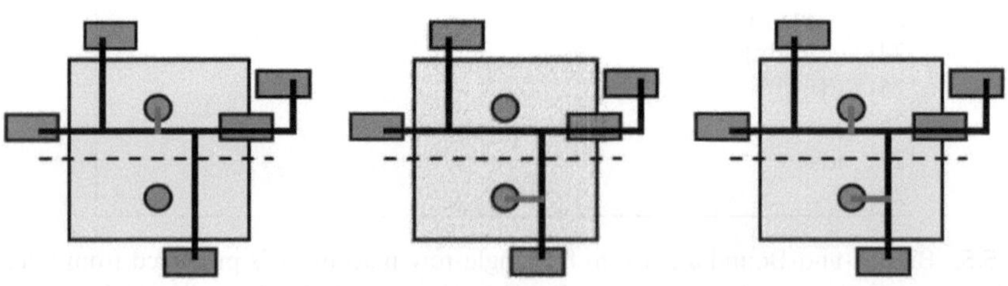

Fig. 5.6. Calculating the three costs for weighted terminal propagation with StWL: w_1 (*left*), w_2 (*middle*), and w_{12} (*right*) [35]. The net has five fixed terminals: four above and one below the proposed cutline. For the traditional HPWL objective, this net would be considered inessential. Note that the structure of the three Steiner trees may be entirely different, which is why w_1, w_2 and w_{12} are evaluated independently. © 2007 IEEE.

cost function for these weighted partitioning problems, this is a framework whereby it can be minimized [35].

The simplicity of this framework for minimizing StWL is deceiving. In particular, the propagation of terminal locations to the current placement bin and the removal of inessential nets [13] – standard techniques for HPWL minimization – cannot be used when minimizing StWL. Moving terminal locations drastically changes Steiner-tree construction and can make StWL estimates extremely inaccurate. Nets that are considered inessential in HPWL minimization (where the x- or y-span of terminals, if the cut is vertical or horizontal respectively, contains the x- or y-span of the centers of child bins) are not necessarily inessential when considering StWL because there are many Steiner trees of different lengths that have the same bounding box. Figure 5.6 illustrates a net that is inessential for HPWL minimization but essential for StWL minimization. Not only computing Steiner trees, but even traversing all relevant nets to collect all relevant point locations can be very time-consuming. Therefore, the main challenge in supporting StWL minimization is to develop efficient data structures and limit additional runtime during placement [35].

Pointsets with Multiplicities

Building Steiner trees for each net during partitioning is a computationally expensive task. To keep runtime reasonable when building Steiner trees for partitioning, the authors of [35] introduce a simple yet highly effective data structure – *pointsets with multiplicities*. For each net in the hypergraph, two lists are maintained. The first list contains all the unique pin locations on the net that are fixed. A fixed pin can come from sources such as terminals or fixed objects in the core area. The second list contains all the unique pin locations on the net that are movable, i.e., all other pins that are not on the fixed list. All points on each list are unique so that redundant points are not given to Steiner evaluators which may increase their runtime. To do so efficiently, the lists are kept in a sorted order. For both lists, in addition to the location of the pin, the number of pins that correspond to a given point is also saved [35].

Maintaining the number of actual pins that correspond to a point in a pointset (the multiplicity of that point) is necessary for efficient update of pin locations during placement. If a pin changes position during placement, the pointsets for the net connected to the pin must be updated. First, the original position of the pin must be removed from the movable pointset. As multiple pins can have the same position, especially early in placement, the entire net would need to be traversed to see if any other pins share the same position as the pin that is moving. Multiplicities allow to know this information in constant time. To remove the pin, one performs a binary search on the pointset and decreases the multiplicity of the pin's position by 1. If this results in the position having a multiplicity of 0, the position can be removed entirely. Insertion of the pin's new position is similar: first, a binary search is performed on the pointset. If the pin's position is already present in the pointset, the multiplicity is increased by 1. Otherwise, the position is added in sorted order with a multiplicity of 1. Empirically, building and maintaining the pointset data structures takes less than 1% of the runtime of global placement [35].

Performance

We compared three Steiner evaluators in terms of runtime impact and solution quality. They chose the FastSteiner [22] evaluator for global placement based on its reasonable runtime and consistent performance on large nets. Empirical results show the use of FastSteiner leads to a reduction of StWL by 3% on average on the IBMv2 benchmarks [40] (with a reduction of routed wirelength up to 7%) while using less than 30% additional runtime [35].

5.6.2 Congestion-Based Cutline Shifting

One of the most important reasons that we use bisection instead of quadrisection is the flexibility that it allows in choosing the cutline of a partitioned bin. Before partitioning, we first choose a direction for the cutline, usually based upon the geometry of the bin. We then choose a tentative cutline in that direction to split the bin roughly in half.

After the partitioner returns a solution, we have the flexibility to keep the cutline as it was chosen before partitioning or to change it to optimize an objective. The WSA [27] technique, applied after placement, geometrically divides the placement area in half and estimates the congestion in both halves of the layout. It then allocates more area to the side with greater routing demand, i.e., shifts the cutline, and proceeds recursively on the two halves of the design. In WSA, cells must be replaced after the whitespace allocation. However, we can avoid this replacement because our cells have not yet been placed and will be taken care of naturally during the min-cut process.

Cutline shifting used to handle congestion necessitates a slicing floorplan. The only work in the literature that describes top-down congestion estimates and uses them in placement assumes a grid structure [8]. Therefore we develop the following technique: Before each round of partitioning, we overlay the entire placement region on a grid. We choose the grid such that each placement bin is covered by 2–4 grid cells. We then build a congestion map using the last updated locations of all pins. We choose the mapping technique from [39] as it shows good correlation with routed congestion.

When cells are partitioned and their positions are changed, the congestion values for their nets are updated. Before cutline shifting, the routing demands and supplies for either side of the cutline are estimated with the congestion map. Given the bounding box of a region, we estimate its demand and supply by intersecting the bounding box with the grid cells of the congestion map. Grid cells that partially overlap with the given bounding box contribute only a portion of their demand and supply based on the ratio of the area of the overlap to the area of the grid cell. Using these, we shift the cutline to equalize the ratio of demand to supply on either side of the cutline.

To show the effectiveness of this dynamic version of WSA, we plot congestion maps of placements of ibm01h produced with and without our technique in Figure 5.7. The left plot illustrates uniform whitespace allocation and the right plot congestion-driven whitespace allocation. Our whitespace allocation technique

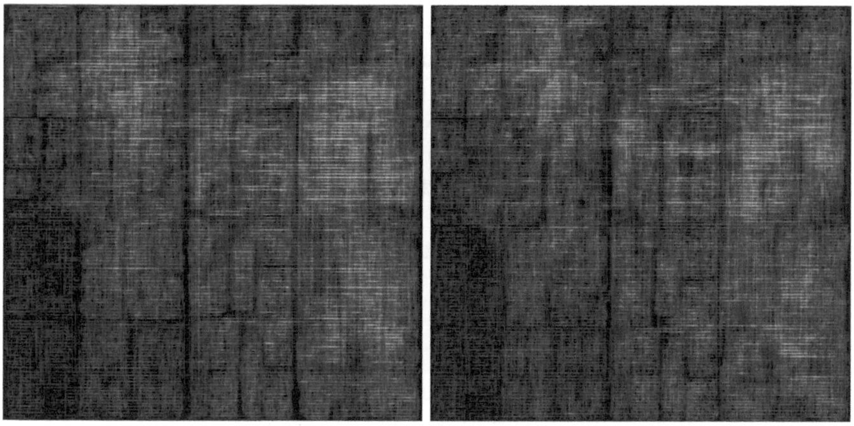

Fig. 5.7. Congestion maps for the ibm01h benchmark: Uniform whitespace allocation (produced with Capo-uniformWS) is illustrated on the *left*, congestion-driven allocation in ROOSTER is illustrated on the *right* [35]. The peak congestion when using uniform whitespace is 50% greater than that for our technique. When routed with Cadence WarpRoute, uniform whitespace produces 3.95% overfull global routing cells and routes in just over 5 h with 120 violations. ROOSTER's whitespace allocation produces 3.18% overfull global routing cells and routes in 22 min without violations. © 2007 IEEE.

reduces the maximum congestion by 50% and the number of overfull global routing cells from 3.9% to 3.18% (as reported by an industrial router).

5.7 Improved RTL Placement

Industrial floorplacement problems are increasingly difficult due to factors such as an increasing number of movable modules and a wide variation of module sizes. There is also insufficient cohesion for whitespace allocation between top-down methods and macro-placement algorithms. For example, a partitioner may misapproximate the area required by a set of macros and incorrectly allocate whitespace. To address these issues, we have integrated into Capo 10 the SCAMPI (SCalable Advanced Macro Placement Improvements) work [31]. The top-down partitioning flow is modified to selectively place large macros, while smaller macros are clustered into soft modules that will be placed later (Figure 5.8). The robustness of the flow is also improved by employing fast *look-ahead* Simulated Annealing on large macros of newly created bins. This allows early detection of bins difficult to floorplan, and alerts the placer to backtrack and seek a different partitioning solution.

5.7.1 Selective Floorplanning for Multimillion Gate Designs

One case that is not considered by either the original floorplacement techniques [34] or those introduced in SCAMPI [31] is where there are an extreme number of movable modules and an extreme ratio between the largest and smallest macro. An example of this is the *newblue1* benchmark from the ISPD'06 placement contest suite. The *newblue1* benchmark contains 64 macros and 330073 standard cells. As we show below, such a configuration is problematic for floorplacement tools.

Recall that a floorplacer utilizes a floorplanner to place macros. As the floorplanner uses Simulated Annealing to pack blocks, clustering is performed on the netlist to improve scalability. However, a very large number of small modules may stress clustering algorithms, which, in the absence of refinement, may undermine the overall solution quality.[3]

```
Variables: queue of placement partitions
Initialize queue with top-level partition
While (queue not empty)
  Dequeue a partition
  If (partition is not marked as merged)
    Perform look-ahead floorplanning on partition
    If look-ahead floorplanning fails
      Undo one partition decision
      Merge partition with sibling
      Mark new partition as merged and enqueue
  Else if (partition has large macros or
      is marked as merged)
    Mark large macros for placement after floorplanning
    Cluster remaining macros into soft macros
    Cluster std-cells into soft macros
    Use fixed-outline floorplanner to pack
      all macros (soft+hard)
    If fixed-outline floorplanning succeeds
      Fix large macros and remove sites beneath
    Else
      Undo one partition decision
      Merge partition with sibling
      Mark new partition as merged and enqueue
  Else if (partition is small enough and
      mostly comprised of macros)
    Process floorplanning on all macros
  Else if (partition small enough)
    Process end case std cell placement
  Else
    Bi-partition netlist of the partition
    Divide the partition by placing a cutline
    Enqueue each child partition
```

Fig. 5.8. Our modified min-cut floorplacement flow [31]. *Bold-faced lines* are new compared to [34].

[3] Refinement algorithms would need to operate on very large netlists and may require long runtimes.

Figure 5.9 shows the *newblue1* benchmark placed with SCAMPI, before and after our most recent improvements. In the original SCAMPI flow, the large block was designated for floorplanning by Parquet at the top level. Parquet precedes annealing with clustering to reduce the size of the netlist. However, given the large number of small modules, the simple-minded clustering algorithm in Parquet ended up taking 16% of total runtime, whereas annealing took only 4%. Additionally, even if clustering were more scalable, clustering such a large number of small macros into large, soft macros can lead to unnatural or unrepresentative netlists. In the original SCAMPI flow, the clusters formed by the standard cells in *newblue1* became large enough to artificially constrain the movement of the large macro during floorplanning. This is mainly a limitation of Simulated Annealing as it becomes impractical in solution quality and runtime for over 100 modules.

Therefore, we propose the following method. Whenever a bin is designated for floorplanning and the largest real module is smaller in area than the largest soft macro built from clustering (this area can be estimated without actually performing clustering), we do not use Simulated Annealing. Instead, a simple analytical placement technique, such as (*Successive Over-Relaxation*) (SOR), is used to determine reasonable locations for the large macros.[4] It has been shown that analytical techniques are good at finding general areas where objects should be placed [6], so this is a reasonable and efficient solution for placing a large macro or macros in this situation. As such, this technique may also be useful in regions with large amounts of whitespace as block-packing often overlooks good solutions in such situations. Objectives other than HPWL, such as routing congestion and timing, are also important, and any

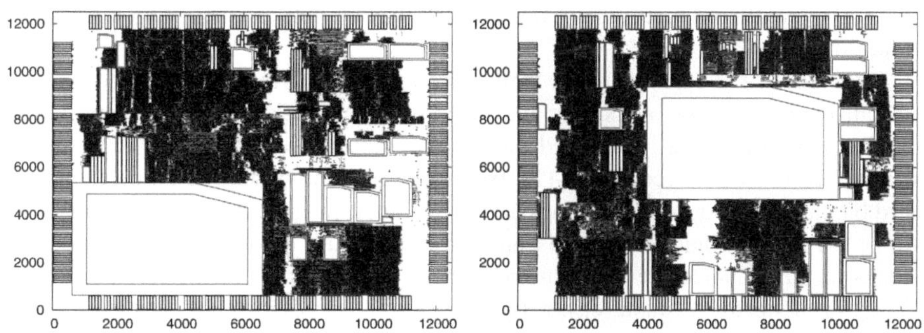

Fig. 5.9. The *newblue1* benchmark placed by SCAMPI before (*left*) and after (*right*) our recent modifications. Before our improvements to SCAMPI, the clusters formed by the smaller modules at the top-level constrain the movement of the largest module and result in it being placed in the *bottom-left corner* of the core. After our improvements, the largest macro is placed using Successive Over-Relaxation (SOR).

[4] Any analytical placement technique can be used, but SOR may be sufficient since we are not necessarily looking for a nonoverlapping placement. For example, we have also used a linearized version of the SOR technique as well and seen improvements in HPWL at the expense of moderately increased runtime.

analytical placer used in this context should place macros with respect to the most relevant objective(s). Our key observation is that placing such macros early is helpful.

When there is only one large macro to be placed, the solution of the analytical tool is used and the macro is fixed in its desired location. To place a small number of large macros with this method, we again compute macro locations with the analytical tool, but must legalize the macro locations to maintain the correct-by-construction paradigm of floorplacement. Overlaps can be legalized in several ways. One way is to use a greedy macro legalization technique such as the macro legalizer described in [34, Sect. 3.3]. Another method for removing macro overlap is the constraint-based floorplan repair algorithm FLOORIST [29]. Following legalization, one can shift the macros so that their center of mass coincides with their center of mass before legalization in keeping with the spirit of the analytical placement. This technique contributed to HPWL improvement over the ISPD 2006 Placement Contest results of Capo by 17% on newblue1, with an overall improvement in the contest score on the ISPD 2006 benchmark suite by 10%, moving Capo three positions higher.

5.7.2 Temporary Macro Deflation

Low-whitespace conditions in block-packing instances formed during floorplacement can worsen solution quality significantly. In such cases, the block-packing engine focuses mainly on finding legal solutions rather than those that have good wire length. In addition, a legal solution may not be found which leads to backtracking and increased runtime as well. To improve the solution quality of block-packing instances created during floorplacement, we prevent these low-whitespace conditions.

To account for standard cells in the floorplacement framework, standard cells are clustered into soft blocks for instances of block-packing [1]. To improve the likelihood of finding a legal fixed-outline solution, these soft blocks representing standard cells are reduced in size [1]. We propose extending this deflation to include hard blocks in addition to soft blocks. When a block-packing instance is formed, we adjust the sizes of hard blocks to maintain a minimum amount of whitespace. All blocks in the instance are sized in the same way and aspect ratios are maintained. The resized instance, made easier by the addition of whitespace, is placed using Simulated Annealing as normal.

Resizing the hard blocks in this way has the positive effect of making fixed-outline block-packing easier, which allows the block-packing engine to focus on HPWL minimization rather than mere legality in cases where whitespace is limited, but removes the correct-by-construction property upon which floorplacement is built. To alleviate this problem, we apply legalization to macros after packing. We use the fast and robust constraint-based floorplan repair algorithm FLOORIST [29] after each layer of placement where block-packing took place. FLOORIST moves macros minimally when repairing macro overlaps, so the reduced HPWL found in easier block-packing instances is preserved.

Empirically we find that the overhead of running FLOORIST for legalization is mitigated by the fact that block-packing is easier and therefore faster. In terms of solution quality, we find that temporary macro deflation reduces HPWL by 2–3%.

5.7.3 Whitespace Reallocation Using Linear Programming and Min-Cost Max-Flow

As we have noted earlier, in order to avoid cases of backtracking which can dramatically increase both HPWL and runtime, Capo allocates whitespace uniformly during partitioning when macros are present. We have shown in Figure 5.10 this whitespace allocation scheme can lead to HPWL that is much larger than a tighter packing. In order to reclaim some of the HPWL lost due to uniform distribution during global placement, we propose a technique to reallocate whitespace during detail placement.

Our technique builds upon the well-known linear programming formulations used, e.g., in [38] and [33] in that we impose linear constraints for movable objects based on their relative positions with respect to core boundaries and other movable objects. More details on the linear programming formulation such as types of constraints and the objective function are given below. We include additional linear inequalities to account for fixed obstacles and region constraints. One major difference from previous work is that we guarantee that the x and y locations found align to legal sites and rows, as explained later in this section.

We handle reallocation of whitespace separately for the horizontal and vertical directions, and preserve local relative ordering of movables in each direction. In other words, movable objects may not jump over each other or any fixed obstacles when whitespace is being reallocated. Unlike in global placement [33], we start with legal or nearly legal locations. This simplifies our selection of relative constraints to include into the LP formulation as follows. In the horizontal case, we examine each row individually. For each cell or macro that intersects the row, we determine its immediate neighbors to the left and to the right (those objects with which the current object could feasibly overlap if it would slide to the left or right). These neighbors can

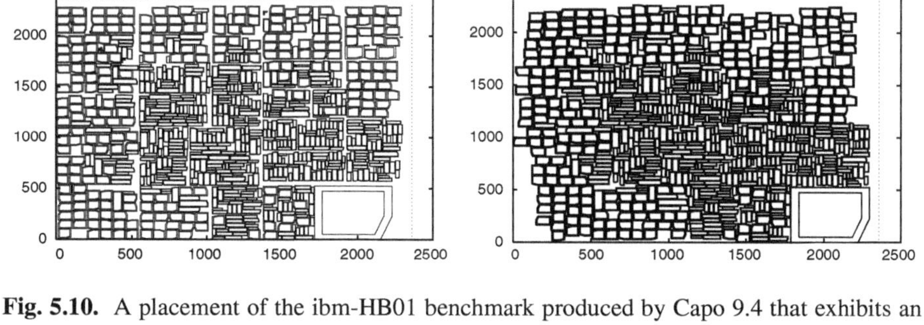

Fig. 5.10. A placement of the ibm-HB01 benchmark produced by Capo 9.4 that exhibits an overly generous whitespace allocation scheme in Capo. After reallocating whitespace with a min-cost max-flow technique, we decrease HPWL by 4.5%.

include movable objects, row or region boundaries as well as fixed obstacles. After the neighborhood relations are determined, we constrain an object to lie between its left- and right-hand neighbors. Construction of constraints for the vertical case is analogous where rows are replaced with columns and site width is replaced by row height. Unlike the formulation from [33], ours guarantees an overlap-free placement and needs to be solved only once. In contrast with [38], we include only several constraints per movable object rather than a quadratic number of constraints read from a sequence-pair. This significantly improves scalability and allows one to pack more tightly.

In addition to the constraints above, we minimize HPWL. This is done by adding $x_{\min}, x_{\max}, y_{\min}, y_{\max}$ variables for each net, and the terms $(x_{\max} - x_{\min})$ and $(y_{\max} - y_{\min})$ to the objective function. To solve the entire LP efficiently, we dualize it as in [38] and cast the dual as a min-cost max-flow instance. The latter is solved using the scaling push-relabeling algorithm of Goldberg [19]. An important feature of our technique is the use of integrality of the solutions found by this algorithm – we scale the coordinates so that integer x values correspond to legal sites and integer y values correspond to standard-cell rows. Figure 5.10 illustrates whitespace reallocation in the horizontal and vertical directions applied to a placement of the ibm-HB01 benchmark. HPWL is improved by 4.5% while runtime of the technique is less than 1% of placement runtime.

5.8 Incremental Placement

To develop a strong incremental placement tool, ECO-system, we build upon an existing global placement framework and must choose between analytical and top down. The main considerations include robustness, the handling of movable macros and fixed obstacles, as well as consistent routability of placements and the handling of density constraints. Based on recent empirical evidence [31,35,37], the top-down framework appears a somewhat better choice. Indeed the 2 out of 9 contestants in the ISPD 2006 Competition that satisfied density constraints were top-down placers. However, analytical algorithms can also be integrated into our ECO-system when particularly extensive changes are required. ECO-system favorably compares to recent detail placers in runtime and solution quality and fares well in high level and physical synthesis.

5.8.1 General Framework

The goal of ECO-system is to reconstruct the internal state of a min-cut placer that could have produced a given placement *without the expense of global placement.* Given this state, we can choose to accept or reject previous decisions based on our own criteria and build a new placement for the design. If many of the decisions of the placer were good, we can achieve a considerable runtime savings. If many of the decisions are determined to be bad, we can do no worse in terms of solution quality

```
   Variables: queue of placement bins
   Initialize queue with top-level placement bin
1  While(queue not empty)
2    Dequeue a bin
3    If(bin not marked to place from scratch)
4      If(bin overfull)
5        Mark bin to place from scratch, break
6      Quickly choose the cutline which has
         the smallest net cut considering
         cell area balance constraints
7      If(cutline causes overfull child bin)
8        Mark bin to place from scratch, break
9      Induce partitioning of bin's cells from cutline
10     Improve net cut of partitioning with
         single pass of Fiduccia–Mattheyses
11     If(% of improvement > threshold)
12       Mark bin to place from scratch, break
13     Create child bins using cutline and partitioning
14     Enqueue each child bin
15   If(bin marked to place from scratch)
16     If(bin small enough)
17       Process end case
18     Else
19       Bi-partition the bin into child bins
20       Mark child bins to place from scratch
21       Enqueue each child bin
```

Fig. 5.11. Incremental min-cut placement [36]. *Bold-faced lines* 3–15 and 20 are different from traditional min-cut placement. © 2007 IEEE.

than placement from scratch. An overview of the application of ECO-system to an illegal placement is depicted in Figure 5.12. See the algorithm in Figure 5.11.

To rebuild the state of a min-cut placer, we must reconstruct a series of cutlines and partitioning solutions efficiently. To extract a cutline and partitioning solution from a given placement bin, we examine all possible cutlines as well as the partitions they induce. We start at one edge of the placement bin (left edge for a vertical cut and bottom edge for a horizontal cut) and move towards the opposite edge. For each potential cutline encountered, we maintain the cell area on either side of the cutline, the partition induced by the cutline and the net cut.

5.8.2 Fast Cutline Selection

For simplicity, assume that we are making a vertical cut and are moving the cutline from the left to the right edge of the placement bin (the techniques necessary for a horizontal cut are analogous). Pseudocode for choosing the cutline is shown in Figure 5.13. To find the net cut for each possible cutline efficiently, we first calculate the bounding box of each net contained in the placement bin from the original

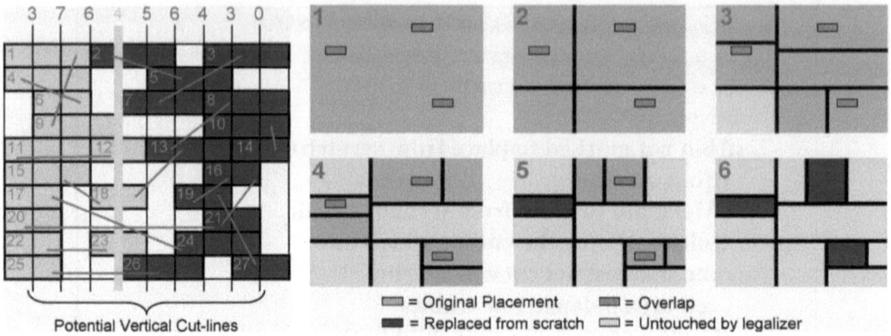

Fig. 5.12. Fast legalization by ECO-system [36]. The image on the *left* illustrates choosing a vertical cutline from an existing placement. Nets are illustrated as *red lines*. Cells are individually numbered and take two or three sites each. Cutlines are evaluated by a left-to-right sweep (net cuts are shown above each line). A cutline that satisfies partitioning tolerances and minimizes cut is found (*thick green line*). Cells are assigned to *left* and *right* according to the center locations. On the right, placement bins are subdivided using derived cutlines until (1) a bin contains no overlap and is ignored for the remainder of the legalization process or, (2) the placement in the bin is considered too poor to be kept and is replaced from scratch using min-cut or analytical techniques. © 2007 IEEE.

placement. We create two lists with the left and right x-coordinates of the bounding boxes of the nets and sort them in increasing x-order. While sliding the cutline from left to right (in the direction of increasing x-coordinates), we incrementally update the net cut and amortize the amount of time used to a constant number of operations per net over the entire bin. We do the same with the centers of the cells in the bin to incrementally update the cell areas on either side of the cutline as well as the induced partitioning. While processing each cutline, we save the cutline with smallest cut that is legal given partitioning tolerances. An example of finding the cutline for a partitioning bin is shown in Figure 5.12.

Once a partitioning has been chosen, we accept or reject it based on how much it can be improved by *a single pass of a Fiduccia–Mattheyses partitioner with early termination* (which takes only several seconds even on the largest ISPD'05 circuit).[5] The intuition is that if the constructed partitioning is not worthy of reuse, a single Fiduccia–Mattheyses pass could improve its cut nontrivially. If the Fiduccia–Mattheyses pass improves the cut beyond a certain threshold, we discard the solution and bisect the entire bin from scratch. If this test passes, we check legality: if a child bin is overfull, we discard the cutline and bisect from scratch.

5.8.3 Scalability

Pseudocode for the cutline location process used by ECO-system is shown in Figure 5.13. The runtime of the algorithm is linear in the number of pins incident to the bin, cells incident contained in the bin, and possible cutlines for the bin. Since

[5] We do not assume that the initial placement was produced by a min-cut algorithm.

```
Input: placement bin, balance constraint
Output: x-coord of best cutline
numCutlines =
   1 + ⌊(rightBinEdgeX−leftBinEdgeX)/cellSpacing⌋
Create three arrays of size numCutlines:
   LEFT, RIGHT, AREA
Set all elements of LEFT, RIGHT, and AREA to 0
Foreach net
  Calculate x-coord of left- and right-most pins
  leftCutlineIndex =
   max(0,⌈(leftPinX−leftBinEdgeX)/cellSpacing⌉)
  rightCutlineIndex =
   max(0,⌈(rightPinX−leftBinEdgeX)/cellSpacing⌉)
  if(leftCutlineIndex < numCutlines)
    LEFT[leftCutlineIndex]+=1
  if(rightCutlineIndex < numCutlines)
    RIGHT[rightCutlineIndex]+=1
Foreach cell
  Calculate x-coord of the center of the cell
  cutlineIndex =
   max(0,⌈(centerX−leftBinEdgeX)/cellSpacing⌉)
  if(cutlineIndex < numCutlines)
    AREA[cutlineIndex]+=cellArea
Set X = leftBinEdge, CURCUT = 0, BESTCUT = ∞
BESTX = ∞, LEFTPARTAREA = 0
For(I = 0;I < numCutlines;I+=1,X+=cellSpacing)
  CURCUT+=LEFT[I]
  CURCUT−=RIGHT[I]
  LEFTPARTAREA+=AREA[I]
  If(CURCUT < BESTCUT and
     LEFTPARTAREA satisfies balance constraint)
    BESTCUT = CURCUT
    BESTX = X
Return BESTX
```

Fig. 5.13. Algorithm for finding the best vertical cutline from a placement bin. Finding the best horizontal cutline is largely the same process. Note that the runtime of the algorithm is linear in the number of pins incident to the bin, cells incident contained in the bin, and possible cutlines for the bin. ©2007 IEEE.

a single Fiduccia–Mattheyses pass takes also takes linear time [18], the asymptotic complexity of our algorithm is linear. If we let P represent the number of pins incident to the bin, C represent the number of cells in the bin, and L represent the number of potential cutlines in the bin, the cutline selection process runs in $O(P + C + L)$ time. In the vast majority of cases, $P > C$ and $P > L$, so the runtime estimate simplifies to $O(P)$.

The number of bins may double at each hierarchy layer, until bins are small enough for end-case placement. End-case placement is generally a constant amount of runtime for each bin, so it does not affect asymptotic calculations. Assume that ECO-system is able to reuse all of the original placement. Since ECO-system performs bisection, it will have $O(\log C)$ layers of bisection before end-case placement. At layer i, there will be $O(2^i)$ bins, each taking $O\left(P/2^i\right)$ time. This gives a total time per layer of $O(P)$. Combining all layers gives $O(P \log C)$. Empirically, the runtime of the cutline selection procedure (which includes a single pass of a Fiduccia–Mattheyses partitioner) is much smaller than partitioning from scratch. On large benchmarks, cutline selection requires 5% of ECO-system runtime time whereas min-cut partitioning generally requires 50% or more of ECO-system runtime.

5.8.4 Handling Macros and Obstacles

With the addition of macros, the flow of top-down placement becomes more complex. We adopt the technique of "floorplacement" which proceeds as traditional placement until a bin satisfies criteria for block-packing [31, 34]. If the criteria suggest that the bin should be packed rather than partitioned, a fixed-outline floorplanning instance is induced from the bin where macros are treated as hard blocks and standard cells are clustered into soft blocks. The floorplanning instance is given to a Simulated Annealing-based floorplanner to be solved. If macros are placed legally and without overlap, they are considered fixed. Otherwise, the placement bin is merged with its sibling bin in the top-down hierarchy and the merged bin is floorplanned. Merging and re-floorplanning continues until the solution is legal.

We add a new floorplanning criterion for our legalization technique. If no macros in a placement bin overlap each other, we generate a placement solution for the macros of the bin to be exactly their placements in the initial solution. If some of the macros overlap with each other, we let other criteria for floorplanning decide. If block-packing is invoked, we must discard the placement of all cells and macros in the bin and proceed as described in [34].

During the cutline selection process, some cutline locations are considered invalid – namely those that are too close to obstacle boundaries but do not cross the obstacles. This is done to prevent long and narrow slivers of space between cutlines and obstacle boundaries. Ties for cutlines are broken based on the number of macros they intersect. This helps to reduce overfullness in child bins allowing deeper partitioning, which reduces runtime.

5.8.5 Relaxing Overfullness Constraints

One of the primary objectives of ECO-system is to reuse as much relevant placement information as possible from a given placement. As described above, it is possible to find a cutline which has a good cut but is not legal due to space constraints. In these cases, ECO-system must discard these good solutions and partition from scratch.

In order to make better use of the given placement, we propose the following addition to ECO-system. In these situations, we allow ECO-system to shift the cutline to legalize the derived partition with respect to area. Cutline shifting is

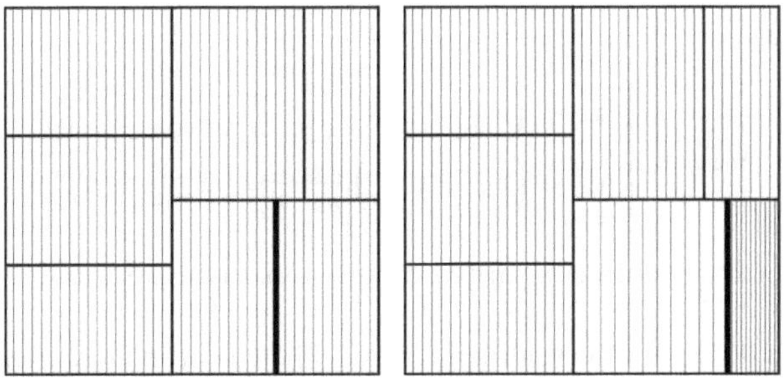

Fig. 5.14. Shifting a cutline chosen during ECO cutline selection. Unlike the WSA technique [27, 28], cutline shifting during ECO is not done on geometric cutlines but instead on those cutlines which are chosen during fast cutline selection. The image on the left shows a placement that has been divided into bins during the course of ECO-system. In the image on the right, the chosen cutline of the bottom-right bin is shifted to the right. The density of vertical lines represents the initial placement and its scaling around the moving cutline (*shown in red*). © 2007 IEEE.

a technique commonly used in the top-down min-cut placement for allocation of whitespace [4, 27, 28, 35, 37]. The cutline is shifted as little as possible to make the derived partitioning legal with respect to area. If it is impossible to find an area-legal cutline, the derived partitioning must be discarded and ECO-system proceeds normally.

If cutline shifting is successful in correcting the illegality, the original placement must be modified for purposes of consistency. To do so, cells are scaled proportionately within the placement bin based on their original positions, the position of the originally chosen cutline and the position of the shifted cutline in a manner similar to that in the WSA technique [27,28]. As the centers of cells are used to determine in what partitions cells belong during fast cutline selection, we shift cell locations based on center locations as well to ensure that cutline shifting will not change derived partitions. We seek to shift cell locations and maintain the following property: The relative position between cells before and after shifting is maintained. Also, if a cell were in the middle of a partition before shifting, it should remain in the middle of a partition after shifting. Let x_{L} and x_{R} represent the x-coordinates of the left and right sides of the placement bin, $x_{\mathrm{orig}}^{\mathrm{cut}}$ and $x_{\mathrm{new}}^{\mathrm{cut}}$ the x-coordinates of the original and new cuts, and, lastly, $x_{\mathrm{orig}}^{\mathrm{cell}}$ and $x_{\mathrm{new}}^{\mathrm{cell}}$ the x-coordinates of the center of a particular cell before and after shifting. We wish to maintain the following ratios (for vertical partitioning):

$$\frac{x_{\mathrm{orig}}^{\mathrm{cell}} - x_{\mathrm{L}}}{x_{\mathrm{orig}}^{\mathrm{cut}} - x_{\mathrm{L}}} = \frac{x_{\mathrm{new}}^{\mathrm{cell}} - x_{\mathrm{L}}}{x_{\mathrm{new}}^{\mathrm{cut}} - x_{\mathrm{L}}}, \quad x_{\mathrm{orig}}^{\mathrm{cell}} \leq x_{\mathrm{orig}}^{\mathrm{cut}},$$

$$\frac{x_{\mathrm{R}} - x_{\mathrm{orig}}^{\mathrm{cell}}}{x_{\mathrm{R}} - x_{\mathrm{orig}}^{\mathrm{cut}}} = \frac{x_{\mathrm{R}} - x_{\mathrm{new}}^{\mathrm{cell}}}{x_{\mathrm{R}} - x_{\mathrm{new}}^{\mathrm{cut}}}, \quad x_{\mathrm{orig}}^{\mathrm{cell}} > x_{\mathrm{orig}}^{\mathrm{cut}}.$$

Solving for x_{new}^{cell} :

$$x_{\mathrm{new}}^{\mathrm{cell}} = \begin{cases} x_{\mathrm{L}} + \left(x_{\mathrm{orig}}^{\mathrm{cell}} - x_{\mathrm{L}}\right) \frac{x_{\mathrm{new}}^{\mathrm{cut}} - x_{\mathrm{L}}}{x_{\mathrm{orig}}^{\mathrm{cut}} - x_{\mathrm{L}}}, & x_{\mathrm{orig}}^{\mathrm{cell}} \leq x_{\mathrm{orig}}^{\mathrm{cut}}, \\ x_{\mathrm{R}} - \left(x_{\mathrm{R}} - x_{\mathrm{orig}}^{\mathrm{cell}}\right) \frac{x_{\mathrm{R}} - x_{\mathrm{new}}^{\mathrm{cut}}}{x_{\mathrm{R}} - x_{\mathrm{orig}}^{\mathrm{cut}}}, & x_{\mathrm{orig}}^{\mathrm{cell}} > x_{\mathrm{orig}}^{\mathrm{cut}}. \end{cases}$$

The new y-coordinates of cells shifted during horizontal partitioning are calculated analogously.

Figure 5.14 illustrates the scaling involved when a cutline is shifted. In the figure, the cutline of the bottom-right bin is shifted to the right. All objects to the left and right of the cutline are scaled appropriately. Objects that were to the left of the original cutline remain to the left and are spread out and objects on the right are packed closer together.

Shifting proportionately in this way maintains the relative ordering of all the cells within the current placement bin. Also the partitioning induced by the cutline remains unchanged so ECO-system can proceed as normal. Shifting the cutline in this manner can allow deeper ECO partitioning which can reduce both runtime and cell displacement.

5.8.6 Satisfying Density Constraints

A common method for increasing the routability of a design is to inject whitespace into regions that are congested [4, 27]. One can also require a minimum amount of whitespace (equivalent to a maximum cell density) in local regions of the design to achieve a similar effect [37]. As one of ECO-system's legality checks is essentially a density constraint (checking to see if a child bin has more cell area assigned to it than it can physically fit), this legality check is easy to generalize. The new criterion for switching from using the initial placement and partitioning from scratch is based on a child bin having less than a threshold percent of relative whitespace, which is controlled by the user.

The cutline shifting feature of ECO-system can also be used to satisfy density constraints. As ECO-system proceeds, cutlines can be shifted as described above to implement a variety of whitespace allocation schemes [27, 28, 35, 37]. Specifically, ECO-system can implement the hierarchical whitespace injection of WSA [27, 28]. WSA chooses cutlines based only on the geometry of a placement bin and shifts these cutlines from the top down. ECO-system chooses cutlines that are more natural to the original placement, shifts cutlines top down, and also supports fixed objects, and movable macros.

5.9 Memory Profile

Capo's nonuniform whitespace allocation techniques tend to produce unbalanced partitionments at the top layers. As peak memory usage grows with partitioning problem size, memory consumption can stay near the peak for longer periods of time during placement. To counteract the increased possibility of thrashing, Capo 10 has several memory improvements which include the slimming down of data structures and carefully choosing the lifetimes of major data structures so that fewer need to be in main memory simultaneously. The most radical of these changes involves removing the netlist hypergraph from main memory during the largest partitioning instances and rebuilding it from scratch afterwards. These changes reduce peak memory consumption by 2× compared to Capo 9.1 but slow down global placement by 10%.

5.10 Performance on Publicly Available Benchmarks

To illustrate Capo's ability to handle a wide range of placement instances, we evaluate Capo on benchmarks with routing information, mixed-size benchmarks and the extremely large benchmarks with generous amounts of whitespace from the ISPD 2005 and 2006 placement competitions.

5.10.1 Routing Benchmarks

To show Capo's performance on placement instances with routing information, we show results for the IBMv2 [40], IWLS [20] and Faraday suites of benchmarks [1]

Table 5.2. A comparison of ROOSTER to the most recent version of mPL-R + WSA and APlace 2.04 on the IBMv2 benchmarks [40]. All routed wirelengths (rWL) are in meters. "Time" represents routing runtime in minutes. Note that while APlace 2.04 achieves overall smaller wire length than ROOSTER, it routes with violations on 2 of the 16 benchmarks. Best legal rWL and via counts are in bold.

	ROOSTER				Latest mPL-R + WSA				APlace 2.04 -R 0.5			
	rWL	#Vias	#Vio.	Time	rWL	#Vias	#Vio.	Time	rWL	#Vias	#Vio.	Time
ibm01e	0.733	**122286**	0	42	**0.718**	123064	0	11	0.790	158646	85	132
ibm01h	0.746	**124307**	0	32	**0.691**	213162	0	11	0.732	161717	2	121
ibm02e	2.059	259188	0	13	**1.821**	**250527**	0	11	1.846	254713	0	9
ibm02h	2.004	262900	0	14	**1.897**	**260455**	0	13	1.973	268259	0	14
ibm07e	4.075	**476814**	0	17	4.130	492947	0	21	**3.975**	500574	0	17
ibm07h	4.329	**489603**	0	19	4.240	516929	0	26	**4.141**	518089	0	23
ibm08e	4.242	**559636**	0	17	4.372	579926	0	23	**3.956**	588331	0	18
ibm08h	4.262	**574593**	0	20	4.280	599467	0	26	**3.960**	595528	0	18
ibm09e	3.165	**466283**	0	11	3.319	488697	0	17	**3.095**	502455	0	11
ibm09h	3.187	**475791**	0	11	3.454	502742	0	19	**3.102**	512764	0	12
ibm10e	6.412	**749731**	0	22	6.553	777389	0	30	**6.178**	782942	0	23
ibm10h	6.602	**775018**	0	27	6.474	799544	0	33	**6.169**	801605	0	28
ibm11e	**4.698**	**605807**	0	15	4.917	633640	0	22	4.755	648044	0	18
ibm11h	**4.697**	**618173**	0	16	4.912	660985	0	25	4.818	677455	0	24
ibm12e	9.289	**918363**	0	36	10.185	995921	0	57	**8.599**	921454	0	32
ibm12h	9.289	**938971**	0	43	9.724	976993	0	50	**8.814**	961296	0	50
Ratio	1.000	1.000			1.007	1.069			0.968	1.073		

Table 5.3. A comparison of Capo with ROOSTER extensions to Cadence AmoebaPlace on the IWLS 2005 Benchmarks [20]. All routed wirelengths (rWL) are in meters. "Time" represents routing runtime in minutes. ROOSTER outperforms AmoebaPlace by 12.0% in rWL and 1.1% in via counts. Best rWL and via counts are in bold.

Benchmark	ROOSTER + NanoRoute				AmoebaPlace + NanoRoute			
	rWL	#Vias	#Vio.	Time	rWL	#Vias	#Vio.	Time
aes_core	**1.339**	**125939**	2	32	1.657	131049	1	28
ethernet	**7.287**	**467777**	1	27	7.745	471800	1	28
mem_ctrl	**1.061**	**87276**	0	22	1.224	90067	0	21
pci_bridge32	**1.336**	**114880**	0	35	1.598	117326	2	35
usb_funct	**0.995**	**84717**	0	19	1.106	85739	0	19
vga_lcd	25.906	1131591	2	57	**25.405**	**1076178**	2	90
Ratio	1.000	1.000			1.120	1.011		

Table 5.4. Routing results on the Faraday benchmarks with movable macro blocks fixed [1]. All routing wirelengths (rWL) are in meters. "Time" represents routing runtime in minutes. Best rWL and via counts are highlighted in bold.

Bench-mark	ROOSTER				Silicon Ensemble Ultra v5.4.126			
	rWL	#Vias	#Vio.	Time	rWL	#Vias	#Vio	Time
DMA	**0.554**	**116414**	0	3	0.644	125328	0	3
DSP1	**1.110**	209274	0	5	1.224	**204863**	0	6
DSP2	**1.067**	**194971**	0	6	1.230	207521	0	6
RISC1	**1.868**	**328699**	5	9	1.957	345615	4	6
RISC2	**1.786**	**324278**	5	7	1.959	347515	2	5
Ratio	1.000	1.000			1.112	1.048		

in Tables 5.2–5.4. Capo with ROOSTER extensions consistently produces routable placements with the best published routed wirelength on several benchmarks and best via counts overall.

5.10.2 Mixed-Size Benchmarks

To show Capo's performance on difficult mixed-size placement instances, we show results on difficult floorplanning instances identified by the authors of [31]. Comparisons of Capo with other tools on two difficult benchmark suites are shown in Tables 5.5 and 5.6. Most other tools are unable to place these benchmarks legally within the time limit, but Capo with SCAMPI extensions completes all of these benchmarks quickly and legally. Considering the designs successfully placed by PATOMA 1.0 and Capo 9.4, Capo with SCAMPI extensions produces placements with smaller HPWL by 31% and 13%.

Table 5.5. Runs of Capo with SCAMPI extensions and other tools on recent designs from Calypto Design Systems, Inc. [31].

cal	PATOMA 1.0			Capo 9.4 -faster			APlace 2.0			FengShui 5.1			SCAMPI				
bench	HPWL (e+04)	ovlp (%)	time (s)	HPWL (e+04)	ovlp (%)	time (s)	HPWL (e+04)	ovlp (%)	time (s)	HPWL (e+04)	ovlp (%)	time (s)	HPWL (e+04)	ovlp (%)	time (s)	vs. PATOMA (HPWL)	vs. CAPO (HPWL)
040	177.2	0.0	9.6	18.7	0.0	45.4	20.7	0.3 ⊗	239.0	20.6	0.0	37.9	**17.7**	0.0	39.5	0.10x	0.94x
098	52.3	0.0	11.2	31.8	1.3	788.2	22.6	0.3	271.6	24.0	0.0 ⊗	6.0	**26.9**	0.0	264.4	0.51x	-
336	2.8	0.0	1.2	3.5	9.1	22.5	2.2	0.1 ⊗	83.5	7.6	0.0	0.2	**2.8**	0.0	11.4	0.99x	-
353	7.6	0.0	1.0	6.5	0.5	52.6	4.6	0.3	211.8	31.5	1.6 ⊗	0.8	**5.5**	0.0	26.0	0.73x	-
523	123.7	0.0	3.4	34.7	0.3	240.2	27.5	0.3	920.3	348.7	0.0	2.8	**30.3**	0.0	157.2	0.24x	-
542	0.9	0.0	0.1	0.8	0.0	3.3	0.7	0.1	42.8	×	×	×	**0.8**	0.0	2.0	0.85x	0.96x
566	83.6	0.0	4.9	63.8	1.9	225.7	46.9	0.5	341.1	493.6	3.8 ⊗	3.2	**71.6**	0.0	188.5	0.86x	-
583	47.0	0.0	2.3	26.1	0.6	190.6	20.6	0.2	421.2	×	×	×	**21.5**	0.0	141.3	0.46x	-
588	8.8	0.0	0.7	6.3	1.1	60.4	4.8	0.5	41.5	×	×	×	**5.6**	0.0	26.4	0.63x	-
643	4.9	0.0	0.6	3.8	0.9	18.8	3.0	0.4	29.3	15.3	0.2 ⊗	0.5	**3.4**	0.0	11.5	0.68x	-
DCT	×	×	×	×	×	>1800	33.1	1.7 ⊗	719.4	184.7	0.0	8.0	**37.2**	0.0	123.5	-	-
													Average			0.51x	0.95x

× indicates time-out, crash, or a run completed without producing a solution; ⊗ indicates an out-of-core solution.

Best legal solutions are emphasized in bold.

Table 5.6. Runs of Capo with SCAMPI extensions and other tools on the IBM-HB$^+$ benchmarks [31].

ibm	PATOMA 1.0			Capo 9.4 -faster			APlace 2.0			FengShui 5.1			SCAMPI				
-HB$^+$ bench	HPWL (e+06)	ovlp (%)	time (s)	HPWL (e+06)	ovlp (%)	time (s)	HPWL (e+06)	ovlp (%)	time (s)	HPWL (e+06)	ovlp (%)	time (s)	HPWL (e+06)	ovlp (%)	time (s)	*vs.* PATOMA (HPWL)	*vs.* CAPO (HPWL)
01	3.9	0.0	5.6	5.4	1.4	651.5	2.7	2.7	68.0	3.0	0.2 ⊗	16.6	**3.2**	0.0	57.6	0.83x	–
02	×	×	×	19.1	0.0	1539.7	5.0	2.6	101.5	8.7	0.9 ⊗	43.6	**6.9**	0.0	185.4	–	0.36x
03	×	×	×	×	×	>1800	7.4	2.1	101.3	×	×	×	**10.1**	0.0	179.9	–	–
04	×	×	×	×	×	>1800	8.2	2.8	113.9	10.8	0.2 ⊗	41.4	**11.1**	0.0	145.8	–	–
06	×	×	×	×	×	>1800	8.2	1.0	122.5	10.7	1.4 ⊗	36.0	**9.3**	0.0	201.7	–	–
07	16.8	0.0	13.6	**15.8**	0.0	115.31	13.7	1.4	218.4	37.1	0.0	5.1	16.1	0.0	90.7	0.96x	1.02x
08	×	×	×	×	×	>1800	16.6	1.0 ⊗	294.2	21.8	0.5 ⊗	60.6	**18.8**	0.0	240.0	–	–
09	×	×	×	20.2	0.2	188.9	15.1	0.9	222.4	20.6	1.2 ⊗	42.9	**20.9**	0.0	185.7	–	–
10	×	×	×	45.9	2.7	263.7	39.9	0.4	544.8	×	×	×	**55.2**	0.0	319.9	–	–
11	**25.3**	0.0	49.2	28.1	0.0	140.5	24.5	1.1	270.3	30.4	0.2 ⊗	63.8	26.9	0.0	137.3	1.06x	0.96x
12	×	×	×	**63.4**	0.0	482.2	×	×	>1800	52.3	0.0 ⊗	39.2	64.0	0.0	397.6	-	1.01x
13	**37.5**	0.0	34.7	39.6	0.0	221.5	31.7	0.5	240.4	×	×	×	39.7	0.0	159.8	1.06x	1.00x
14	68.7	0.0	70.9	68.2	0.0	320.7	57.1	1.0 ⊗	392.9	74.0	2.7	89.7	**63.8**	0.0	238.8	0.93x	0.94x
15	×	×	×	×	×	>1800	87.5	1.5	422.2	90.6	0.0 ⊗	100.3	**86.4**	0.0	508.3	–	–
16	**100.3**	0.0	74.4	106.9	0.0	431.5	89.8	0.3	528.1	×	×	×	101.8	0.0	254.2	1.01x	0.95x
17	**141.4**	0.0	95.9	152.6	0.1	397.1	133.9	0.5	799.3	×	×	×	146.3	0.0	380.0	1.03x	–
18	**72.6**	0.0	67.2	75.9	0.7	220.1	69.1	0.6	344.0	×	×	×	74.7	0.0	181.9	1.03x	–
													Average			0.99x	0.85x

× indicates time-out, crash, or a run completed without producing a solution; ⊗ indicates an out-of-core solution

Best legal solutions are emphasized in bold.

5.10.3 ISPD Contest Benchmarks

The ISPD 2005 and 2006 Placement Contests introduced 16 new benchmarks into the public domain based on industrial designs. These designs have many movable objects, an abundance of fixed obstacles and relatively low utilizations. Tables 5.7 and 5.8 compare Capo's performance at the contests to Capo's current performance on the 2005 and 2006 contest benchmarks, respectively. Since the contests, Capo

Table 5.7. Comparison of Capo's current performance to that at the ISPD 2005 Placement Contest. Capo was run using the commandline options "-ispd05" and "-tryHarder." The results for Capo at the ISPD 2005 Placement Contest were the best placements produced over the period of 1 week. Current Capo results are the best of three independent runs of Capo.

Benchmark	ISPD 2005	Current		
	HPWL (e8)	HPWL (e8)	Runtime (m)	HPWL Ratio
adaptec1	-	0.863	95	-
adaptec2	0.997	1.001	128	1.004
adaptec3	-	2.340	274	-
adaptec4	2.113	2.071	257	0.980
bigblue1	1.082	1.071	152	0.990
bigblue2	1.723	1.624	291	0.943
bigblue3	3.826	4.006	984	1.047
bigblue4	10.988	9.470	1335	0.862
Average				0.969

Table 5.8. Comparison of Capo's current performance to that at the ISPD 2006 Placement Contest. "Overflow" represents the HPWL penalty for not effectively enforcing density constraints on the benchmarks. Results at the ISPD06 contest were the result of a single run of Capo. Current results are the median of three independent runs of Capo. Using the SCAMPI improvements, Capo's HPWL is reduced by 6.3% overall.

Benchmark	ISPD 2006			Current			
	HPWL (e8)	Over-flow%	Runtime (m)	HPWL (e8)	Over-flow%	Runtime (m)	HPWL Ratio
adaptec5	4.916	0.62	162	4.836	0.42	153	0.984
newblue1	0.984	0.13	43	0.850	0.12	47	0.864
newblue2	3.086	0.29	94	2.866	0.21	125	0.929
newblue3	3.612	0.01	101	3.299	0.01	92	0.913
newblue4	3.583	1.15	115	3.512	0.83	96	0.980
newblue5	6.574	0.33	348	6.391	0.26	212	0.972
newblue6	6.683	0.05	308	6.522	0.05	251	0.976
newblue7	15.185	0.02	916	13.482	0.01	525	0.888
Average							0.937

has been able to improve its solution quality by 3.1% on the ISPD 2005 benchmarks (while using considerably less runtime than the week allowed for the original contest) and 6.3% for the ISPD 2006 benchmarks.

Since the ISPD 2005 and 2006 contests, variants of the contest benchmarks have been proposed with known optimal or near-optimal wire lengths in order to gauge how much room for improvement is left with state-of-the-are placement methods. In the original work on placements with known optimal solutions, Capo 8.6 placements had nearly twice the HPWL of optimal placements [16]. As shown in Table 5.9, Capo placements are less than 60% from optimal which represents a significant

Table 5.9. Comparison of Capo's current performance on the PEKO-ISPD 2005 benchmarks to optimal results. Capo results are the best of three independent runs of Capo. These results represent an improvement in Capo's performance vs. optimal since the original work on placements with know optimal solutions where Capo placements had nearly twice optimal wire length [16].

Benchmark	Optimal HPWL (e8)	Capo HPWL (e8)	Capo Runtime (m)	Capo HPWL Ratio
adaptec1	0.201	0.301	35	1.498
adaptec2	0.250	0.401	42	1.604
adaptec3	0.410	0.657	376	1.602
adaptec4	0.394	0.578	455	1.467
bigblue1	0.209	0.296	51	1.416
bigblue2	0.423	0.664	141	1.570
bigblue3	0.944	1.898	321	2.011
bigblue4	1.714	2.533	889	1.478
Average				1.572

Table 5.10. Comparison of Capo's current performance on the PEKO-ISPD 2006 benchmarks to optimal results. Capo results are the best of three independent runs of Capo.

Benchmark	Optimal HPWL (e8)	Capo HPWL (e8)	Capo Over-flow%	Capo Runtime (m)	Capo HPWL Ratio
adaptec5	0.611	1.295	4.97	322	2.119
newblue1	0.195	0.563	1.53	29	2.887
newblue2	0.273	0.910	1.17	46	3.333
newblue3	0.303	1.210	1.72	136	3.993
newblue4	0.436	0.792	6.43	196	1.817
newblue5	0.858	1.679	6.33	615	1.957
newblue6	0.800	1.952	2.30	578	2.440
newblue7	1.510	4.196	2.11	1439	2.779
Average					2.580

improvement, especially on such challenging benchmarks as the ISPD 2005 contest benchmarks. The focus of the ISPD 2006 benchmarks is less on HPWL and more on satisfying the required density constraints, and on these benchmarks Capo achieves the density constraint requirements but, as Table 5.10 shows, at the expense of HPWL were Capo produces solutions with more than twice the optimal wire length on average.

5.11 Conclusions

In this chapter, we have described in detail the workings of the robust and scalable academic placement tool Capo. Capo is a min-cut floorplacer that provides (1) scalable multiway partitioning, (2) routable standard-cell placement, (3) integrated mixed-size placement, (4) wire length-driven fixed-outline floorplanning as well as (5) incremental placement. Capo produces best published results on several publicly available benchmark suites for routability as well as difficult instances of floorplacement. Capo has been used as part of Synplicity's Amplify ASIC product and is freely available for all uses as part of the UMpack (`http://vlsicad.eecs.umich.edu/BK/PDtools/`).

References

1. Adya SN, Chaturvedi S, Roy JA, Papa DA, Markov IL (2004) Unification of partitioning, placement and floorplanning. In Proc ICCAD 550–557
2. Adya SN, Markov IL (2003) Fixed-outline floorplanning: enabling hierarchical design. IEEE Trans on VLSI 11(6):1120–1135
3. Adya SN, Markov IL (2005) Combinatorial techniques for mixed-size placement. ACM Trans on Design Auto of Elec Sys 10(5)
4. Adya SN, Markov IL, Villarrubia PG (2006) On whitespace and stability in physical synthesis. Integration: the VLSI J 25(4):340–362
5. Agnihotri A et al. (2003) Fractional cut: improved recursive bisection placement. In Proc ICCAD 307–310
6. Alpert CJ, Nam G-J, Villarrubia PG, (2003) Effective free space management for cut-based placement via analytical constraint generation. IEEE Trans on CAD 22(10):1343–1353
7. Brenner U, Vygen J (2000) Faster optimal single-row placement with fixed ordering. In Proc DATE 117–121
8. Brenner U, Rohe A (2003) An effective congestion driven placement framework. IEEE Trans. on CAD 22(4):387–394
9. Caldwell AE, Kahng AB, Mantik S, Markov IL, Zelikovsky A (1999) On wirelength estimations for row-based placement. IEEE Trans on CAD 18(9):1265–1278
10. Caldwell AE, Kahng AB, Markov IL (2000) Improved algorithms for hypergraph bipartitioning. In Proc ASPDAC 661–666
11. Caldwell AE, Kahng AB, Markov IL (2000) Can recursive bisection alone produce routable placements? In Proc DAC 477–482

12. Caldwell AE, Kahng AB, Markov IL (2000) Design and implementation of move-based heuristics for vlsi hypergraph partitioning. ACM J of Experimental Algorithms 5
13. Caldwell AE, Kahng AB, Markov IL (2000) Optimal partitioners and end-case placers for standard-cell layout. IEEE Trans on CAD 19(11):1304–1314
14. Caldwell AE, Kahng AB, Markov IL. VLSI cad bookshelf. `http://vlsicad.eecs.umich.edu/BK/`. See also Caldwell AE, Kahng AB, Markov IL (2002) Toward cad-ip reuse: the marco gsrc bookshelf of fundamental cad algorithms. IEEE Design and Test 72–81
15. Caldwell AE, Kahng AB, Markov IL (2003) Hierarchical whitespace allocation in top-down placement. IEEE Trans on CAD 22(11):716–724
16. Chang C-C, Cong J, Romesis M, Xie M (2004) Optimality and scalability study of existing placement algorithms. IEEE Trans on CAD 23(4):537–549
17. Chen TC, Chang YW, Lin SC (2005) IMF: interconnect-driven multilevel floorplanning for large-scale building-module designs. In Proc ICCAD 159–164
18. Fiduccia CM, Mattheyses RM (1982) A linear-time heuristic for improving network partitions. In Proc DAC 175–181
19. Goldberg AV (1997) An efficient implementation of a scaling minimum-cost flow algorithm. ACM J. Algorithms 22:1–29
20. IWLS 2005 Benchmarks, `http://iwls.org/iwls2005/benchmarks.html`
21. Kahng AB, Mantik S, Markov IL, (2002) Min–max placement For large-scale timing optimization. In Proc ISPD 143–148
22. Kahng AB, Mandoiu II, Zelikovsky A (2003) Highly Scalable Algorithms for rectilinear and octilinear steiner trees. In Proc ASPDAC 827–833
23. Kahng AB, Wang Q (2005) Implementation and extensibility of an analytic placer. IEEE Trans on CAD 25(5):734–747
24. Kahng AB, Reda S (2004) Placement feedback: a concept and method for better min-cut placement. In Proc DAC 143–148
25. Karypis G, Aggarwal R, Kumar V, Shekhar S (1997) Multilevel hypergraph partitioning: applications in vlsi domain. In Proc DAC 526–629
26. Khatkhate A, Li C, Agnihotri AR, Yildiz MC, Ono S, Koh C-K, Madden PH (2004). Recursive bisection based mixed block placement. In Proc ISPD 84–89
27. Li C, Xie M, Koh C-K, Cong J, Madden PH (2004) Routability-driven placement and whitespace allocation. In Proc ICCAD 394–401
28. Li C, Koh C-K, Madden PH (2005) Floorplan management: incremental placement for gate sizing and buffer insertion. In Proc ASPDAC 349–354
29. Moffitt MD, Ng AN, Markov IL, Pollack ME (2006) Constraint-driven floorplan repair. In Proc DAC 1103–1108
30. Nam G-J, Alpert CJ, Villarrubia P, Winter B, Yildiz M (2005) The ISPD 2005 placement contest and benchmark suite. In Proc ISPD 216–220
31. Ng AN, Markov IL, Aggarwal R, Ramachandran V (2006) Solving hard instances of floorplacement. In Proc ISPD 170–177
32. Papa DA, Adya SN, Markov IL (2004) Constructive benchmarking for placement. In Proc GLSVLSI 113–118 `http://vlsicad.eecs.umich.edu/BK/FEATURE/`
33. Reda S, Chowdhary A (2006) Effective linear programming based placement methods. In Proc ISPD 186–191
34. Roy JA, Adya SN, Papa DA, Markov IL (2006) Min-cut floorplacement. IEEE Trans on CAD 25(7):1313–1326
35. Roy JA, Markov IL (2007) Seeing the forest and the trees: steiner wirelength optimization in placement. To appear in IEEE Trans on CAD

36. Roy JA, Markov IL (2007) ECO-system: embracing the change in placement. To appear IEEE Trans on CAD
37. Roy JA, Papa DA, Ng AN, Markov IL (2006) Satisfying whitespace requirements in top-down placement. In Proc ISPD 206–208
38. Tang X, Tian R, Wong MDF (2005) Optimal redistribution of whitespace for wirelength minimization. In Proc ASPDAC 412–417
39. Westra J, Bartels C, Groeneveld P (2004) Probabilistic congestion prediction. In Proc ISPD 204–209
40. Yang X, Choi B-K, Sarrafzadeh M (2002) Routability driven whitespace allocation for fixed-die standard-cell placement. IEEE Trans on CAD 22(4):410–419

6

Congestion Minimization in Modern Placement Circuits

Taraneh Taghavi[1], Xiaojian Yang[2], Bo-Kyung Choi[3], Maogang Wang[4] and Majid Sarrafzadeh[1]
[1]Computer Science Department, UCLA, {taghavi, majid@cs.ucla.edu}@cs.ucla.edu
[2]Magma Design Automation Inc., {bkchoi}@magma-da.com
[3]Synplicity Inc., {xiaojian}@synplicity.com
[4]Blaze DFM, Inc., {mwang}@blaze-dfm.com

6.1 Introduction

In this chapter, we propose a placement tool called Dragon which deploys hierarchical techniques to place large-scale mixed size designs that may contain thousand of macro blocks and millions of standard cells [1–3]. Min-cut-based top-down approach is taken to handle the large complexity of designs and simulated annealing is used to minimize the total wire length. Min-cut partitioning should be aware of large macro cells and may result in bins with different sizes. During simulated annealing, different bin sizes have to be considered. The techniques discussed in this work can be easily incorporated into any hierarchical placement flow and effectively produce legal final layouts with a short runtime.

As VLSI circuits are growing in both size and complexity, not only the half-perimeter wire length but also *congestion* need to be emphasized at the placement stage. Congestion is one of the main optimization objectives in global routing. However, the optimization performance is constrained because the cells are already fixed at this stage. A highly congested region in the placement often leads to routing detors around the region, in turn results in a larger routed wire length. Congested areas can also downgrade the performance of global router and, in the worst case, create an unroutable placement in the *fix-die* regime [12].

Congestion can be modeled as the summation of linear [15] or quadratic [10] function of difference between *routing demand* and *routing resource*. Existing congestion reduction techniques include incorporating congestion into cost function of simulated annealing [10], combining a regional router into placement tool [26] and performing a post placement processing step [15, 16]. While congestion reduction at late or post placement stage is empirically effective, congestion estimate achieved at early placement stages would be equally valuable. First, a congestion driven placement tool guided by early congestion information might be more powerful. Such a tool could use techniques like white space allocation to relieve layout congestion. Second, early congestion estimates could be utilized by combined logic and layout optimization to improve design convergence. For example, when logic designers are given a number of different netlists, they can estimate the congestion by running only several steps of placement. The netlists with bad estimated congestion will be discarded much earlier.

The use of white space in fixed-die standard-cell placement is an effective way to improve routability. In this work, we present a white space allocation approach that dynamically assigns white space according to the congestion distribution of the placement. Experimental results show that the proposed allocation approach, combined with a multilevel placement flow, significantly improves placement routability and layout quality.

Target utilization is another metric to estimate the capability of a placement tool to generate routable designs. Target utilization is defined as a constraint to force the placement tool to produce designs with utilization close to the design utilization which is an inherit feature of the design.

To tackle the problem of controlling the target utilization, we incorporated two methods into our placement tool, Dragon. Dragon beats all the other contestants with quite a big margin on generating placements with very low penalty of violating the design utilization in ISPD2006 Placement Design Contest which was held by International Symposium on Physical Design in conjunction with IBM Corporation.

6.2 Overview of Dragon [1]

A typical top-down hierarchical placement approach can be generalized as follows: At a given hierarchical level, the layout area is partitioned into several global bins. All the cells of the circuit are distributed into these global bins to minimize a certain placement objective. This cell distribution problem is called a hierarchical placement problem. If a cell is distributed into a particular global bin, it will be placed within the area of this bin in the final layout. As we proceed to more refined levels, the number of global bins increases and the physical size of global bins decreases. Thus we can get more and more detailed information about physical locations of cells as we proceed. The top-down approach terminates when there are only a few cells in each global bin. Dragon is divided into two phases, global placement (GP) and detailed placement (DP). A top-down hierarchical approach is used in the GP phase. We recursively solve the hierarchical placement problem and quadrisect each

[1] Portions reprinted, with permission, from [1–3, 5]. © [2006, 2005, 2001, 2000] IEEE.

global bins at each level. Overlap between cells are allowed in the GP phase. The DP phase takes the output from GP and produces an overlap free layout. Then it iteratively improves the legal layout using a greedy heuristic. Due to the computational complexity, the DP heuristic is only capable of performing optimization locally. Thus it is expected that the top-down hierarchical GP phase should finish the majority of work in placement.

6.2.1 Framework of Dragon

Figure 6.1 shows our placement flow. The circuit is recursively partitioned alternatively along horizontal and vertical cut lines. The subcircuits after partitioning are assigned to rectangular bins. At some points a bin-based simulated annealing where the objects that are moved are the subcircuits in the bins, is performed to improve the current placement. Such a procedure terminates when a certain stop criteria (e.g. average number of cells per bin is less than a given number) is met. An adjustment step is then executed to fit the current bin-based placement into row structures. The next step is a cell-based simulated annealing. The bin structure still exists and the cells are moved between the centers of bins. The locations of these centers can be changed during the annealing procedure. The final step simply spreads overlapped cells and makes local improvements to obtain the detailed placement.

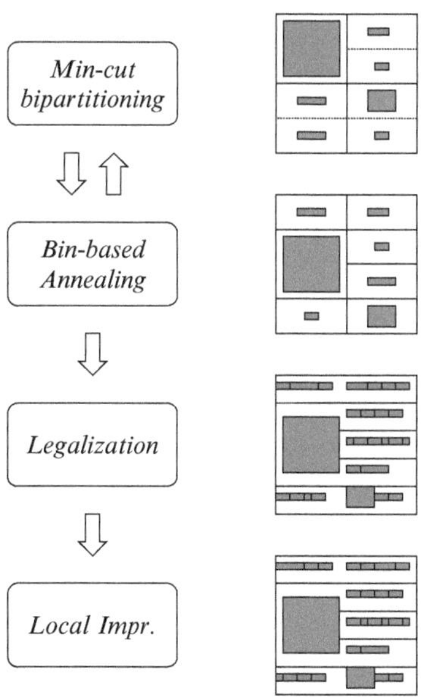

Fig. 6.1. Overall flow of proposed mixed-size placement.

Partitioning

To handle the high complexity of the problem, the input netlist is recursively divided into two partitions using a state-of-the-art min-cut partitioner, hMetis [23]. Two things have to be considered during partitioning. One is the number of cuts across the partitions and the other is the balance in the sizes of two partitioned sets. hMetis is shown to be able to get very good solutions in terms of both the cutsize and the balance [27].

Simulated Annealing

A weakness of pure min-cut type placement is its irreversibility. Once a cell is assigned to one side of the cut line, it will never move to the other side to improve the placement. Combining simulated annealing in this flow helps placements move out of the local minima. We use multilevel simulated annealing in this placement flow. The key idea is to reduce the number of movable objectives in annealing. The difference between our flow and hierarchical annealing is instead of using a single cooling schedule through three hierarchical levels, we use low temperature annealing at each level and do not fix the number of levels. Moreover, we avoid using simulated annealing at the final placement stage and use a fast greedy improvement instead. Both bin annealing and cell annealing use total wire length as the cost function. Also they adopt the same cooling schedule. Swapping is the main move in both types of annealing, and shifting is used a little bit in cell annealing. The disadvantage of simulated annealing is its expensive runtime cost. Although our flow tries to reduce this cost by bin-based approach, annealing is still the most time consuming part.

6.3 Mixed-Size Placement [2]

In this section, we propose hierarchical techniques to place large-scale mixed size designs that may contain thousand of macro blocks and millions of standard cells. Min-cut-based top-down approach is taken to handle the large complexity of designs and simulated annealing is used to minimize the total wire length. Min-cut partitioning should be aware of large macro cells and may result in different size bins. During simulated annealing, different bin sizes have to be considered.

To build a mixed-size placement tool that can handle macro cells as well as standard cells, we follow the basic flow a min-cut and simulated annealing-based placer [3], that is believed to be very successful [28], and add functionalities that can resolve the problems caused by the presence of macro cells of various sizes.

The input circuit netlist is recursively bipartitioned, and bin-based simulated annealing is done after each bipartitioning. Once the number of cells in a bin is less than some threshold, we stop partitioning and annealing processes and the detailed placement starts. During detailed placement, legalization has to be done to remove overlaps between the cells. After legalization, a greedy local improvement step is performed by switching adjacent standard cells or flipping them.

[2] Portions reprinted, with permission, from [1–3,5]. © [2006, 2005, 2001, 2000] IEEE.

6.3.1 Macro-Aware Partitioning

To handle the high complexity of the problem, the input netlist is recursively divided into two partitions using a state-of-the-art min-cut partitioner, hMetis [23]. Two things have to be considered during partitioning. One is the number of cuts going across the partitions and the other is the balance in the sizes of two partitioned sets. hMetis is shown to be able to get very good solutions in terms of both the cutsize and the balance [27].

Min-cut-based hierarchical approaches run into trouble in mixed-size placement, when there is a large macro cell that is bigger than the bin size at a certain hierarchical level. Figure 6.2 illustrates this problem. In Figure 6.2 (a), we are trying to vertically cut the bin. The size of both sub-bins have to be equal to have a regular bin structure. However, the macro is too large to fit into any of the sub-bins, even though the actual area of the macro is equal to the half of the bin are that is being cut. Since each cell has to be assigned to only one bin, we have to put the macro either the left of the right sub-bin. If we put the macro into the one sub-bin and the rest standard cells into the other sub-bin, the resulting layout will be extremely illegal. Figure 6.2 (b) shows a possible placement solution, which can never be obtained by traditional min-cut-based approaches.

In order to deal with macro cells, we give up the regularity of bin structure, meaning that bins can have different sizes. For example, we are now vertically partitioning the bin in Figure 6.3. By scanning all the cells in the bin, we can identify the largest cell in the bin that has the width which is larger than the half of the current bin width. We cannot partition this bin into equal size sub-bins. Instead, we do unbalanced partitioning. Let α and β be the ratio of the macro width and height, to the bin width and height, respectively. (In case of horizontal partitioning, α and β are defined conversely.)

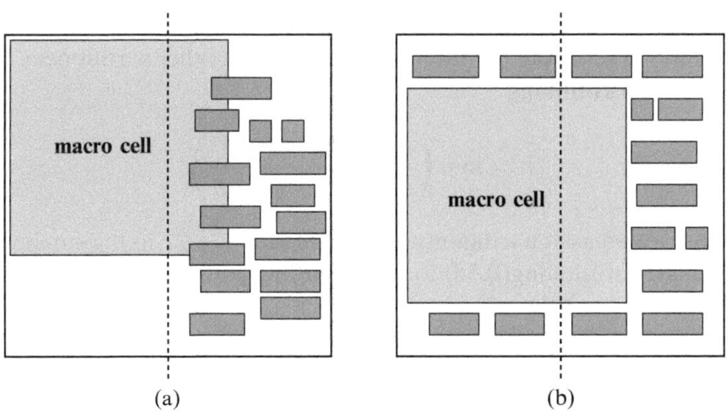

Fig. 6.2. (**a**) A macro cell is too large to fit into a sub-bin. (**b**) Possible placement solution for the bin, which can never be obtained by traditional approaches.

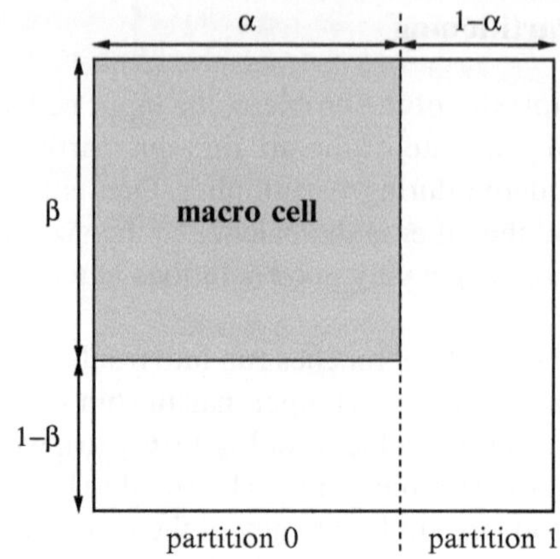

Fig. 6.3. Partitioning when a bin contains a large macro.

$$\alpha = \frac{\text{width of macro}}{\text{width of bin}} \quad (0 < \alpha < 1)$$
$$\beta = \frac{\text{height of macro}}{\text{height of bin}} \quad (0 < \beta < 1)$$

Let us denote the partition that the macro will be assigned to as partition 0 and the other as partition 1. The ratio in partition sizes in terms of area will be $\alpha(1-\beta)$: $(1-\alpha)$, excluding the macro. Therefore, we perform min-cut partitioning such that the ratio of areas is

$$\alpha(1-\beta) : (1-\alpha) = \frac{\alpha(1-\beta)}{1-\alpha\beta} : \frac{1-\alpha}{1-\alpha\beta}$$

and assign the resulting partitions to the sub-bins.

We use hMetis [23] as the partitioner. hMetis does not support the function to control the ratio of resulting partitions, neither do most other partitioners. In order to make $r : (1-r)$ partitioning

$$r = \min\left(\frac{\alpha(1-\beta)}{1-\alpha\beta}, \frac{1-\alpha}{1-\alpha\beta}\right)$$

using hMetis, we preassign a dummy node with area $1-2r$ to the smaller partition, and do balanced partitioning(0.5:0.5). The resulting partitions excluding the dummy node will be $r : (1-r)$.

If there are more than one macros in the bin being partitioned, we preassign macros so that they can fit in the sub-bins they will belong to, and perform partitioning for the rest of standard cells. If a macro can fit in neither of the sub-bins, it is preassigned to a sub-bin that minimizes the violation. When a bin contains only one cell/macro, the bin is no longer partitioned but still can move around during simulated annealing to minimize wire length.

6.3.2 Bin-Based Simulated Annealing

After each bipartition, bin-based simulated annealing takes place to find a good location for each partition to be placed in, minimizing the total wire length.

Because the bin structure is irregular due to unbalance partitioning, we have to take care of different bin sizes during simulated annealing. There are three types of moves in bin-based simulated annealing: horizontal switch, vertical switch, and diagonal switch. These moves switch two adjacent bins. If the bin structure is regular, we can freely choose any type of move. However, we now have constraints for these moves: Diagonal switches are allowed only when the two bins have the same size (both width and height), vertical switches are allowed only when the widths of both bins are the same, and horizontal switches occur only when the heights are the same. The moves that do not satisfy these constraints are automatically rejected.

When we accept a move, the size and the position of bins have to be updated accordingly to keep the bin structure correct. Figure 6.4 shows an example of horizontal switching. We keep the center position, the width and the height of bins to maintain the irregular bin structure. The coordinates of the center and the widths have to be updated as

$$x'_0 = x_0 - w_0/2 + w_1/2$$
$$x'_1 = x_1 + w_1/2 - w_0/2$$
$$w'_0 = w_1$$
$$w'_1 = w_0$$

(assuming that $h_0 = h_1, y_0 = y_1, x_0 < x_1$).

By restricted bin-based simulated annealing, macro cells can still move around to find a better location to improve the quality. Also, the solution space is limited because of the constraints, resulting in speed-up. The speed-up of simulated annealing by limiting the moves within some boundary is well discussed in [29].

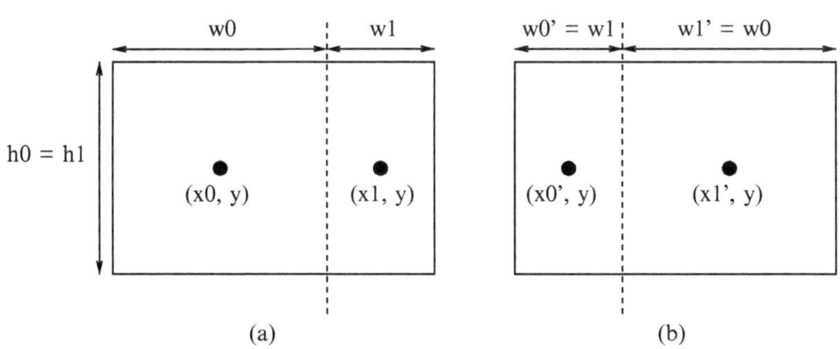

Fig. 6.4. Example of horizontal switch in irregular bin structure.

6.3.3 Legalization

Once average number of cells in a bin gets less than a certain number, recursive partitioning and simulated annealing is stopped and we proceed to the detailed placement step.

First, overlaps between cells have to be resolved to get a legal placement. This step is called legalization. Without macro cells, this stage is very simple. All that has to be done is just to place cells next to each other in a row from left to right. However, with the presence of macro cells that spans through multiple rows, the problem is no longer straightforward. Since we have tried to put macro cells inside bin boundaries during the previous steps, we expect that macro cell do not cause much problem, but simply placing the cells next to each other will easily end up with having cells outside of the chip boundary. When placing cell to remove overlaps, we have to consider two factors: The degradation in placement quality and the legality of result, which usually conflict with each other. To address this conflict, we use a cost function for placing each cell that combines the two factors. For macro cells, the cost function of putting cell c into row r is :

$$\text{cost}(c,r) = \alpha \cdot (\text{position offset}) + (1-\alpha) \cdot x_{\text{final}}$$

and for standard cells

$$\text{cost}(c,r) = \alpha \cdot (\text{wirelength change}) + (1-\alpha) \cdot x_{\text{final}}$$

where position offset is the distance from the original position to the final position, *wire length change* is the change in wire length caused by moving the cell (that can be negative when wire length decreases), and x_{final} is the x-coordinate of the cell after legalization. α is a coefficient to control the importance of each term. By selecting the row that minimizes $cost(c,r)$, we try to minimize both the wire length (displacement in case of macros) and x-coordinate of cells. Minimizing x_{final} helps increase the chance to get a legal solution, leaving room for the rest of cells to be placed.

6.4 Congestion Estimation [3]

In this section we are going to estimate both peak congestion and congestion distribution at early top-down placement stages. Specifically, we quantitatively estimate the maximum congestion prior to placement stage. Also we give a congestion distribution picture of the chip layout at coarse levels of hierarchical placement flow. Both estimates are made based on Rent's rule – a well-known stochastic model for real circuits.

---

[3] Portions reprinted, with permission, from [5,11]. © [2001, 2000] IEEE.

Congestion is a function of routing demand and routing resource. Once the technology feature and chip characteristics (die size, number of layers, position of preplaced macros) are fixed, the routing resource is roughly determined [4] Congestion and routing demand are so closely related that it is straightforward to convert one to the other. In this work, we will focus on the estimation for routing demand.

During the global routing, the chip is divided into *bins*. The bin is small enough that each placement region covers an integral number of bins. All the nets will be routed by connecting the cells of each net using grid wires. For each *boundary* of the bins b, the *routing demand* $d(b)$ is the number of wires that cross this boundary; the *routing supply* $s(b)$ is the number of wires that are allowed to cross the boundary. The *overflow* of a boundary $c(b)$ is $\max(d(b) - s(b), 0)$. The congestion of a placement region is the summation of the overflow over all the boundaries within this placement region. The *peak congestion* of a placement is the maximum overflow over all the bin boundaries.

6.4.1 Rent's Rule

Rent's rule is an empirical observation first described by Landman and Russo [17]. It states the relationship between the number of elementary blocks G in a subcircuit of a partitioned design, and the number of external connections T of the subcircuit. Specifically,

$$T = tG^p \tag{6.1}$$

where t is the average number of interconnections per block, and p is the Rent exponent ($0.4 < p < 0.8$ in real circuits). The Rent exponent p can be computed by plotting the T vs. G relation in a log–log diagram for every value of G in a top-down partitioning process, and then fitting a line on the plotted points. The slope p of this line represents the Rent exponent.

Rent's rule has been widely used to estimate interconnect wire length [18,20,21]. In general, a higher Rent exponent will result in a longer average wire length, which in turn implies a larger wiring area and more congested layout [22].

6.4.2 Peak Congestion Analysis

Cut Ratio in Recursive Bipartitioning

In order to analyze peak congestion over all the bin boundaries of the layout, we assume that the circuit is an *ideal circuit* which strictly obeys Rent's rule. This ideal circuit is placed using a hierarchical placement flow which is based on recursively bipartitioning. On each hierarchical level of the top-down placement, each subcircuit

---

[4] Accurate available routing resource can only be obtained after placement and global routing with the consideration of the layer area occupied by placed cells and the number of routing layers.

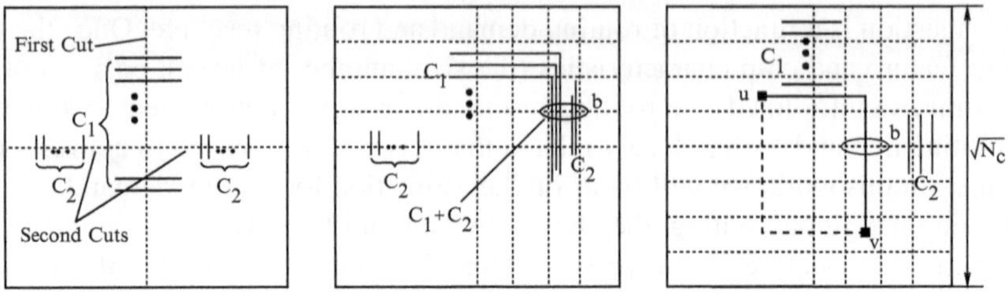

(a) (b) (c)

Fig. 6.5. Relationship between net cut and congestion. (**a**) Recursive bipartitioning and cut. (**b**) Worst case routing demand analysis. (**c**) Average case routing demand analysis.

is quadrisectioned into four smaller subcircuits. A quadrisection step consists of a vertical bipartitioning followed by a horizontal one [5].

Let C_1 be the net cut of the first bipartitioning by a vertical cut line. Let $C_{2,1}$ and $C_{2,2}$ be the net cuts of the second horizontal bipartitioning (Figure 6.5(a)). Similarly, the net cuts of the ith bipartitioning are $C_{i,1}, \ldots, C_{i,2^{i-1}}$. For an ideal circuit,

$$C_{i,1} \approx C_{i,2} \approx C_{i,3} \approx \ldots \quad \text{for } i = 1, \ldots 2H$$

where H is the number of hierarchical levels in the top-down recursive quadrisection placement. For simplicity, all the cuts on the same level i are denoted as C_i for $i = 1, \ldots 2H$.

Theorem 6.1. *In a recursive bipartitioning approach on an ideal circuit, the ratio between the net cut of the* $(i + 1)$*th bipartitioning* C_{i+1} *and the net cut of the* i*th bipartitioning* C_i *is* 2^{-p}, *where* p *is the Rent exponent of the circuit.*

Proof. Consider the subcircuits to be bipartitioned at each hierarchical level. Let G_i be the size of subcircuit at the ith level. Thus the size of subcircuit at the $(i + 1)$th bipartitioning is $G_{i+1} = G_i/2$. For and ideal circuit, all the subcircuits have the same Rent exponent p and Rent coefficient t. According to equation (6.1), the number of external interconnects for subcircuits at ith level is $T_i = tG_i^p$, and the number of external interconnects for subcircuits at $(i + 1)$th level is $T_{i+1} = tG_{i+1}^p$. At a given level i, the interconnects between two subcircuits that are split from one of the ith bipartitionings, are a subset of the external interconnects of each of these two subcircuits. Let k $(0 < k < 1)$ be the ratio between the number of cut nets of the ith bipartitioning and the number of external nets for subcircuit at ith level. We have $C_i = kT_i$. For an ideal circuit, assuming k is fixed through all the hierarchical levels, we have,

$$\frac{C_{i+1}}{C_i} = \frac{kT_{i+1}}{kT_i} = \frac{kt(G_{i+1})^p}{kt(G_i)^p} = \frac{(G_i/2)^p}{(G_i)^p} = 2^{-p}$$

□

Worst Case Analysis

The top-down placement flow terminates at the $H = \log_4 N_c$ level where N_c is the number of cells of the circuit. In the final placement each cell occupies one bin.[5] The global router uses L-shape routing model, in which a net is routed using either the upper or the lower part of the bounding box of this net. This is not a good routing method but it gives a general picture of wire distribution.

Now we want to find out the maximum routing demand over all the bin boundaries without placing the circuit. First we discuss the worst case. Let us denote the maximum routing demand of a bin boundary as C_{max}.

Theorem 6.2. *In a recursive bipartitioning approach on an ideal circuit, the maximum routing demand over all the bin boundaries*

$$C_{max} < C_1 \frac{1 - \alpha^{2H}}{1 - \alpha}$$

where C_1 is the net cut of the first bipartitioning and $\alpha = C_{i+1}/C_i$ is the ratio between net cuts of two consecutive partitionings.

Proof. In Figure 6.5(b), the circuit is partitioned into two parts with a net cut C_1. It follows that there are C_1 nets crossing between left half and right half. Let us look at a bin boundary located at the right half. In the worst case, all these C_1 nets pass this specific boundary. Hence the first bipartitioning contributes C_1 to the routing demand of this boundary. Similarly, the ith bipartitioning contributes C_i to the routing demand. Thus, for any boundary, the upper bound of the routing demand is

$$\sum_{i=1}^{2H} C_i = C_1 \sum_{i=0}^{2H-1} \alpha^i$$
$$= C_1 \frac{1 - \alpha^{2H}}{1 - \alpha}$$

□

Uniform Distribution of Cut Nets

In the previous discussion we assume that all the nets that are cut in a bipartitioning cross a particular bin boundary. That is, obviously, not the general case. However, once we construct a framework like the model in the earlier subsection, we can study the congestion behavior using different cut net distribution models.

We continue the analysis using a *uniform distribution model*, in which the cut nets of a bipartitioning are uniformly distributed over all the subcircuit area. In other words, the cells in the partitioned subcircuit have equal probabilities to be connected to a cut net.

[5] A bin has unit width and height.

Theorem 6.3. *In a recursive bipartitioning approach on an ideal circuit, assuming cut nets are uniformly distributed, the expected maximum routing demand over all the bin boundaries is,*

$$C_{\max} = \frac{C_1}{\sqrt{N_c}}\left(\frac{1}{2} + 2\alpha\right)\frac{\sqrt{N_c}\alpha^{2H} - 1}{2\alpha^2 - 1}$$

where C_1 is the net cut of the first bipartitioning and $\alpha = C_{i+1}/C_i$ is the ratio between net cuts of two consecutive partitioning operations.

Proof. In Figure 6.5(c), the first bipartitioning result C_1 means that there are C_1 nets connecting the left half and right half of the design. We know that the number of final bins on either half is $N_c/2$. Since the cut nets are uniformly distributed, for each final bin, the average number of cut nets connected to this bin is $2C_1/N_c$. Among all the horizontal bin boundaries, the ones on the center horizontal line of the chip accommodate the maximum number of net crossing caused by those C_1 nets. Note that for a specific horizontal bin boundary b at the center line, only the nets which connect to the bins at the same column could cross b. This is because we are using the L-shape routing model. There are $\sqrt{N_c}$ bins at the same column with boundary b. For each of them, if the connected bin in the other half is located at the different upper/lower part (half of the connections of this bin have this property), like bin v and u in Figure 6.5(c), the probability that this route crosses boundary b is 1/2 (because of L-shape routing). Therefore the first cut contributes on average $C_1\sqrt{N_c}/(2N_c)$ crossings to b. The case for the second bipartitioning is relatively simple. Since there are C_2 nets crossing $\sqrt{N_c}/2$ boundaries, on average each boundary is crossed by $C_2/(\sqrt{N_c}/2)$ nets. Using the same approach for the third, fourth, etc. bipartitioning, we have the expected maximum number of crossings over all the boundaries:

$$C_{\max} = \frac{1}{2}\frac{C_1}{\sqrt{N_c}} + 2\frac{C_2}{\sqrt{N_c}} + \frac{C_3}{\sqrt{N_c}} + 4\frac{C_4}{\sqrt{N_c}} + \cdots \tag{6.2}$$

According to Theorem 6.1 we have $C_i = C_1\alpha^{i-1}$, plug it into (6.2), we obtain

$$\begin{aligned} C_{\max} &= \frac{C_1}{\sqrt{N_c}}\left(\frac{1}{2} + 2\alpha\right)\sum_{i=0}^{H-1}(2\alpha)^{2i} \\ &= \frac{C_1}{\sqrt{N_c}}\left(\frac{1}{2} + 2\alpha\right)\frac{\sqrt{N_c}\alpha^{2H} - 1}{2\alpha^2 - 1} \end{aligned}$$

□

6.4.3 Regional Congestion Estimation

In Sect. 6.4.2 we have discussed the peak congestion estimation problem. Another estimation requirement, regional congestion estimation, appears at early placement stages. In this section, we propose a routing demand estimation approach in the context of top-down placement.

Definition 1 *For a given region r in a globally routed design, the routing demand $D(r)$ is the summation of the number of net crossings over all the bin boundaries within region r.*

Note that the *region* in the definition is a common term in top-down placement. It is called *placement region* in [12] or *global bin* in [3]. A region contains a number of adjacent global routing bins. Estimating the routing demand for all the regions during the placement will give us a rough congestion map, which is valuable for early design evaluation.

Given a placement region, the nets which cause the edge crossings can be classified into two types: the *internal nets* which connect cells within this region and the *external nets* which span toward other regions or cross this region while connecting no cells. Thus we give the following terms:

Definition 2 *For a given region r in a design, the internal routing demand $ID(r)$ is the summation of the number of crossings caused by internal nets for all bin boundaries; the external routing demand $ED(r)$ is the summation of the number of crossings caused by external nets for all bin boundaries.*

The total routing demand $D(r)$ for a region r can be calculated by:

$$D(r) = ID(r) + ED(r)$$

Figure 6.6 shows the concepts of internal routing demand and external routing demand at a top-down placement stage. The original circuit is divided into subcircuits and each subcircuit is assigned into a region. The dashed lines are internal nets. The thicker, solid lines represent external nets. The routing demand in a region consists of two parts: net crossings caused by the internal nets and those by the external nets.

At the very coarse placement stage, the subcircuits are loosely coupled, i.e. the number of external nets is much smaller than the number of internal nets. The routing demand of a region is primarily determined by the interconnect complexity of the subcircuit which belongs to the region. As the top-down placement flow goes into

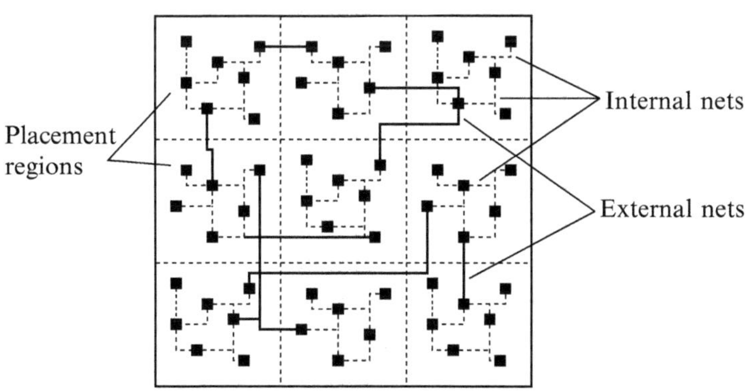

Fig. 6.6. Internal and external routing demand.

deeper levels, the routing demand of a placement region is determined by not only the internal complexity of the subcircuit in this region, but also the geometrical locations of other subcircuits and the interconnects between them.

Internal Routing Demand

In a typical top-down placement scheme, e.g., min-cut placement, the cells of a partitioned subcircuit will eventually be placed within the area that is assigned for this subcircuit. Therefore, estimating the internal routing demand becomes feasible. For a certain region in a top-down placement, the internal routing demand is proportional to the total routed wire length after global routing [15]. [6]

The wire length estimation problem has been studied for many years. There are several successful estimation techniques based on Rent's rule: Donath's classical method [18], its extension [20] and a more recent model [21]. In these methods the wire length distribution of the entire design is predicted before place and route.

Most of the research done on wire length estimation is based on regularly placed circuits such as standard cell designs. With the trend toward IP-block-based design, macro cells as blockage (sometimes referred to as obstacle), are more likely to be present in the circuit. The presence of the IP blocks may significantly increase wire length and cause congestion [19]. Since the presence of blockage makes the traditional wire length estimations far from reality, new techniques should be derived to address the problem of wire length estimation. In this section, starting from Donath's hierarchical technique [18] , his approach is extended to be able to consider obstacles in the placement area.

Similar to [18], our technique to estimate the average wire length is based on a top-down hierarchical placement of the circuit into a square Manhattan grid in the presence of obstacles [4, 6]. The circuit is partitioned hierarchically into four subcircuits. This hierarchical partitioning is continued until the number of the standard cells in all of the subcircuits is equal or less than β, where β is a predefined constant. At each level of hierarchy, we deduce the average number n_h of interconnections and the average length L_h of interconnections between each two subcircuits belonging to the same $h+1$ level of hierarchy, but different h level of hierarchy. Given the above model for the circuit, the feature parameter of the circuit P which is given by Rent's rule, and the above partitioning scheme, we want to estimate the total interconnection length of the circuit in the presence of obstacles. This is done by calculating the average number of interconnections n_h and the average length of the interconnections L_h at every hierarchical level h. The total interconnection length over all hierarchical levels is then obtained from

$$L = \sum_{h=0}^{H} n_h L_h \tag{6.3}$$

where H is the finest level of hierarchy. Since at every step of partitioning, each subcircuit is divided by four, and in the last level of hierarchy the number of cells

[6] We assume that the global routing bins are square.

inside each subcircuit is less than the factor β, the number of levels can be calculated from

$$H = \log_4\left(\frac{C}{\beta}\right)$$

The average number of interconnections between the subcircuits in each level of hierarchy is extracted based on Rent's exponent which is experimentally proven to be a good indicator of the complexity of the circuit. Using a similar type of analysis as [19], the average length of interconnection between the subcircuits is calculated in each level of hierarchy. Then, using formula 6.3, we estimate the total wire length by multiplying the average number of interconnections by the average length of interconnections for each level of hierarchy and summing out all these values over all the hierarchical levels.

The average number of interconnections at each hierarchical level can be calculated using Rent's rule as in [18] to be

$$n_h = \alpha AC(1 - 4^{p-1})4^{L(p-1)} \tag{6.4}$$

The Average Length of Interconnections at Each Level of Hierarchy

To start analyzing average wire length at each level of hierarchy, we need to define some terminologies and deploy some assumptions at first.

Assumption 1 *To compute L_h we assume that all of the nets have two terminals. This simplification is based on the knowledge that these nets are much more than all the other nets in the circuit and that multiterminal nets can be modeled as a collection of two-terminal nets.*

The effect of multiterminal nets is incorporated into our estimation by using higher values for in the calculation of the average number of interconnections n_h as shown in the earlier section.

Assumption 2 *We assume that the available routing layers are such that the blockages are obstructions for both placement and routing. This model is based on what commercial tools support for placement and routing of large-scale circuits.*

Definition 3 *In level h of hierarchy an intrabin wire is the wire that its terminals belong to the same bin, i.e. same part of the chip area.*

Definition 4 *An interbin wire is the wire that its terminals belong to different bins in level h of hierarchy, but to one bin in the level $(h + 1)$ of hierarchy.*

Definition 5 *In the presence of the obstacles, the transparent-block wire length, L_{TB}, is defined as the wire length when the obstacle is assumed to be transparent and wires can pass through it. For two-terminal nets, transparent-block wire length is the Manhattan distance between them.*

Definition 6 *Detor wire length, L_{DT}, is the detor length needed in a routing wire in the presence of the obstacles.*

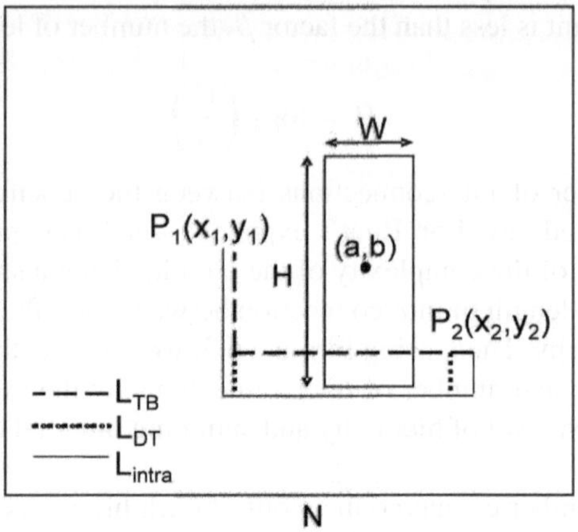

Fig. 6.7. Definition of intrabin transparent-block and detor wire length.

In other words, $L_{\text{TB}} = L - L_{\text{DT}}$, where L is the steiner minimal length of the net such that no part of the wire is routed inside any of the obstacles. Figure 6.7 shows the intrabin transparent-block and detor definitions.

Assumption 3 *We assume that the entire possible terminal pairs $P_1 = (x_1, y_1)$ and $P_2 = (x_2, y_2)$ for a two-terminal net have equal probability to occur. It means that the probability distribution of the terminals is uniform inside the nonblocked area of the chip.*

To obtain the average wire length, we decompose it into three parts, namely i transparent-block and detor in X and Y directions such that

$$\overline{L} = \overline{L}_{\text{TB}} + \overline{L}_{\text{DT}}^{h} + \overline{L}_{\text{DT}}^{v} \tag{6.5}$$

where $\overline{L}_{\text{DT}}^{h}$ and $\overline{L}_{\text{DT}}^{v}$ are the average detor in X and Y direction and $\overline{L}_{\text{TB}}$ is the average transparent-block wire length.

If there is no obstacle in the bin area, the average intrabin wire length can be easily obtained from 6.6, considering a uniform probability distribution for all the terminals [19].

$$\overline{L}_{\text{intra}} = \frac{\int_0^M \int_0^M \int_0^N \int_0^N (|x_1 - x_2| + |y_1 - y_2|)\mathrm{d}x_1\mathrm{d}x_2\mathrm{d}y_1\mathrm{d}y_2}{\int_0^M \int_0^M \int_0^N \int_0^N \mathrm{d}x_1\mathrm{d}x_2\mathrm{d}y_1\mathrm{d}y_2} = \frac{N+M}{3} \tag{6.6}$$

where N is width and M is height of a bin. Moreover, subscript intra denotes that the average is taken over all intrabin nets (nets that both of their terminals are in one bin), in contrast with the interbin nets which will be discussed later. In the presence of an obstacle, however, the average intrabin wire length would be different. To obtain the average wire length in this case, let us assume the obstacle's center is at position (a, b) and its width and height are, respectively, W and H (see

Figure 6.7). It should be noticed that in this case P_1 and P_2 must be placed outside of the obstacle, i.e. , where A is the set of all the points inside the bin and S is the set of all the points inside the obstacle. With the same type of analysis as for formula 6.6, we can calculate the average transparent-block wire length [19], $\overline{L}_{\mathrm{TB,intra}}$, $\psi(N, M, W, H, a, b)/(NM\text{-}WH)^2$ where

$$\psi(N, M, W, H, a, b) = \int\int\int\int_{P_1, P_2 \in A-S} (|x_1 - x_2| + |y_1 - y_2|)\mathrm{d}x_1\mathrm{d}x_2\mathrm{d}y_1\mathrm{d}y_2 \quad (6.7)$$

Basically, in (6.7), we get the integral over the nonblocked area of the bin. By that, we mean that the terminals can be everywhere except the blocked i part of the bin, but the interconnections can pass through the blockage. Analyzing the intrabin average detor wire length [19] is more complicated and is omitted here for brevity. It can be shown that the average detor wire length in X direction is

$$\overline{L}^h_{\mathrm{DT,intra}} = \frac{2}{3}W\frac{W(M - b - H/2)W(b - H/2)}{(MN - WH)^2} \quad (6.8)$$

Similar formula holds for the average detor wire length in Y direction.

In the presence of an obstacle in the adjacent bins, the formulas for calculating wire length is complicated. The wire length in this case, consists of both transparent-block and detor parts.

Horizontally adjacent bins are shown in Figure 6.8. In this case, the average transparent-block wire length can be computed as:

$$\overline{L}_{\mathrm{TB,inter}} = \frac{\psi(2N, M, W, H, a, b) - \psi(N, M, W_1, H, a_1, b_1) - \psi(N, M, W_2, H, a_2, b_2)}{(NM - W_1H)(NM - W_2H)} \quad (6.9)$$

where $W = W_1 + W_2$, $b_1 = b_2 = b$, $a_1 = a - W/2 + W_1/2$, and $a_2 = W_2/2$ and ψ is the same function as in (6.7). The average detor wire length in Y direction can be expressed as

$$\overline{L}^v_{\mathrm{DT,inter}} = Pr^v_{\mathrm{DT}}.L^v_{\mathrm{DT,inter}} \quad (6.10)$$

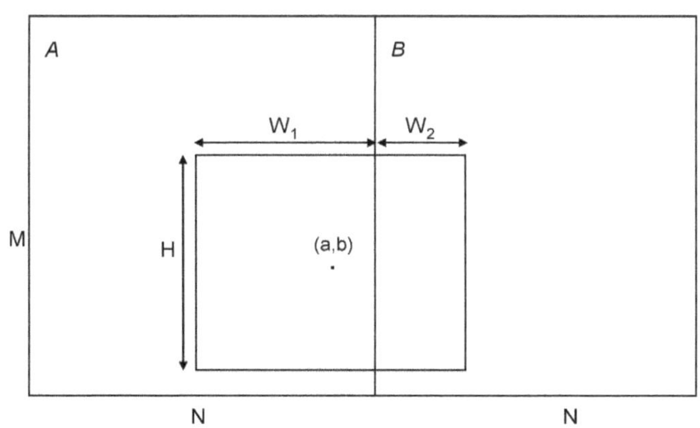

Fig. 6.8. Two horizontal adjacent bins.

where $\Pr^{v}_{\mathrm{DT}}$ is the probability of occurring a detor in Y direction. Moreover, $L^{v}_{\mathrm{DT,inter}}$ is the average detor length in Ydirection given that a detor occurred in this direction. Similar to the analysis in (6.8), $L^{v}_{\mathrm{DT,inter}}$ equals to $\frac{1}{3}H$. The average detor wire length in X direction can be found in similar way.

Having had $\overline{L}_{\mathrm{TB}}$, $\overline{L}^{h}_{\mathrm{DT}}$ and $\overline{L}^{v}_{\mathrm{DT}}$ as the average transparent-block and detor wire length of horizontal adjacent bins A and B, $L_h(A, B)$ in this case can be extracted from (6.5).

Calculating the average interbin wire length in the case of two diagonally adjacent bins is i somewhat similar is omitted here for brevity.

Having had the average interbin wire length for horizontally, vertically and diagonally adjacent bins, the average interbin wire length can be obtained for every level of hierarchy h. For every level of hierarchy, the average interbin wire length can be written as

$$\overline{L}_{\mathrm{inter}} = \frac{1}{6}(L^{h}_{\mathrm{inter}}(A, B) + L^{h}_{\mathrm{inter}}(C, D) + L^{v}_{\mathrm{inter}}(A, C) + L^{v}_{\mathrm{inter}}(B, D) + L^{d}_{\mathrm{inter}}(A, D) + L^{d}_{\mathrm{inter}}(B, C))$$

where h, v, and d, respectively, denote that the corresponding bins are horizontally, vertically, or diagonally adjacent.

The analysis of interbin average wire length in the presence of multiple blockages can be performed by using the analysis for average wire length in the presence of a single blockage. The interbin average wire length for vertically and diagonally adjacent bins use the same type of analysis as horizontally adjacent bins and are omitted here for brevity.

As shown in Figure 6.9, in the presence of multiple blockages, the average transparent-block wire length can be calculated from:

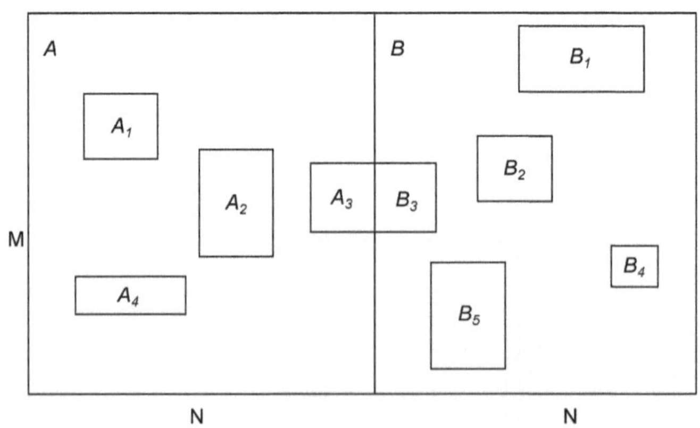

Fig. 6.9. Horizontal adjacent bins with multiple blockages.

$$\psi\left(A - \sum_i A_i, B - \sum_i B_i\right)$$
$$= \psi(A, B) - \left(\sum_i \psi(A_i, B) + \sum_i \psi(A, B_i) - \sum_j \sum_i \psi(A_i, B_j)\right) \quad (6.11)$$

It is shown by Cheng et al. [19] that if the obstacles do not overlap neither in X span nor in Y span, the effect of the obstacles on the average detor wire length is additive and this problem can be treated as a combination of single blockage problems. This assumption, however, seems to be too simplifying for real circuits. So, instead of making an assumption on the geometry of the blockages, we use the look-up table (LUT) estimation method based on numerous experiments. This LUT is indexed by two parameters; number of blockages and percentage of blocked area over the whole chip area where the indices of the latter are from 5 to 70% in the steps of 5%. The indices of the number of blockage are 4^n where $1 \leq n \leq 6$. Each entry of the LUT is obtained by generating some random circuit instance and performing the GP and routing for each instance. The total detor wire length for each instance is measured as the difference between the HPWL and the routed wire length.

6.5 Congestion Removal [7]

Achieving autoroutability is one of the main goals in modern standard-cell placement. Total estimated wire length, or bounding box wire length, was widely used as an objective function to optimize routability. It is commonly believed that a shorter total wire length implies better routability. Wire length optimization has been extensively studied during the past two decades. Successful placement techniques include min-cut [8, 9], simulated annealing [13], and analytical approach [14, 24].

Congestion is another important indicator of routability in placement, and it has become dominant for large, tight designs. Previous works address the congestion problem using various methods. Mayrhofer and Lauther [25] combine congestion function in cut minimization. Cheng [10] employs a congestion model in simulated annealing approach. Wang et al. [11] propose a postprocessing step to remove congestion for a wire length optimized placement. These methods attempt to reduce congestion by obtaining a placement with less gathered wires and are successful for improving routability.

White space allocation is another way to alleviate congestion in placement orthogonal to above congestion management techniques. White space is a term associated with *fixed-die* placement, which is the common design style in current industry practice. For fixed-die designs, chip area, core area, rows and available sites are given before placement and routing. White space, or the empty space that is not occupied by the standard-cells, varies from 0.1 to 30% for real designs. In fixed-die placement with large white space, purely minimizing wire length tends to place all the

[7] Portions reprinted, with permission, from [7]. © [2002] IEEE.

cells close to each other. But the congestion of this "packed" placement is worse than a spread-out placement. Fixed-die placement tool has to take white space into consideration to improve routability.

Since congestion originates from the discrepancy between routing demand and routing supply, increasing supply, as well as reducing demand, is a natural way to reduce congestion. However, the problem of allocating white space without much loss of placement quality (e.g., wire length) is not trivial.

6.5.1 Problem Formulation

In top-down placement flow, white space is allocated at later levels, where the congestion information acquired from the current placement can be used to guide allocation. Basically, we tend to allocate more white space to congested areas [7].

In an $m \times n$ grid mesh, assuming that the congestion of each grid is known, we want to determine the white space of each grid. Let c_{ij} be the congestion of the grid at column i and row j, and w_{ij} be the white space to be assigned into this grid. Assuming that the total white space of the design is W, we have,

$$\sum_{i=1}^{m}\sum_{j=1}^{n} w_{ij} = W \tag{6.12}$$

In addition, the total white space assigned to each row should be balanced, i.e., there is no row containing too much or too little white space. Let $w_{\min}$ and $w_{\max}$ be the minimum and maximum total white space for rows, respectively. We then have the following constraints:

$$w_{\min} \leq \sum_{i=1}^{m} w_{ij} \leq w_{\max} \qquad j = 1, ..., n \tag{6.13}$$

The problem of allocating white space is to find a mapping function f from congestion to white space for each grid. The function should be monotone, i.e.,

$$\begin{aligned} w_{ij} &= f(c_{ij}) && \text{and} \\ w_{ij} &\leq w_{uv} && \text{if} \quad c_{ij} \leq c_{uv} \\ & 1 \leq i, u \leq m, && 1 \leq j, v \leq n \end{aligned} \tag{6.14}$$

It is not hard to find a feasible solution (function f) to meet constraint (6.12)–(6.14). However, we require a *smooth* mapping function for a reasonable allocation. In this work, instead of directly solving this problem, we solve two alternative problems. We first allocate white space to each row of the grid mesh, then allocate white space to each grid within rows. We will describe them in Sects. 6.5.2 and 6.5.3.

6.5.2 Row White Space Allocation

To allocate white space for each row, we are dealing with a similar problem to the global white space allocation. Assume that there are n rows in the design and the total congestion for row j is c_j. Let w_j be the white space to be allocated to row j. The capacity constraint is,

$$\sum_{j=1}^{n} w_j = W \tag{6.15}$$

Without loss of generality, we assume that the congestions of the rows are in a nondecreasing order, i.e.,

$$c_1 \leq c_2 \leq \cdots \leq c_n$$

Hence,

$$w_{\min} \leq w_1 \leq w_2 \leq \cdots \leq w_n \leq w_{\max}$$

If we fix $w_1 = w_{\min}$ and $w_n = w_{\max}$, the problem can be understood as finding an increasing curve on a plane such that point $(c_1, w_{\min})$ and $(c_n, w_{\max})$ are on the curve, and the integral of the function on $[c_1, c_n]$ is W. We have the following constraints:

$$f(c_1) = w_{\min} \tag{6.16}$$

$$f(c_n) = w_{\max} \tag{6.17}$$

$$\int_{c_1}^{c_n} f(x)dx = W \tag{6.18}$$

where $f(x)$ is the function of the curve.

The linear function determined by the two points does not meet the integration constraint (6.18). A quadratic curve fits well in this situation. Let $f(x)$ be in the form $f(x) = a_1x^2 + a_2x + a_3$. According to the constraints (6.15)–(6.17),

$$a_1c_1^2 + a_2c_1 + a_3 = w_{\min} \tag{6.19}$$

$$a_1c_n^2 + a_2c_n + a_3 = w_{\max} \tag{6.20}$$

$$a_1\sum_{i=1}^{n} c_i^2 + a_2\sum_{i=1}^{n} c_i + a_3n = W \tag{6.21}$$

where $c_i, w_i, i = 1, ..., n$ are known and a_1,a_2 and a_3 are unknown parameters. The unique solution of equation (6.19)–(6.21) determines function $f(x)$.

However, the curve $f(x)$ may have extremum inside $[c_1, c_n]$, as shown in Figure 6.10(a), (b). This happens when $c_1 < -a_2/(2a_1) < c_n$. In these cases, the curve is no longer monotone within $[c_1, c_n]$. We need to relax either $w_1 = w_{\min}$ or $w_n = w_{\max}$ to satisfy the capacity constraint.

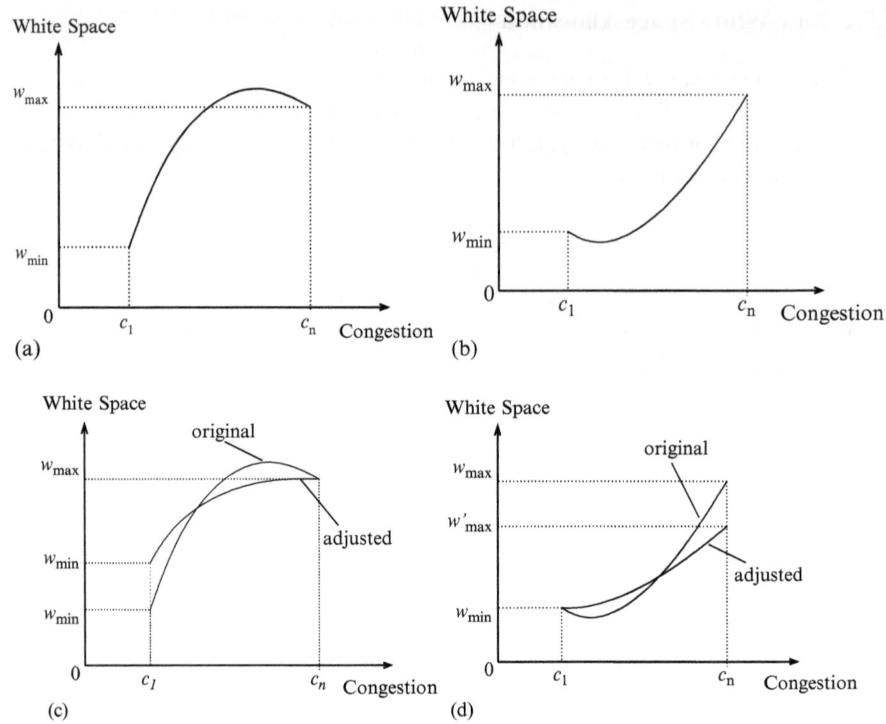

Fig. 6.10. The quadratic curves with extremum inside $[c_1, c_n]$ ((**a**), (**b**)), and adjusted curves that are strictly increasing within $[c_1, c_n]$ (**c, d**).

In the case $a1 < 0$ (Figure 6.10(a)), we relax the constraint $w_1 = w_{min}$ to $w_1 \geq w_{min}$ and take the point (c_n, w_{max}) as the extremum of the quadratic curve. Therefore, the following equation is used to replace (6.20),

$$c_n = -\frac{a_2}{2a_1} \tag{6.22}$$

The solution of (6.19), (6.21), and (6.22) corresponds to the adjusted curve shown in Figure 6.10(c). In the white space allocation problem, w_{max} is the maximum white space for a row, i.e., $w_{max} > W/n$. Thus there exists a w'_{min} such that point (c_1, w'_{min}) is on the adjusted curve ($w'_{min} = f(c_1)$) and $w'_{min} < w_{max}$, i.e., the new curve is monotone within $[c_1, c_n]$.

Similarly, in the case $a1 > 0$ (Figure 6.10(b)), we relax the constraint $w_n = w_{max}$ to $w_n \leq w_{max}$ and take the point (c_1, w_{min}) as the extremum of the quadratic curve. The following equation is used to replace (6.19),

$$c_1 = -\frac{a_2}{2a_1} \tag{6.23}$$

The solution of (6.20), (6.21), and (6.23) corresponds to the adjusted curve shown in Figure 6.10(d). Since $w_{\min} < W/n$, there exists a $w'_{\max}$ such that point $(c_n, w'_{\max})$ is on the adjusted curve ($w'_{\max} = f(c_n)$) and $w'_{\max} > w_{\min}$, i.e., the new curve is monotone within $[c_1, c_n]$.

The obtained quadratic function is used to compute white space for each row according to the congestion of this row. The white space of the row is then assigned to each grid.

6.5.3 Grid White Space Allocation

Unlike row white space allocation, there is no maximum or minimum white space limitation for grid white space allocation. The white space for a grid can be zero, if the grid is not congested. If a grid is highly congested, its neighbor grid is likely to be congested as well. This prevents one congested grid from being assigned too much white space.

For each grid, it is reasonable to allocate white space proportional to the ratio of the congestion to the total congestion, i.e., $w_{ij} = w_j c_{ij}/c_j$. Other ratios can be used, for instance, the ratio of the grid congestion square to the total square of the grid congestion. The specific model used to allocate grid white space varies and should takes congestion model into consideration.

6.5.4 Placement Flow

In general, placement quality (e.g. wire length) degrades after white space allocation, since the locations of the cells are changed without considering the wire length. Therefore, the allocation must not be the final step of the placement. A detailed placement optimization, usually a low temperature annealing step, is a good solution to the loss of quality in white space allocation. On the other hand, if the placement flow does not contain such a detailed optimization step, it is better to start allocating white space at earlier top-down levels and do multiple allocations. This is because the loss of wire length quality by allocating white space will be less severe by multiple uses of allocation.

For our top-down hierarchical placement flow, we use the white space allocation two times in the placement, both are in the detailed placement stage. The congestion information at this stage is more accurate. The first allocation is made after grid adjustment. At this moment, the number of rows in the grid mesh is the same as the number of standard-cell rows for the design. The second allocation is made after the annealing improvement step, and before the overlap removal. The entire placement flow with white space allocation steps is shown in Figure 6.11.

6.5.5 Post-Allocation Optimization

The optimization steps after two allocations are crucial for the successful allocation, as white space allocation changes the current placement and results in loss of placement quality. We use a simulated annealing-based step after the first allocation, and

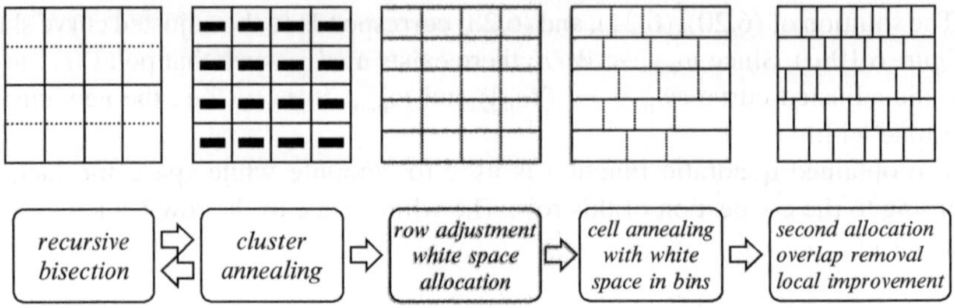

Fig. 6.11. Our placement flow with two white space allocations.

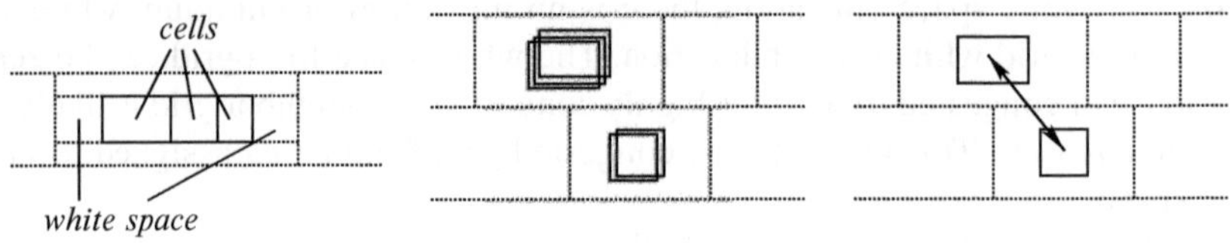

(a) Bin width is determined by cell width and white space (b) Cells are placed at the bin center with overlap (c) Swapping cells between two bins

Fig. 6.12. Simulated annealing after white space allocation.

a fast greedy algorithm after the second allocation. Both steps improve the total wire length by swapping or moving cells. We describe the first approach in this section.

As shown in Figure 6.12, after the white space allocation, each grid is assigned some white space. The width of a grid is determined by the white space of this grid, and the total widths of cells in this grid. A grid spreading step is performed before simulated annealing to determine the location of every grid. This step will be repeated periodically in simulated annealing. All the cells are placed at the centers of the grids with overlaps. For each move in annealing, cells are either swapped between two grids or are moved from one grid to another. The calculation of row overflow penalty should take the white space of this row into consideration, i.e., penalizing the rows for which the total cell width plus white space is greater than the row capacity.

6.6 Target Utilization Control

Another metric for estimationg the routability of a circuit is by measuring its target utilization. The target utilization (or density) can be defined as a constraint for the placement tool. The target utilization should be set higher than the design utilization which is a characteristic of the designed circuit. In ISPD 2006 contest, there was a general placement solution scoring function defined which considered target utilization along with wire length to emphasis on routability as well as optimizing the wire length. That scoring function was defined as

$$HPWL \times (1 + ScaledOverflowFactor + CPUFactor) \qquad (6.24)$$

To compute the Scaled Overflow Factor, a bin grid is imposed over the whole circuit, with each bin width and height equal to 10 circuit row height. The bin overflow is defined as:

$$BOF = \Sigma Movable\ Area\ Bin\ \text{-}\ Bin\ Free\ Space \times Target\ Density \tag{6.25}$$

and total overflow is defined as ΣBOF. Thus, the Scaled Overflow Factor is defined as

$$Scaled\ Overflow\ Factor = \left(\frac{TOF \times Bin\ Area \times Target\ Density}{\Sigma Movable\ Object\ Area} \right)^2 \tag{6.26}$$

We used two methods to control the target utilization of the placed circuits generated by Dragon considering the above cost function. In the first method, the key idea is to redistribute cells inside a region around a highly utilized bin to decrease the utilization level of each bin inside the region to the target utilization constraint. This region should be determined big enough so that its utilization is not more than the target utilization constraint. To redistribute cells inside that region we use the min-cut partitioning algorithm. The details of this algorithm is given in the Figure 6.13.

In the second method, the key idea is to move cells out of a bin which is highly utilized. This move would be accepted if it helps in reducing the total cost function composed of wire length and density factors. This process continues till there is no more bins left which is highly utilized. Figure 6.14 describes this method in details.

Input: Placement of $m \times n$ bin structure with some over-utilized bins, Target utilization constraint TUC

Output: Placement of $m \times n$ bin structure; Utilization of all bins are less than target utilization

---

repeat
  Pick the bin B with highest density
  Initial working region $WR = B$
  repeat
    Add low-utilized bins to WR in an $m \times n$ structure
  until WR utilization $\leq TUC$
  Collect all the cells in the WR into CellSet
  repeat
    Alternatively do horizontal and vertical min-cut on CellSet considering cells current location
    Balance the cell area in the cut according to the cut area
  until Each cut size = bin size
  Assign the cells to their bins
  Update each cell coordinate
until All bins utilization $\leq TUC$

Fig. 6.13. Utilization Control by cell redistribution.

Input: Placement of $m \times n$ bin structure with some over-utilized bins, Target utilization constraint TUC

Output: Placement of $m \times n$ bin structure; Utilization of all bins are less than target utilization

```
repeat
  Pick the bin B with highest density
  Sort all cells attached to B by their area in descending order
  repeat
    Pick the first cell C from the sorted list
    Move C to the closed bin with utilization ≤ TUC
    Compute the scoring function from (6.24)
    if scoring function is ≤ previous scoring function then
      accept the move
    else
      Reject the move
    end if
  until Bin B utilization ≤ TUC
until All bins utilization ≤ TUC
```

Fig. 6.14. Utilization Control by cell migration.

Table 6.1. Pure Wire length Optimization; ISPD 2005 Suite.

Circuit	Wire length	Run Time (h)
Adaptec1	83.2768	2.05
Adaptec2	94.7201	5.5
Adaptec3	231.0787	4.7
Adaptec4	200.8822	8.8
Bigblue1	102.3929	10.6
Bigblue2	159.7095	19.5
Bigblue3	380.4462	20
Bigblue4	903.9639	41.1

6.7 Experimental Result

Our proposed placement tool, Dragon, has been implemented in C under Linux environment. We have conducted two sets of experiments to show the capabilities of Dragon for minimizing the wire length and congestion. To verify our theoretical results on real-world circuits, all the experiments are done on ISPD 2005, ISPD 2006, Peko 2005, and Peko 2006 Suites [30].

In the first set of experiments, to emphasis the strength of Dragon to minimize the wire length, pure wire length minimization has been used on all the benchmarks from ISPD 2005 and Peko 2005 suites. The results are shown in Tables 6.1 and 6.2.

Since pure wire length minimization cannot necessarily generate routable designs which is a key concern in real world, the second set of experiments uses the scoring function presented in ISPD 2006 design contest as the metric. For ISPD 2006 suite,

Table 6.2. Wire length Optimization; Peko 2005 Suite.

Circuit	Design Util (%)	WL	Run Time (h)
Adaptec1	60.02	52.754	1.0
Adaptec2	60.04	76.666	1.2
Adaptec3	60.11	120.988	4.5
Adaptec4	60.19	87.1832	4.5
Bigblue1	60.01	57.1388	1.3
Bigblue2	59.15	200.7992	2.6
Bigblue3	56.84	635.1079	1
Bigblue4	60	520.0035	15.9

Table 6.3. Concurrent Wire length and Congestion Optimization; Peko 2006 Suite.

Circuit	Design Util (%)	Target Util (%)	WL	Overflow (%)	Run Time (h)	Scaled WL
Adaptec5	51.27	51.27	242.02	0.29	2691	242.72
newblue1	87.00	87.00	63.75	0.008	1413	63.76
newblue2	91.27	91.27	137.99	0.13	2470	138.17
newblue3	81.23	81.23	256.75	0.14	2228	257.11
newblue4	51.22	51.22	129.84	0.29	1862	130.22
newblue5	51.24	51.24	300.04	0.3	3988	300.94
newblue6	81.15	81.15	254.07	0.17	8328	254.50
newblue7	80.00	81.00	776.07	0.11	10556	776.96

Table 6.4. Concurrent Wire length and Congestion Optimization; ISPD 2006 Suite.

Circuit	Design Util (%)	Target Util (%)	WL	Overflow (%)	Run Time (h)	Scaled WL
Adaptec5	49.98	60	435.9676	0.04	5.2	436.1419
newblue1	39.83	80	79.8244	0.01	2.1	79.8323
newblue2	57.58	90	251.6102	0.14	1.8	251.9624
newblue3	26.31	80	437.5762	0.01	1.8	437.6199
newblue4	46.45	50	336.9413	0.06	2.3	337.1434
newblue5	49.56	50	609.2434	0.07	4.5	609.6698
newblue6	38.78	80	562.8104	0.03	5.7	562.9792
newblue7	49.31	80	377.5833	0.05	13.2	1378.2720

all the target densities are picked to be the same as or close to the ISPD 2006 design contest

As it can be seen in Tables 6.3 and 6.4, the overflow penalty used in design contest is very low in all of the designs, even the ones with very close target and design utilizations. The results on these tables verify that the methods presented in Sects. 6.5 and 6.6 are very successful in alleviating the congestion over the whole design.

References

1. Taghavi T, Yang X, Choi B.K, Wang M, Sarrafzadeh M (2006) Dragon2006: Blockage-Aware Congestion-Controlling Mixed-Sized Placer. International Symposium on Physical Design 209–211
2. Taghavi T, Yang X, Choi B.K, Wang M, Sarrafzadeh M (2005) Dragon2005: Large-Scale Mixed-Sized Placement Tool. International Symposium on Physical Design 245–247
3. Wang M, Yang X, Sarrafzadeh M (2000) Dragon2000: Fast standard-cell placement for large circuits. International Conference on Computer-Aided Design 260–263
4. Taghavi T, Amelifard B, Sarrafzadeh M (2006) Hierarchical Wirelength Estimation for Large-Scale Circuits in the Presence of IP Blocks. Submitted to IEEE Transaction on Very Large Scale Integration Systems, Special Section on System Level Interconnect Prediction
5. Yang X, Kastner R, Sarrafzadeh M (2001) Congestion Estimation during Top-Down Placement International Symposium on Physical Design(ISPD) 164–169
6. Taghavi T, Sarrafzadeh M (2006) Blockage-Oriented Placement. IEEE Electronic Design Process Workshop
7. Yang X, Choi B.K, Sarrafzadeh M (April 2002) Routability-Driven White Space Allocation for Fixed-Die Standard-Cell Placement. ACM International Symposium on Physical Design 42–47
8. Breuer M.A (1977) A Class of Min-cut Placement Algorithms. IEEE/ACM Design Automation Conference 284–290
9. Dunlop A.E, Kernighan B.W (Jan. 1985) A Procedure for Placement of Standard Cell VLSI Circuits. IEEE Transactions on Computer Aided Design 4(1):92–98
10. Cheng C.E (1994) RISA: Accurate and Efficient Placement Routability Modeling. International Conference on Computer-Aided Design 690–695
11. Wang M, Yang X, Sarrafzadeh M (2000) Congestion Minimization During Placement. IEEE Transactions on Computer Aided Design 19(10):1140–1148
12. Caldwell A.E, Kahng A.B, Markov I.L (June 2000) Can Recursive Bisection Alone Produce Routable Placements?. IEEE/ACM Design Automation Conference 477–482
13. Sechen C, Sangiovanni-Vincentelli A (1986) TimberWolf3.2: A New Standard Cell Placement and Global Routing Package. IEEE/ACM Design Automation Conference 432–439
14. Sigl G, Doll K, Johannes F.M (1991) Analytical Placement: A Linear or a Quadratic Objective Function. IEEE/ACM Design Automation Conference 427–432
15. Wang M, Sarrafzadeh M (April 1999) "Behavior of Congestion Minimization During Placement". ACM International Symposium on Physical Design pages 145–150
16. Wang M, Yang X, Eguro K, Sarrafzadeh M (April 2000) Multi-Center Congestion Estimation and Minimization During Placement. ACM International Symposium on Physical Design 147–152
17. Landman B, Russo R. (1971) On a Pin Versus Block Relationship for Partitions of Logic Graphs. IEEE Transactions on Computers c-20:1469–1479
18. Donath W.E (April 1979) Placement and Average Interconnection Lengths of Computer Logic. IEEE Transactions on Circuits and Systems 26(4):272–277
19. Cheng C.-K, Kahng A.B, Liu B. L, Stroobandt D (Feb 2001) Toward Better Wireload Models in the Presence of Obstacles. Asia and South Pacific Design Automation Conf. 527-532.
20. Stroobandt D, Campenhout J.V (1999) Accurate Interconnection Length Estimations for Predictions Early in the Design Cycle. VLSI Design, Special Issue on Physical Design in Deep Submicron 10(1):1–20

21. Davis J.A, De V.K, Meindl J (March 1998) A Stochastic Wire-Length Distribution for Gigascale Integration(GSI) - Part I: Derivation and Validation. IEEE Transactions on Electron Devices 45(3):580–589
22. Hagen L, Kahng A.B, Kurdahi F.J, Ramachandran C (Jan 1994) On the Intrinsic Rent Parameter and Spectra-Based Partitioning Methodologies. IEEE Transactions on Computer Aided Design 13(no.1):27–37
23. Karypis G, Aggarwal R, Kumar V, Shekhar S (1997) Multilevel Hypergraph Partitioning: Application in VLSI Domain. IEEE/ACM Design Automation Conference 526–529
24. Kleinhans J.M, Sigl G, Johannes F.M, Antreich K.J (1991) GORDIAN: VLSI Placement by Quadratic Programming and Slicing Optimization. IEEE Trans. on Computer Aided Design 10(3):365–365
25. Mayrhofer S, Lauther U (1990) Congestion-Driven Placement Using a New Multi-partitioning Heuristic. International Conference on Computer-Aided Design 332–335
26. Parakh P.N, Brown R.B, Sakalleh K.A (June 1998) Congestion Driven Quadratic Placement. IEEE/ACM Design Automation Conference 275–278
27. Cong J, Romesis M, Xie M (Apr 2003) Optimality, scalability and stability study of partitioning and placement algorithms. International Symposium on Physical Design 88–94
28. Madden P.H. (Apr. 2001) Reporting of standard cell placement results. ACM International Symposium on Physical Design 30–35
29. Xu H, Wang M, Choi B.-K., Sarrafzadeh M (Nov. 2003) Toop: A trade-off oriented placement tool. International Conference on Computer-Aided Design 467–471
30. http://www.ispd.cc

Part IV

Multilevel Placement Techniques

7

APlace: A High Quality, Large-Scale Analytical Placer

Andrew B. Kahng[1], **Sherief Reda**[2] **and Qinke Wang**[1]
[1]Univeristy of California, San Diego
[2]Brown University, Division of Engineering
{abk, qinke}@cs.ucsd.edu, {Sherief_Reda}@brown.edu

Modern design requirements have brought additional complexities to netlists and layouts. Millions of components, whitespace resources, and fixed/movable blocks are just a few to mention in the list of complexities. With these complexities in mind, placers are faced with the burden of finding an arrangement of placeable objects under strict wirelength, timing, and power constraints. In this chapter, we describe the architecture and novel details of our high quality, large-scale analytical placer APlace2 (and the subsequent APlace3) [26–28]. The performance of APlace2, has been recognized in the recent ISPD-2005 placement contest, and in this paper we disclose many of the technical details that we believe are key factors to its performance. We describe (1) a new clustering architecture, (2) a dynamically adaptive analytical solver, and (3) better legalization schemes and novel detailed placement methods. We also provide extensive experimental results on a number of benchmark sets, including the IBM ISPD'04, IBM-PLACE 2.0, ICCAD'04, ISPD'05, PEKO'05, ISPD'06, PEKO'06 as well as using the zero-change netlist transformation benchmarking framework.

7.1 Introduction

Beside enormous sizes, modern VLSI circuits exhibit a wide range of features that require careful handling by physical design tools. These features include thousands of fixed as well as movable blocks, a large number of I/O pads that are not necessary on the peripheral of the layout, millions of standard cells, and a large amount of whitespace for routing and timing requirements. These features are challenging to handle with existing placers, as has been demonstrated in the recent ISPD-2005 placement contest [36].

To tackle these challenges, modern placers combine a wide range of techniques and components. For instance, (1) clustering is routinely used to cut down runtime and enable scalable implementations [5,19,42], (2) core placement engines are based on min-cut [4,10,48] and/or analytical solvers [11,16,46], (3) legalization component [7,21,33] (4) iterative improvement heuristics [17,40], and (5) detailed placers and whitespace distributers [8, 15, 21, 25, 47]. All these techniques must readily handle the presence of blocks – whether fixed or movable – and whitespace. Furthermore, all components have to be carefully tuned to squeeze out every possible increment of Quality of Result (QOR).

In this paper we describe the architecture and details of our placer APlace [26–28]. The outline architecture of our placement tool is given in Figure 7.1. Clustering is used as an initial pre-processing step to condense the netlist in a multi-level fashion to just around a couple of thousand components. Global placement works on the clustered netlist until a "decent" spreading is achieved. At that point, unclustering

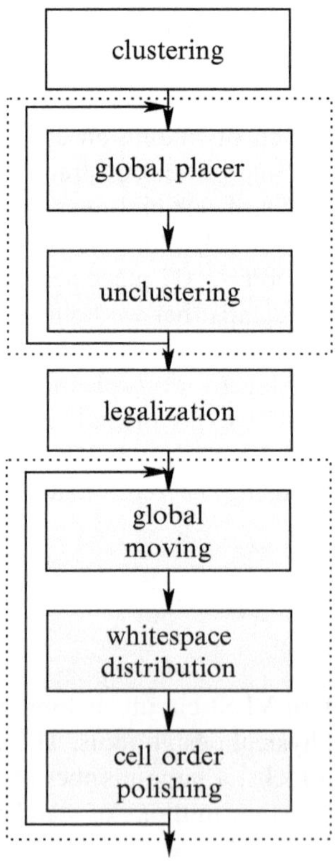

Fig. 7.1. The outline of our placement flow.

breaks down the clusters to reveal the next level of clustering. The process of global placement and unclustering is iterated until the original flat netlist is well spread. Then, legalization assigns valid positions for all movable components with no overlaps. Legalization typically incurs an increase in wirelength of the placement. The ensuing detailed placement phase attempts to recover any loss of quality due to legalization. Detailed placement is comprised of three phases (1) global moving where cells are moved globally to reduce wirelength, (2) whitespace distribution where whitespace is optimally distributed to minimize wirelength while maintaining the relative cell ordering in every layout row, and (3) cell order polishing where successive small windows of cells are optimally re-ordered. The three phases of detailed placement may be iterated until negligible improvements in wirelength are observed.

The organization of this paper is as follows. Section 7.2 gives the details of the clustering and unclustering phases. Section 7.3 discusses various global placement ideas and details. Section 7.4 provides the details of our legalization scheme and various phases of detailed placement. Section 7.5 gives experimental results for various benchmark sets.

7.2 Clustering and Unclustering

Executing an analytical global placer on a flat cell design might give the best placement results – depending on how scalable the placer is – but nevertheless can incur extremely long runtimes. Clustering offers an attractive choice to reduce runtime, and with careful tuning this can have no impact on placement quality. Our clustering approach can be viewed as a middle-ground between the top-down multi-level paradigm of MLPart [6] and hMetis [31] on one side, and fine-grain clustering [19] and semi-persistent clustering [5] on the other side.

Our clustering pre-processes the input netlist to reduce its size to only a couple of thousand clusters. However, this clustering is executed in a multi-level paradigm, where each clustering level is about tenth the size of the previous clustering level. For example, if the input size is around two million objects – roughly the size of the largest circuit in the IBM ISPD'05 benchmark set – then the clustering hierarchy is around four levels with vertex cardinalities: 2M, 200k, 20k, and 2k. After clustering, the pre-processed netlist is given to the global analytical placer to operate on. The global placer keeps on solving the netlist until it achieves a non-overlapping, or a "sufficiently" small overlapping placement. At this point, unclustering is triggered and the components of the next clustering hierarchy level replace the existing components. The components of an unclustered object are initially placed at the center location of their component with a slight random perturbation.

To move from one clustering level to the next during the initial pre-processing step, we use the *best choice* heuristic [5] with tight control on cluster area and using lazy updates. This can be summarized as follows. Initially, the *affinity* of every object u to its neighbors is calculated and the neighbor object with largest affinity is declared the *closest*; its affinity becomes the *score* of node u. The affinity between a pair of objects u and v is the total weight of the hyperedges joining them divided by

Input: Flat netlist.
Output: Clustered netlist.

until number of clusters < 2000:
  target number of clusters = $\frac{\text{current number of clusters}}{\text{clustering ratio=10}}$.
  target cluster area (CA) = $\frac{\text{total cell area}}{\text{target number of clusters}} * 1.5$.
  for each object u:
    calculate the most affine neighbor to u and u's score.
  sort all objects by their score descendingly using a heap.
  until the target number of clusters is met:
    if (i) top of the heap u is not marked invalid **and**
      (ii) clustering does not violate CA
      then cluster u with its most affine neighbor.
    else if u is marked invalid
      then recalculate its score, insert in heap and mark valid.
    else remove u from the heap and continue.
    update netlist and calculate the new clustered object score.
    insert the new object into the heap.
    mark the neighbors of the new object invalid.

Fig. 7.2. Clustering algorithm.

their area (similar to first choice clustering [32]), where the weight of a hyperedge is inversely proportional to its cardinality. After the scores of all nodes are calculated, they are inserted in a priority queue that is sorted in a descending order. Clustering then proceeds as follows: (1) cluster the best node – essentially the one with highest score – with its closest neighbor, unless the node is marked "invalid" or clustering violates the area constraints, (2) update the netlist and insert the score of the new clustered node in the proper position in the priority queue, and (iii) mark the neighbors of the new node as invalid. The clustering algorithm is given in Figure 7.2.

The interaction between the best choice clustering heuristic with the multi-level paradigm in the presence of different cell areas creates an unbalanced cluster hierarchy at each level, and the boundaries of such a hierarchy must be clearly marked to allow correct unclustering. For example, Figure 7.3 gives a possible clustering hierarchy, where we label each node with its level. We can clearly see that a node might take part in a number of clusterings during the same level – as long as area constraints are not violated. Thus, during clustering it is important to remember the boundaries of the clustering hierarchy to allow exact reversal of the clustering process.

Another concern during clustering is what we call *clustering saturation*. In netlists with large numbers of fixed blocks and I/O pads, it is possible that clustering is not able to meet its final target number of clusters since a large number of clusters are just connected to fixed components. These fixed components slow down the clustering, causing saturation. In this case, we can *bypass* the fixed objects – especially the small ones – to allow further clustering. This bypassing can be achieved by adding an artificial net connecting all the neighbors of a fixed object together.

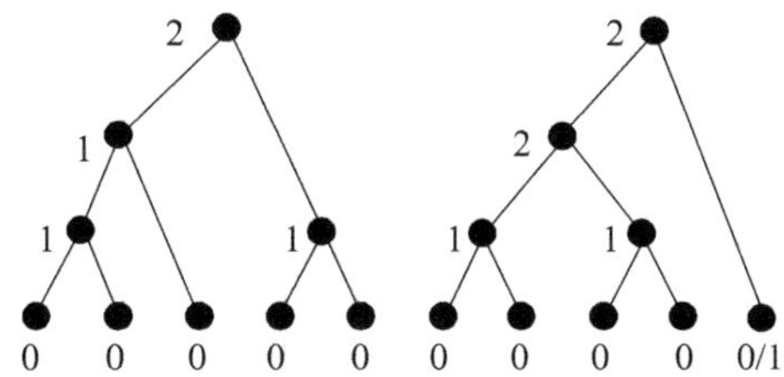

Fig. 7.3. A multi-level clustering hierarchy, with a clustering ratio of 2 and a required final target number of clusters equal to 2. Each node is labelled with its position in the clustering hierarchy.

7.3 Global Placement

7.3.1 Constrained Minimization Formulation

We regard global placement as a *constrained nonlinear optimization problem*: We divide the placement area into uniform grids, and seek to minimize total half-perimeter wirelength (HPWL) under the constraint that total module area in every grid is equalized. The problem is expressed using the following formulation:

$$\begin{aligned} &min\ HPWL(\mathbf{x}, \mathbf{y}) \\ &s.t.\ \ D_g(\mathbf{x}, \mathbf{y}) = D_g \text{ for each grid } g \end{aligned} \tag{7.1}$$

where $(\mathbf{x}, \mathbf{y})$ is the center coordinates of modules, $HPWL(\mathbf{x}, \mathbf{y})$ is the total HPWL of the current placement, $D_g(\mathbf{x}, \mathbf{y})$ is a density function that equals the total module area in grid g and D_g is the expected total module area in grid g, which is usually a constant denoting the average module area over all grids.

To solve the problem using nonlinear optimization techniques, first we need to have smooth wirelength and density functions.

LOG–SUM–EXP Wirelength Function

While wirelength and overall placement quality is typically evaluated according to HPWL, this “linear wirelength” function can not be efficiently minimized. In our placer, we use a *log–sum–exp* method to capture the linear HPWL while simultaneously obtaining the desirable characteristic of continuous differentiability. The log–sum–exp formula picks the most dominant terms among pin coordinates; it is proposed for wirelength approximation in [37] and applied in recent academic placers [13,22,24]. For a net e with pin coordinates $\{(x_1, y_1),\ (x_2, y_2),\ \dots\ (x_n, y_n)\}$, the smooth wirelength function is

$$\begin{aligned} \mathrm{WL}(e) = &\ \alpha \cdot (\log(\textstyle\sum e^{x_i/\alpha}) + \log(\textstyle\sum e^{-x_i/\alpha})) \\ &+ \alpha \cdot (\log(\textstyle\sum e^{y_i/\alpha}) + \log(\textstyle\sum e^{-y_i/\alpha})) \end{aligned} \tag{7.2}$$

where α is a smoothing parameter. $\mathrm{WL}(e)$ is strictly convex, continuously differentiable and converges to $HPWL(e)$ as α converges to 0 [37].

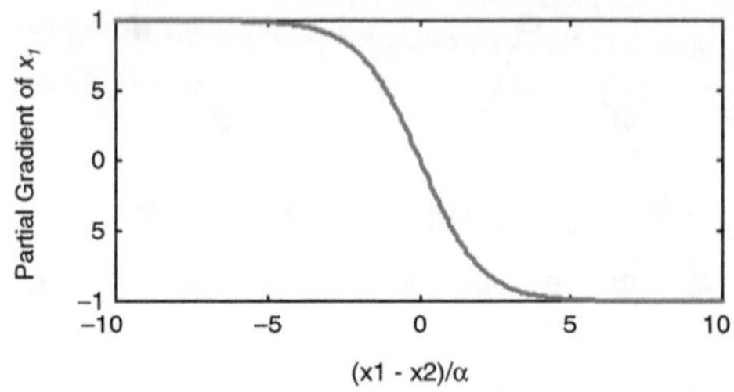

Fig. 7.4. Partial wirelength gradient for x_1 as a function of $(x_1 - x_2)/\alpha$.

Intuitively, for the overall placement problem, the smoothing parameter α can be regarded as a "significance criterion" for choosing nets with large wirelength to minimize. For example, for a two-pin net with pin coordinates $\{(x_1, y_1), (x_2, y_2)\}$, the partial gradient of the wirelength function WL for x_1 is

$$\frac{\partial \text{WL}}{\partial x_1} = 1/(1 + e^{(x_1 - x_2)/\alpha}) - 1/(1 + e^{(x_2 - x_1)/\alpha}) \tag{7.3}$$

As shown in Figure 7.4, when the net length $|x_1 - x_2|$ is relatively small compared to α, the partial gradient is close to 0; otherwise, the gradient is close to 1 or -1. It means that the length of long nets (relative to α) will be minimized more efficiently than short nets when optimizing the wirelength function for the whole netlist. Our placer uses this important characteristic to facilitate the multi-level algorithm that will be described in Sect. 7.3.3.

Bell-Shaped Potential Function

The density function $D_g(\mathbf{x}, \mathbf{y})$ in (7.1) is also not smooth or differentiable. Function $D_g(\mathbf{x}, \mathbf{y})$ can be expressed as the following form:

$$D_g(\mathbf{x}, \mathbf{y}) = \textstyle\sum_v P_x(g, v) \cdot P_y(g, v) \tag{7.4}$$

where functions $P_x(g, v)$ and $P_y(g, v)$ denote the overlap between the grid g and module v along the x and y directions, respectively. For example, suppose we have a grid g with width w_g and a standard cell c. Since the size of cell c is usually small relative to the grid size, we ignore the cell size and assume it to be a dot with unit area. Then function $P_x(g, c)$ is a 0/1 function as shown in Figure 7.5(a): $P_x(g, c)$ is 1 when the horizontal distance between grid g and cell c, $d_x = |x_c - x_g|$ is less than $w_g/2$, and is 0, otherwise.

Naylor et al. [37] propose to replace the above "rectangle-shaped" function with a "bell-shaped" function $p_x(g, c)$ as shown in Figure 7.5(b)

$$p_x(g, c) = \begin{cases} 1 - 0.5d_x^2/w_g^2 & (0 \leq d_x \leq w_g) \\ 0.5(d_x - 2w_g)^2/w_g^2 & (w_g \leq d_x \leq 2w_g) \end{cases} \tag{7.5}$$

This function is implemented in our original placer and has been proved effective. Since the function decides an "area potential" exerted by a cell to its nearby grids, we call it *area potential function*.

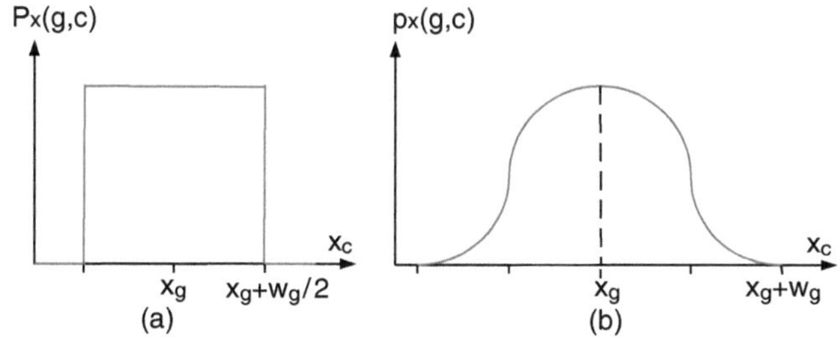

Fig. 7.5. (**a**) "Rectangle-shaped" function $P_x(g, v)$; and (**b**) "Bell-shaped" smooth function $p_x(g, v)$.

We follow the above idea and apply a similar "bell-shaped" area potential function in our current placer. Unlike the above potential function, our potential function also takes care of large blocks, as well as standard cells, and extends the scope of area potential according to the block size so that a larger block will have non-zero potential with respect to more nearby grids.

Suppose a module v has a large width w_v. The scope of this module's x-potential is $w_v/2 + 2w_g$, i.e., every grid within horizontal distance of $w_v/2 + 2w_g$ from the module's center has a non-zero x-potential from this module. Therefore, the area potential function for the x-direction $p_x(g, v)$ becomes

$$p_x(g, v) = \begin{cases} 1 - a * d_x^2 & (0 \le d_x \le w_v/2 + w_g) \\ b * (d_x - 2w_g)^2 & (w_v/2 + w_g \le d \le w_v/2 + 2w_g) \end{cases} \tag{7.6}$$

where

$$\begin{aligned} a &= 4(w_v + 4w_g^2)/((w_v + 8w_g^2)(w_v + 2w_g)^2) \\ b &= 4/(w_v + 8w_g^2) \end{aligned} \tag{7.7}$$

so that the function is continuous when $d_x = w_v/2 + w_g$.

Similarly, we define a smooth y-potential function $p_y(g, v)$ and the non-smooth function $D_g(\mathbf{x}, \mathbf{y})$ in (7.1) is replaced by a continuous function:

$$SD_g(\mathbf{x}, \mathbf{y}) = \textstyle\sum_v C_v \cdot p_x(g, v) \cdot p_y(g, v) \tag{7.8}$$

where C_v is a normalization factor so that $\sum_g C_v \cdot p_x(g, v) \cdot p_y(g, v) = A_v$, i.e., each module v has a total area potential equal to its area A_v.

Congestion-Directed Placement

To improve routability of placement results, we have integrated congestion information into the objective functions to direct cell distribution. We use Kahng and Xu's accurate bend-based congestion estimation method [29] in our placer. The blockage-aware method takes into account (1) the impact of the number of bends in a routing path on the probability of the path's occurrence; and (2) the impact of neighboring nets on a path's probability. If a particular grid is determined to be congested (respectively, uncongested), the expected total cell potential of the grid in (7.1) is reduced (respectively, increased) accordingly. The sum of expected area potential over all grids is kept constant, and equal to the total cell area. Specifically, expected cell potential is adjusted as follows:

$$D_g \propto 1 + \gamma \left(1 - 2\frac{\text{Congestion}(g)}{\max_g\{\text{Congestion}(g)\}}\right) \tag{7.9}$$

where γ is the congestion adjustment factor and decides the extent of congestion-directed placement.

7.3.2 Quadratic Penalty Method and Conjugate Gradient Solver

In the current version of our placer, we solve the constrained optimization problem in (7.1) using the simple *quadratic penalty method*. That is, we solve a sequence of unconstrained minimization problems of the form

$$\min \text{WL}(\mathbf{x}, \mathbf{y}) + \tfrac{1}{2\mu} \textstyle\sum_g (SD_g(\mathbf{x}, \mathbf{y}) - D_g)^2 \tag{7.10}$$

for a sequence of values $\mu = \mu_k \downarrow 0$ and use the solution of the previous unconstrained problem as an initial guess for the next one.

Empirical studies show that the values of μ is very important to the solution quality. Theoretically, when the optimal solution of the unconstrained problem in (7.10) is reached, the gradients derived from the wirelength term are opposite to those derived from the density penalty term. Therefore, we decide the initial μ according to the absolute values of wirelength and density gradients

$$\mu_0 = \frac{1}{2}\frac{\sum_{x_i,y_j}\sum_g |SD_g - D_g| \cdot \left(|\frac{\partial SD_g}{\partial x_i}| + |\frac{\partial SD_g}{\partial y_j}|\right)}{\sum_{x_i,y_j}\left(|\frac{\partial WL}{\partial x_i}| + |\frac{\partial WL}{\partial y_j}|\right)} \tag{7.11}$$

After that, μ decreases by half: $\mu_{k+1} = 0.5\mu_k$.

We solve the unconstrained problem in (7.10) using the *Conjugate Gradient* (CG) method, as shown in Figure 7.6. The conjugate gradient method is quite useful in finding an unconstrained minimum of a high-dimensional function, even though the function is not convex. Also the memory required is only linear in the problem size, which makes it adaptable to large-scale placement problems.

7.3.3 Multi-Level Algorithm

Our placer applies a multi-level algorithm to improve scalability in a similar way as in [11,13]. We use two different multi-level methods in the placer: (1) multiple levels of placeable objects and (2) multiple levels of grids.

Multiple Levels of Clusters

Before global placement, our placer builds up a hierarchy of clusters as described in Sect. 7.2, performs placement for each level of clusters and use the solution of

Conjugate Gradient Algorithm

Input:
A high dimensional function $f(x)$
Initial solution x_0
Minimum step length ϵ
Initial maximum step length γ_0
Maximum number of iterations N

Output:
Local minimum x^*

Algorithm:
01. Initialize # iterations $k = 1$, step length $\alpha_0 = \infty$
gradients $g_0 = 0$ and conjugate directions $d_0 = 0$
02. **For** ($k < N$ and step length $\alpha_{k-1} > \epsilon$)
03. Compute gradients $g_k = \nabla f(x_k)$
04. Compute Polak–Ribiere parameter $\beta_k = \frac{g_k^{\mathrm{T}}(g_k - g_{k-1})}{||g_{k-1}||^2}$
05. Compute conjugate directions $d_k = -g_k + \beta_k d_{k-1}$
06. Compute step length α_k within γ_{k-1}
using Golden Section line search algorithm
07. Update new solution $x_k = x_{k-1} + \alpha_k d_k$
08. Update maximum step length γ_k =MAX$\{\gamma_0, 2\gamma_{k-1}\}$
17. Return minimum $x^* = x_k$

Fig. 7.6. Conjugate Gradient Algorithm.

the current level cluster placement as an initial guess for the next level placement problem.

Clustering reduces the number of placeable objects and thus speeds up the calculation of density penalty. For each level in the cluster hierarchy, we compute the density penalty by regarding a cluster as a square block with area equal to the total module area of the cluster. Moreover, the decrease of the number of variables also greatly reduces the number of conjugate gradient iterations required to obtain a good solution of the unconstrained optimization problem. For wirelength calculation, we assume modules to be located at the center of the cluster and only consider the inter-cluster parts of nets, which speeds up the wirelength calculation.

Multiple Levels of Grids

Beside the commonly used method of multiple cluster levels, our placer also employs multiple levels of grids to achieve better scalability and global optimization.

Various grid sizes provide different levels of relaxation for the constrained wirelength minimization problem in (7.1). For example, in the optimal solution of the constrained minimization problem with a larger grid size, modules in the same grid are expected to cluster together instead of spread evenly over the grid in order to reduce total wirelength, although total module area in each grid is equal. However, this solution can be used as the initial solution for the placement problem constrained with finer grids, to obtain a more even module placement.

We adaptively modify the smoothing parameter α according to the grid size, instead of using a small constant value. For a wirelength minimization problem constrained with coarser grids, minimization of short nets (relative to the grid size) leads to undesirably clustered cells. Therefore, the value of α should be comparable to the grid size, so that only long nets (probably connecting modules in different grids) are "chosen" to be minimized and short nets (probably connecting modules in the same grid) are "ignored". Empirical studies show that better placement quality is obtained by setting α to half of the grid size.

Using an initial larger grid size and wirelength smoothing parameter in our placer not only leads to better global optimization, but also greatly speeds up the placer. As shown in (7.6), the scope of modules' potential is proportional to the grid size. Therefore, a larger grid size helps to spread cells faster than a smaller grid size.

Top-Down Multi-Level Algorithm

Combining the two methods discussed in Sects. 7.3.3 and 7.3.3, our top-down multi-level algorithm is described in Figure 7.7. Notations used are summarized as follows:

α	wirelength smoothing parameter
ϵ	minimum step length of CG solver
f	unconstrained objective function
N_l	the number of clusters at level l
L	the number of cluster levels
$\{Gradient_l(i)\}$	a vector of conjugate gradients
$\{ClusterPosition_l(i)\}$	a vector of cluster positions

Subscript ranges, where not explicit, are: $l = 0, \ldots, L$; and $i = 1, \ldots, N_l$.

Initially, the global placements of all modules is initialized to be at the center of the placement area. Unlike most analytical placers, our placer can also place circuits without fixed pads, or simultaneously place modules and peripheral/area I/O pads. In this case, the placeable objects are initially placed randomly close to the center.

For each level in the cluster hierarchy, the grid size is determined according to the number of clusters, assuming the total module area of each cluster is similar. We then decide most important control parameters according to the grid size.

After that, the global placer basically uses the CG optimizer to solve the constrained wirelength minimization problem. Note that when α is small, the wirelength approximation in Sect. 7.3.1 is close to the HPWL. Thus for flat placement, we use the actual HPWL instead of the approximation in the line search algorithm, in order to reduce runtime. We define *discrepancy* within a window of area A_w as the maximum ratio of total module area within the window to the window area over all windows of area A_w. We stop global placement when the discrepancy is less than a user specified target value, which is 1.0 at default.

Top-Down Multi-Level Algorithm

Input:
User-defined target density discrepancy $TargetDisc$
User-defined max #iterations per optimization $MaxIters$

Output:
Global placement

Algorithm:
Construct a hierarchy of clusters
For (each cluster level l from top to down)
  Set initial placement $\{ClusterPosition_l(i)\}$
  $GridSize \propto 1/sqrt(N_l)$
  $\alpha = 0.5 \cdot GridSize$
  $\epsilon = 0.1 \cdot GridSize$
  $\mu = 0.5 \frac{TotalAbsoluteDensityGradient}{TotalAbsoluteWirelengthGradient}$
  While ($Discrepancy > TargetDisc$)
   While ($\#Iter < MaxIters$)
    $f = WL + \frac{1}{2\mu} \cdot QuadraticPenalty$
    Compute conjugate gradients $Gradient_l$
    $StepLength = LineSearch(f, Gradient_l)$
    $ClusterPosition_l += StepLength * Gradient_l$
    If ($StepLength < \epsilon$)
     $\mu = 0.5\mu$
     break
Return module placement $\{ClusterPosition_0(i)\}$

Fig. 7.7. Top-down multi-level algorithm.

7.4 Legalization and Detailed Placement

Our legalization scheme is based on the schemes of [18] and [33] with few modifications. In the basic scheme, cells are first sorted according to their horizontal locations, and then they are processed in order from left to right, where each cell is assigned to the closest available position. We then repeat the above procedure except that cells are processed in the reverse order from right to left this time. We pick the better of the two legalization results.

Detailed placement is composed of three phases that can be iterated a number of times until a negligible threshold of improvement is attained. The three stages are (1) global cell moving, (2) whitespace distribution, and (3) cell order polishing. We start by describing global cell moving.

7.4.1 Global Moving

The objective of global moving is to move each cell to the optimal location among available whitespace without changing other modules' positions. Global moving is applied in our placer to improve placement quality for designs with a low utilization ratio. However, for designs without plenty of whitespace, since the global placer is already quite strong, the effect of global moving could be negligible.

We design an efficient heuristic to find a suboptimal available location for each cell. For each cell, we first traverse all the nets connected to the cell, and decide the optimal region for the cell's placement based on the median idea of [17]. Then we search for an available location in the optimal region, if the current placement is not already in it.

If the optimal region is already filled up, we divide the placement area into uniform bins, choose a "best" bin according to available whitespace in the bin and the cost (wirelength difference) of moving the module to the center of the bin, and then search for a best available location in the candidate bin. To quickly estimate if it is possible for a bin to have a continuous whitespace wider than the cell, we assume a normal distribution of whitespace with respect to its width, and obtain the average μ and standard deviation σ at the beginning of global moving. Therefore the number of whitespaces in a bin with total whitespace s that can hold a cell with width w is $s \cdot \frac{1}{2} \cdot \operatorname{erfc}\left(\frac{w-u}{\sqrt{2}\sigma}\right)$.

7.4.2 Whitespace Distribution

The objective of whitespace distribution is to optimally place the whitespace within each row to minimize the wirelength while reserving the cell order [21, 25]. We briefly sketch our procedure[1]. We define a *subrow* S_i as sequence of ordered sites $S_i = \{s_1, s_2, \ldots, s_n\}$ starting from a left fixed boundary – layout periphery or a fixed object – and ending at a right fixed boundary. Let $C_i = \{c_1, c_2, \ldots, c_m\}$ denote the set of m cells residing at subrow S_i, $x(.)$ indicates the leftmost site occupied by a cell and $w(\cdot)$ the width of a cell. To optimally redistribute the whitespace in subrow S_i, we construct a directed acyclic graph $G = (V, E)$ as shown in Figure 7.8 with vertex set $V = \{0, \ldots, n\} \times \{0, \ldots, m\}$, and edge set

$$E = \{(j, k-1) \to (j, k) \mid 0 \le j \le m, 1 \le k \le n\} \cup \{(j-1, k) \\ \to (j, k + w(c_j)) \mid 0 \le j \le m, 1 \le k \le n - w(c_j)\}.$$

The set of edges E is composed of the union of horizontal edges and diagonal edges in G. A diagonal edge indicates the placement of a cell at its tail, while a horizontal edge indicates no placement. Thus to minimize HPWL, each diagonal edge starting at $(j-1, k)$ is labelled by the cost (wirelength difference) of placing a cell c_j in position k, and all horizontal edges are labelled by zero. Finding an arrangement of the cells that optimally distributes whitespace corresponds to calculating the shortest path in G from the *start* node to the *end* node. Since G is a directed acyclic graph, the shortest path can be calculated using topological traversal of G in $O(mn)$ steps.

[1] While it is possible to use faster methods such as the CLUMPING algorithm [25], we use the method described since it is more convenient with cell order polishing described in next Sect. 7.4.3.

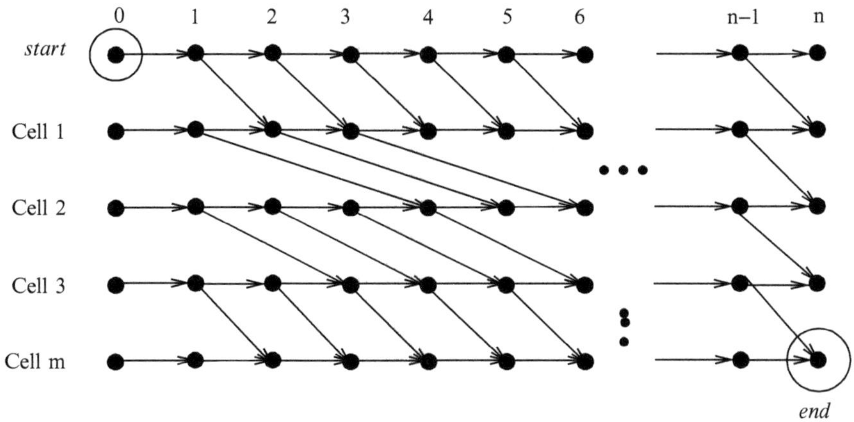

Fig. 7.8. The directed acyclic graph G for finding the optimal whitespace distribution.

A dynamic programming algorithm is applied to find the shortest path from the *start* node to the *end* node in G as shown in Figure 7.8. The algorithm uses a table of size mn, computes the shortest distance from the *start* node to the node (j, k) row by row from left to right and marks for each node whether the shortest path comes from the left node $(j, k-1)$ in the same row or the node $(j-1, k-w(c_j))$ in the upper row. In the table, the element at (j, k) is the minimum cost (total HPWL difference) of placing the first j cells in the first k sites. After the calculation is finished for all the nodes, the shortest path can be traced from the *end* node back to the *start* node, and the optimal placement is obtained.

We also speed up the algorithm by constantly comparing the total size of non-placed cells to available whitespace. Suppose the algorithm is currently computing the element (j, k). Let t be the total size of the remaining $m - j$ cells. If $t < n - k$, i.e., there is no enough whitespace left for placing the rest of the cells, the dynamic programming algorithm stops calculating the remaining elements for the jth row of the table, meaning that the remaining $n - k$ sites cannot be occupied by the first j cells in a valid placement.

7.4.3 Cell Order Polishing

The idea of cell order polishing is to permute a small window of cells in order to improve wirelength. Similar techniques are commonly applied in academic placers. For example, Capo applies a detailed placement improvement technique based on the optimal placement [9], RowIroning, which permutes several cells in one row assuming equal whitespace distribution between cells. FengShui's cell ordering technique [2] permutes six objects in one or more rows regarding whitespace as pseudo cells.

In this section, we present a branch-and-bound algorithm that permutes the order of a few nearby cells in one row or multiple rows, and simultaneously considers the optimal placement for each permutation in a small window. Thus, our algorithm allows more accurate, overlap-free permutations and does not have to shift other cells.

Ordering for Intra-Row Cells

For a few consecutively placed cells in one row, we define a small window of available sites, which includes all the sites occupied by the cells and neighboring whitespace. The algorithm permutes the cells, finds the optimal placement for each permutation within the window using the dynamic programming algorithm described in Sect. 7.4.2, and selects the best permutation and the correspondent optimal placement as the solution.

An important fact is that the cost of placing the first j cells of a permutation is not related to the order of the rest of the cells, because they will be definitely placed to the right of the first j cells. E.g., given a net incidental to the jth cell, it is clear whether the cell is the rightmost or leftmost terminal of the net and the cost in x-wirelength can be accurately calculated.

Therefore, the dynamic programming algorithm can be easily combined with permutation of cells to speed up the process. We construct the permutations of cells in lexicographic order, so that the next permutation has a same prefix as the previous one, and thus the beginning rows in the table calculated by the dynamic programming algorithm for the previous permutation can be reused for the next one as possible. When constructing a new permutation that has the same prefix of $j - 1$ cells as the previous one and selecting a different jth cell, we keep the first $j - 1$ rows of the table in the dynamic programming algorithm and recompute the jth row for the costs of placing the new cell.

Note that in the dynamic programming algorithm, after we select the first j cells for a permutation, we already know the minimum cost of placing the first j cells within the window. Therefore, if the best placement for the first j cells of the permutation is already worse than the current best solution of the cell order polishing problem, we will discard all the permutations that have the same prefix of length j. This can be easily implemented in our algorithm.

Ordering for Inter-Row Cells

Similarly, we can design a dynamic programming algorithm for permutating a few nearby cells in multiple rows. Here, the window of available sites will include neighbor sites among multiple rows. The algorithm first decides how many cells will be assigned to each row from up to down, and then finds the optimal placement within the window for each permutation of cells, which also tells which row that each cell is assigned to.

We use the same method as in Sect. 7.4.3 to combine the dynamic programming algorithm with the lexicographic way of constructing permutations. Note that when we construct a new permutation that has the same prefix of $j - 1$ cells as the previous one by selecting a jth cell, and compute the jth row of the table, we only know that the rest of the cells will be placed in the same row as the jth cell or in lower rows. However, whether the remaining cells will be placed to the left or right of the jth cell is not clear. Therefore, the cost in y-wirelength can be accurately calculated, and if there is a net connecting to the jth cell and another non-placed cell, the cost in x-wirelength for the net is inaccurate. Because of the exponential time complexity of the algorithm, in practice, cell order polishing can only be applied to a small subset of cells at one time, and empirical studies show that the method is still very effective.

7.5 ISPD'06 Contest and APlace3.0

The ISPD'06 contest [35] emphasized the important of placement runtime. While APlace2 produces the highest quality results for the ISPD'05 benchmarks, it was not necessarily the fastest. Thus, we have developed APlace3 that is 2–2.5× faster than APlace2. We also investigated the use of alternative formulations for the wirelength objective function and the density constraints. In this section, we summarize our technical efforts in APlace3.0.

7.5.1 Exploring Alternative Wirelength Functions

While placement quality is typically evaluated according to HPWL, this "linear wirelength" function can not be efficiently minimized. Smooth linear wirelength objectives have been proposed in many works. In this section, we examine three important wirelength approximation functions, LOG–SUM–EXP [37], GORDIAN-L [41] and L_p-NORM [30], and propose to use them in a hybrid way in order to speed up the placer without losing wirelength quality.

The LOG–SUM–EXP approximation of HPWL was first proposed in [37] and recently applied in mPL [13] and APlace. The horizontal wirelength of a net e is written as

$$\mathrm{WL}(e) = \alpha \cdot \log(\textstyle\sum_i e^{x_i/\alpha}) + \alpha \cdot \log(\textstyle\sum_i e^{-x_i/\alpha}) \tag{7.12}$$

where α is a smoothing parameter: The LOG–SUM–EXP approximation is strictly convex, continuously differentiable and converges to HPWL as α converges to 0. In APlace, we adaptively modify the smoothing parameter α according to the grid spacing W_g, instead of using a small constant value. Thus, long nets (probably connecting cells in different bins) are "chosen to be minimized" and short nets (probably connecting cells in the same bin) are "ignored." Empirical results show that not only speedup of the placer, but also better global optimization and hence better placement quality, can be achieved in this way.

We are now able to compare the other two wirelength functions against the LOG–SUM–EXP approximation within the APlace framework. The method of GORDIAN-L is proposed in [41] to minimize a linear wirelength objective using iterated quadratic minimizations. The horizontal wirelength of a net e is formulated as

$$\mathrm{WL}(e) = \textstyle\sum_{i,j} (x_i - x_j)^2 / \gamma_{i,j} \tag{7.13}$$

where the factor $\gamma_{i,j} = max\{r_0, |x_i - x_j|\}$, which is constantly updated, and r_0 is the minimum value that $\gamma_{i,j}$ can take in order to prevent overflow. However, in our implementation, we use r_0 to choose long nets for more accurate minimization, and we decide the value of r_0 according the grid spacing W_g.

Our empirical studies show that the value of r_0 significantly affects the placer performance. We have implemented GORDIAN-L using the clique net model with edge weight equal to $1/q^2$ for a q-pin net, and obtained results on, e.g., the IBM ISPD04 benchmark set. We observe that when the value of r_0 is small, although the GORDIAN-L function is closer to linear wirelength, this does not necessarily lead to a better result. A proper value of r_0 can dramatically reduce average WL increase from 25.9% when $r_0 = 0.1W_g$ to only 4.1% when $r_0 = 2W_g$. We have also compared the two functions using the more realistic IBM ISPD05 benchmark circuits, and have found that the smallest average wirelength increase is 3.6%.

Last, we have also implemented the L_p-NORM approximation of HPWL that was proposed in [30]. The horizontal wirelength of a net e is expressed as

$$\mathrm{WL}(e) = \left(\textstyle\sum_{i,j}(x_i - x_j)^p + \beta\right)^{1/p} \tag{7.14}$$

where β is the smoothing parameter and the approximation converges to HPWL as β converges to 0 and p converges to ∞. In APlace3, similar to with the other wirelength functions, we adaptively increase p with cluster and grid levels during the placement and decide the value of β according to grid spacing. The average wirelength increase of the L_p-NORM function is 1.5% on the IBM ISPD05 circuits.

The very similar wirelength quality of the three approximation functions suggests that we may alternate them to achieve specific advantages. For a faster implementation of APlace, we may choose to apply GORDIAN-L with star net model during cluster-level placement to speed up the placer, but then apply the LOG–SUM–EXP approximation during flat-level placement to maintain the placement quality.

We also find out that most of the runtime of APlace is consumed by the line search algorithm of the CG solver. The wirelength function needs to be executed several times per CG iteration to decide the step length in the conjugate direction. Since the LOG–SUM–EXP approximation is very close to HPWL with fine grids and small smoothing parameters during flat-level placement, we only apply the LOG–SUM–EXP approximation to compute the wirelength derivatives for each iteration, but apply the actual HPWL function during line search, in order to reduce runtime.

7.5.2 Exploring Alternative Density Functions

Density function $D_g(\mathbf{x}, \mathbf{y})$ in (7.1), which is the total cell area in bin g, can be expressed in the form

$$D_g(\mathbf{x}, \mathbf{y}) = \textstyle\sum_v P_x(g, v) \cdot P_y(g, v) \tag{7.15}$$

where functions $P_x(g, v)$ and $P_y(g, v)$ denote the overlap between the grid bin g and cell v along the x and y directions, respectively. Figure 7.5 shows the trapezoidal-shaped $P_x(g, v)$ when the cell width W_v is larger than the bin width W_b and when $W_v < W_b$, as a function of the horizontal distance between cell and bin.

Naylor et al. [37] proposed to use a bell-shaped function $p_x(d)$ in quadratic form to approximate the overlap function as

$$p_x(d) = \begin{cases} W_v(1 - 2d^2/W_b^2) & (0 \leq d \leq W_b/2) \\ 2W_v(d - W_b)^2/W_b^2 & (W_b/2 \leq d \leq W_b) \end{cases} \tag{7.16}$$

for standard-cell placement. APlace2 follows this idea, but extends the function to handle large blocks for the purpose of mixed-size placement.

We compare the quadratic function with two other approximations with Gaussian function and ERFC function, as captured, respectively, by

$$p_x(d) = \min\{W_b, W_v\} \cdot e^{-4d^2/(\max\{W_b, W_v\})^2} \tag{7.17}$$

and

$$\begin{aligned} p_x(d) = (d - a) \cdot \operatorname{erfc}((d - a)/\theta)/2 \\ -(d - b) \cdot \operatorname{erfc}((d - b)/\theta)/2 \end{aligned} \tag{7.18}$$

where $0 \leq d \leq W_b$, $a = |W_v - W_b|/2$ and $b = (W_v + W_b)/2$. The second function provides a more flexible approximation of the step function in Figure 7.5, with smoothness controlled by the factor θ: The approximation converges to the step function when θ converges to 0. In our implementation, the value of θ is set to grid spacing W_g, which gives a smoother approximation for standard cells but a more accurate approximation for large macro blocks.

Benchmarking with IBM ISPD04 circuits shows roughly equal (within 1% error) wirelength quality for standard-cell placement using each of the three approximation functions. However, the Gaussian approximation leads to an average of 7.4% wirelength increase on IBM ISPD05 circuits, compared to APlace2, while the ERFC function slightly reduces the wirelength results.

For a faster implementation, we apply the ERFC approximation for flat-level placement; i.e., we only use the ERFC approximation to compute the partial derivatives related to density terms for each CG iteration, but apply the actual cell density function during line search, in order to speed up the placer.

7.6 Experimental Results

In this section, we report the performance of our placer on the different benchmark sets: IBM ISPD'04, ICCAD'04, ISPD'05, and ISPD'06. We use APlace2.0 to report the results of the IBM ISPD'04, ICCAD'04, ISPD'05 benchmarks, and use APlace3.0 to report the results of ISPD'06 benchmarks. We also compare the performance of various placers and APlace2.0 using the Zero-Change Netlist Transformations (ZCNT) benchmarking framework [23]. These benchmarks feature different characteristics, all of which are helpful in testing the placer's capabilities: IBM ISPD'04 benchmarks [44] are composed of standard cells and test basic placer performance without worry about other "extras" such as whitespace distribution, movable blocks and fixed blocks. IBM-PLACE 2.0 benchmarks [49] are also composed of standard cells, but designed to evaluate routability of placements. IBM ICCAD'04 benchmarks [1] contain large movable macros and assess the performance of a placer in simultaneous floorplanning and standard cell placement. IBM ISPD'05 benchmarks [36] contain large amount of whitespace, and fixed blocks and I/Os, as well as designs with over two million components. These benchmarks are directly derived from industrial ASIC designs, and preserve the physical structure of the designs, unlike other benchmark suites. Thus, they test a placer's ability to handle modern layout features such as whitespace and fixed blocks, and represent the current and future physical design challenges.

In all of our experiments, we use a Linux machine with 1.6 GHz CPU and 2 GB of memory. In the first set of experiments, we report our results for the ISPD'05 benchmarks in Table 7.1. We report the results of other placers as published in the contest results [36]. The first column of the table shows the nine placers participating in the ISPD-2005 placement contest [36]. The HPWL for each of the six benchmarks obtained by the placers are shown in the next six columns.[2] We normalize each wirelength result based on the HPWL obtained by our placer. The last column in Table 7.1 shows the average normalized ratio for each placer. Our placer gives the best results on all six benchmarks and on average is better than the best of all other placers by 6%. The entire benchmark set takes 113.2 h of runtime to complete. Executing Capo v9.1 [1] on our machines takes around 37.8 h to complete. Thus, on the average, our placer is 3× slower than Capo.

In the second set of experiments, we evaluate the performance of our placer on the IBM ICCAD'04 mixed-size benchmarks [1]. These recent mixed-size circuits contain large movable blocks with non-ignorable aspect ratios and I/Os placed at the blocks' periphery, and thus are more realistic than the previously widely used IBM ISPD'02 mixed-size benchmarks. Our results are summarized in Table 7.2 and compared to FengShui and Capo. The first five columns of Table 7.2 show the

Table 7.1. Results of all placers on the ISPD-2005 contest benchmarks. The results of other placers are from the ISPD 2005 paper by Nam et al.

Placer	Benchmark						Av.
	adaptec2	adaptec4	bigblue1	bigblue2	bigblue3	bigblue4	
Ours	87.31	187.65	94.64	143.82	357.89	833.21	1.00
mFAR [20]	91.53	190.84	97.70	168.70	379.95	876.28	1.06
Dragon [43]	94.72	200.88	102.39	159.71	380.45	903.96	1.08
mPL [12]	97.11	200.94	98.31	173.22	369.66	904.19	1.09
FastPlace [45]	107.86	204.48	101.56	169.89	458.49	889.87	1.16
Capo [39]	99.71	211.25	108.21	172.30	382.63	1098.76	1.17
NTUP [14]	100.31	206.45	106.54	190.66	411.81	1154.15	1.21
FengShui [3]	122.99	337.22	114.57	285.43	471.15	1040.05	1.50
Kraftwerk+Domino [38]	157.65	352.01	149.44	322.22	656.19	1403.79	1.84

[2] Due to the contest setup, the results are obtained after five days of tuning of the placers with the circuits.

Table 7.2. Results on the IBM ICCAD'04 mixed-size benchmarks. Results of Capo and FengShui are from the ICCAD 2004 paper by Adya et al.

bench	Placer						
	Ours					FS	Capo
	gpWL (e6)	CPU (s)	legWL (e6)	dpWL (e6)	CPU (s)	dpWL (e6)	dpWL (e6)
ibm01	2.17	436	2.20	2.14	28	2.56	2.67
ibm02	4.83	949	4.73	4.61	57	6.05	5.54
ibm03	6.94	1078	6.93	6.72	67	8.77	8.67
ibm04	7.70	1169	7.83	7.60	73	8.38	9.79
ibm05	9.82	915	9.90	9.70	61	9.94	10.82
ibm06	6.31	988	6.17	5.99	87	6.99	7.35
ibm07	10.04	1445	10.35	10.02	124	11.37	11.23
ibm08	12.65	1328	12.64	12.34	149	13.51	16.02
ibm09	12.56	2515	12.63	12.15	170	14.12	15.51
ibm10	30.32	3518	29.82	28.55	354	41.96	34.98
ibm11	19.62	4253	19.41	18.67	236	21.19	22.31
ibm12	34.51	3598	34.56	33.51	314	40.84	40.78
ibm13	24.28	4869	24.07	23.03	308	25.45	28.70
ibm14	37.51	4878	36.87	35.90	479	39.93	40.97
ibm15	49.97	5337	48.93	46.82	708	51.96	59.19
ibm16	57.15	6244	57.02	54.58	905	62.77	67.00
ibm17	67.39	6495	69.01	66.49	834	69.38	78.78
ibm18	44.40	9159	43.11	42.14	797	45.59	50.39
Average				0.86		1.00	1.05

HPWL after global placement, runtime of global placement, HPWL after legalization, HPWL after detailed placement and runtime of detailed placement, respectively, for our placer. According to the table, the legalization and detail placement steps reduce the wirelength by 4% on average, which indicates a strong global placement and effective post-processing. We report the results of FengShui v2.6 and Capo v9.0 as recently published in [1] in the last two columns. The last row in Table 7.2 shows the average normalized wirelength ratio based on FengShui's results. The results show that our placements are better than FengShui and Capo for all the circuits, and the average improvement is around 14% over FengShui and 19% over Capo. We also believe it is possible to further improve our results if cell flipping is applied – an improvement executed by Capo. The runtime of the entire benchmark set takes 18.0h of runtime. The total runtime of FengShui and Capo on a Linux machine with 2.4 GHz CPU are 8.7 h and 14.0 h respectively, as reported in [1].

The focus of our third set of experiments is on the IBM ISPD'04 standard cell benchmark suite. Wirelength after global placement, wirelength after detail placement, and total runtime of our placer are shown in the second to fourth columns of Table 7.3. We also report the latest wirelength results of mPL v5.0 (in the fifth column), as well as the normalized wirelength ratio with respect to mPL's results of Capo v9.0, Dragon v3.01, FastPlace v1.0 and FengShui v5.0 (in the sixth to ninth

Table 7.3. Results on the IBM ISPD'04 benchmarks. Results of Capo, Dragon, FastPlace, FengShui and mPL are from the ISPD 2005 paper by Chan et al.

bench	Placer							
	Ours			mPL5	Capo	Dragon	FP	FS
	gpWL (e6)	dpWL (e6)	CPU (s)	dpWL (e6)	nWL	nWL	nWL	nWL
ibm01	1.60	1.63	333	1.67	1.08	1.02	1.09	1.08
ibm02	3.54	3.48	649	3.62	1.09	1.02	1.06	1.02
ibm03	4.46	4.51	874	4.57	1.10	1.05	1.12	1.03
ibm04	5.56	5.61	996	5.75	1.06	1.00	1.04	1.05
ibm05	9.63	9.49	1245	9.92	1.02	0.98	1.05	1.00
ibm06	4.73	4.78	951	5.10	1.11	0.98	1.04	1.02
ibm07	7.97	7.90	1892	8.23	1.11	1.04	1.08	1.09
ibm08	9.16	9.46	1296	9.38	1.05	0.96	1.02	-
ibm09	8.84	8.93	2104	9.33	1.08	1.07	1.12	1.06
ibm10	17.20	16.95	3089	17.3	1.10	1.04	1.07	1.07
ibm11	13.22	13.38	2936	14.0	1.09	1.03	1.09	1.04
ibm12	21.83	21.47	3124	22.3	1.11	1.03	1.08	1.07
ibm13	16.46	16.60	3702	16.6	1.10	1.05	1.11	1.09
ibm14	30.55	30.76	4648	31.6	1.10	1.05	1.11	1.04
ibm15	38.38	38.81	7364	38.5	1.09	1.04	1.13	1.07
ibm16	41.36	41.32	7181	43.0	1.10	1.05	1.07	1.09
ibm17	60.82	59.22	10261	61.3	1.09	1.08	1.08	1.08
ibm18	39.32	38.98	10127	41.0	1.09	1.02	1.10	1.04
Average		0.97		1.00	1.09	1.03	1.08	1.06

columns, respectively), as published in [13]. The last row in Table 7.3 shows the average normalized ratio with respect to mPL's results for each placer. The results show that our placements are better than other placers for most of the circuits. On average, our placer is better than the best of all other placers, mPL, by 3%, and better than Capo, Dragon, FastPlace, and FengShui by 11%, 6%, 10%, and 8% respectively. The entire benchmark set takes 17.4h to place. The total runtime of mPL5, Capo, Dragon, FastPlace, and FengShui on a Linux machine with 2.4 GHz CPU are 3.2, 7.2, 39.2, 0.6, and 6.4 h respectively, as reported in [13].

Our fourth set of experiments is on the IBM-PLACE 2.0 standard cell benchmark suite for routability. Wirelength after global placement, wirelength after detail placement and the runtime of our placer are shown in the second to fourth columns of Table 7.4. After placement, we use WarpRoute (v2.4.44) to perform routing with the existing grids for global routing (-grouteGrid existing). Routed final wirelength, the number of vias, the number of violations, the over capacity gcells in percentage and total runtime of WarpRoute are shown in the fifth to ninth columns. We also report the latest routed wirelength results of mPL-R+WSA (in the last column), as published in [34]. We observe that almost all of our placements are successfully routable with good wirelength; finished routings with a small number of violations can be manually fixed. The last row in Table 7.4 shows the average normalized ratio with respect to mPL-R+WSA's results for our placer. On average, our placer is better than mPL-R+WSA by 12%.

The fifth set of experiments report the results of APlace3.0 on the ISPD'06 benchmarks [35]. The reports results is calculated by using a weighted combination of the wirelength, runtime, and the final density of the layout. Table 7.5 gives the results normalized to the best placer in the ISPD'06 contest.

Finally, we conduct a sixth experiment on benchmarks with known pre-calculated wirelength placements. We tabulate our results for the PEKO-MS'05 benchmarks in Table 7.6 and the PEKO-MS'06 benchmarks in Table 7.7. In Table 7.7, SOV/bin overflow per bin (i.e., penalty %). SHPWL is the scaled bin overflow adjusted HPWL, i.e., SHPWL = HPWL $\times(1 + \text{SOV/bin})$. We also compare the results of

Table 7.4. Results on the IBM-PLACE 2.0 benchmarks. Results of mPL-R+WSA are from the ICCAD 2004 paper by Li et al. **vio.** give the number of routing violations.

bench	Placer								
	Ours								mPL-R+WSA
	gpWL (e6)	dpWL (e6)	CPU (s)	route WL (e6)	vias	vio.	over (cap%)	CPU (mm:ss)	route WL (e6)
ibm01-e	0.480	0.509	650	0.700	125062	1	1.66%	19:57	0.772
ibm01-h	0.476	0.515	434	0.721	126655	0	2.05%	35:18	0.751
ibm02-e	1.355	1.389	1097	1.804	243595	0	0.43%	11:03	1.890
ibm02-h	1.330	1.373	917	1.855	251958	0	1.60%	13:39	1.940
ibm07-e	3.083	3.182	2857	3.746	478611	1	0.72%	22:57	4.290
ibm07-h	2.977	3.223	2916	3.903	498805	1	1.92%	35:00	4.430
ibm08-e	3.240	3.330	3286	3.980	576366	1	0.20%	27:05	4.580
ibm08-h	3.074	3.222	2327	3.953	574481	1	0.23%	24:08	4.490
ibm09-e	2.658	2.809	3901	3.023	495073	2	0.01%	18:16	3.500
ibm09-h	2.606	2.757	2112	3.027	503410	2	0.02%	17:15	3.650
ibm10-e	5.193	5.340	7529	5.977	758598	3	0.07%	28:37	6.840
ibm10-h	4.889	5.258	5471	5.931	772744	1	0.09%	28:12	6.760
ibm11-e	4.151	4.295	3064	4.577	638523	3	0.05%	22:58	5.160
ibm11-h	4.044	4.218	3645	4.654	656525	4	0.17%	23:31	5.150
ibm12-e	7.203	7.299	7816	8.337	892915	2	0.09%	44:06	10.520
ibm12-h	7.089	7.210	8640	8.317	902465	0	0.15%	42:51	10.130
Average				0.88					1.00

Table 7.5. APlace3 results using the ISPD 2006 contest scoring function (combining wirelength, runtime, and utilization). Results are normalized to the best scoring placer.

Placer	Benchmark								Av.
	adaptec5	nblue1	nblue2	nblue3	nblue4	nblue5	nblue6	nblue7	
APlace3	1.26	1.20	1.05	1.13	1.35	1.21	1.06	1.05	1.16

Table 7.6. Results of APlace on Peko-MS'05 benchmarks with known optimal HPWL.

PEKO-MS-05	Peko opt	APlace	Ratio
adaptec1	20056216	22.64	1.13
adaptec2	24969764	27.90	1.12
adaptec3	40954784	46.20	1.13
adaptec4	39391712	44.38	1.13
bigblue1	20858240	27.24	1.31
bigblue2	42256768	54.41	1.29
bigblue3	94399040	112.11	1.19
bigblue4	171477120	221.41	1.29

Table 7.7. Results of APlace on Peko-MS'06 benchmarks with known optimal HPWL and with constrained bin densities. SOV/bin is the scaled bin overflow per bin (i.e., penalty %). SHPWL is the scaled bin overflow adjusted HPWL, i.e., SHPWL=HPWL $\times$ (1 + SOV/bin).

PEKO MS'06	Peko opt	HPWL	SOV /bin	SHPWL	APlace HPWL	SOV /bin	SHPWL	Hratio	Sratio
adaptec5	81893792	81.89	9.99	90.07	92.37	117.16	200.59	1.13	2.23
newblue1	20500032	20.50	1.73	20.86	27.63	89.69	52.41	1.35	2.51
newblue2	32869280	32.87	10.29	36.25	47.09	162.49	123.61	1.43	3.41
newblue3	73514272	73.51	9.55	80.54	88.18	132.61	205.11	1.20	2.55
newblue4	49143583	49.14	9.26	53.69	55.19	75.07	96.62	1.12	1.80
newblue5	102083104	102.08	9.58	111.87	266.11	217.56	845.07	2.61	7.55
newblue6	90657856	90.66	8.36	98.24	104.22	48.37	154.63	1.15	1.57
newblue7	206175072	206.18	7.07	220.74	269.08	119.79	591.40	1.31	2.68

different placers using the Zero-Change Netlist Transformation (ZCNT) benchmarking approach [23]. In the ZCNT framework, given a circuit and a placer, the placer is executed on the circuit to get a initial placement with some wirelength. Then the given circuit and the initial placement are used to produce a new circuit that is structurally different from the original circuit but yet has two key properties: (1) both circuits have the same wirelength with the respect to the given placement; and (2) the unknown optimal wirelength for the new circuit is greater than or equal to the original circuit. By executing the placer on the new circuit, we can interpret any deviation in wirelength (with respect to the initial placement) as a measure of suboptimality. Figure 7.9 gives the performance of various placers on the IBM (version 1) benchmarks in response to the ZCNTs. Clearly, APlace displays the least amount of suboptimality.

The excellent performance of our placer on all benchmark sets clearly show that our placement methods are: (1) scalable, (2) deliver high quality placements, and (3) capable of handling various netlist and layout features such as movable/fixed blocks and whitespace.

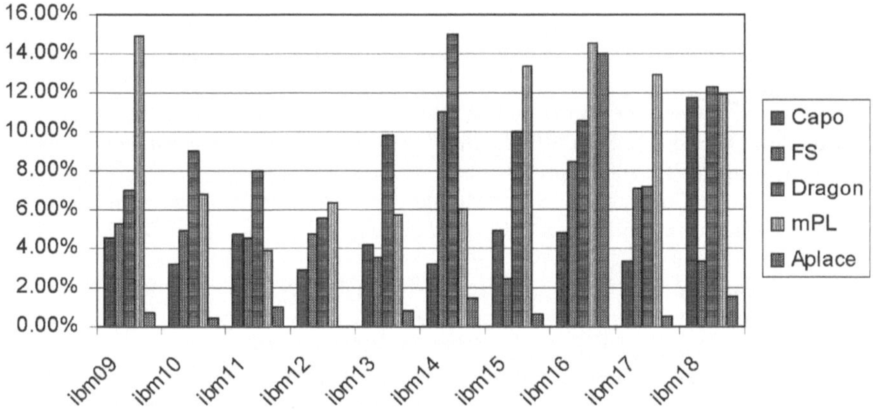

Fig. 7.9. Deviations in wirelength when benchmarking the different placers using the Zero-Change Netlist Transformations (ZCNT) approach [23]. We report results for the largest circuits of the IBM version 1.0 benchmarks.

References

1. S. N. Adya, S. Chaturvedi, J. A. Roy, D. A. Papa and I. L. Markov, "Unification of Partitioning, Placement and Floorplanning," in *Proc. IEEE International Conference on Computer-Aided Design*, 2004, pp. 550–557
2. A. R. Agnihotri, S. Ono, C. Li, M. C. Yildiz, A. Khatkhate, C.-K. Koh and P. H. Madden, "Mixed Block Placement via Fractional Cut Recursive Bisection," *IEEE Transactions on Computer-Aided Design*, vol. 24(5), 2005
3. A. Agnihotri, S. Ono and P. Madden, "Recursive Bisection Placement: Feng Shui 5.0 Implementation Details," in *Proc. ACM/IEEE International Symposium on Physical Design*, 2005, pp. 230–232
4. A. Agnihotri, M. Yildiz, A. Khatkhate, A. Mathur, S. Ono and P. Madden, "Fractional Cut: Improved Recursive Bisection Placement," in *Proc. IEEE International Conference on Computer-Aided Design*, 2003, pp. 307–310
5. C. Alpert, A. Kahng, G.-J. Nam, S. Reda and P. Villarrubia, "A Semi-Persistent Clustering Technique for VLSI Circuit Placement," in *Proc. ACM/IEEE International Symposium on Physical Design*, 2005, pp. 200–207
6. C. J. Alpert, J. H. Huang and A. B. Kahng, "Multilevel Circuit Partitioning," in *Proc. ACM/IEEE Design Automation Conference*, 1997, pp. 530–533
7. U. Brenner, A. Pauli and J. Vygen, "Almost Optimum Placement Legalization by Minimum Cost Flow and Dynamic Programming," in *Proc. ACM/IEEE International Symposium on Physical Design*, 2004, pp. 2–9
8. U. Brenner and J. Vygen, "Faster Optimal Single-Row Placement with Fixed Ordering," in *Proc. Design, Automation and Test in Europe*, 2000, pp. 117–121
9. A. E. Caldwell, A. B. Kahng and I. L. Markov, "Optimal Partitioners and End-Case Placers for Standard-Cell Layout," in *Proc. ACM/IEEE International Symposium on Physical Design*, 1999, pp. 90–96
10. A. E. Caldwell, A. B. Kahng and I. L. Markov, "Can Recursive Bisection Alone Produce Routable Placements?" in *Proc. ACM/IEEE Design Automation Conference*, 2000, pp. 477–482

11. T. F. Chan, J. Cong, T. Kong and J. R. Shinnerl, "Multilevel Optimization for Large-Scale Circuit Placement," in *Proc. IEEE International Conference on Computer-Aided Design*, 2000, pp. 171–176
12. T. F. Chan, J. Cong, M. Romesis, J. R. Shinnerl, K. Sze and M. Xie, "mPL6: A Robust Multilevel Mixed-Size Placement Engine," in *Proc. ACM/IEEE International Symposium on Physical Design*, 2005, pp. 227–229
13. T. F. Chan, J. Cong and K. Sze, "Multilevel Generalized Force-directed Method for Circuit Placement," in *Proc. ACM/IEEE International Symposium on Physical Design*, 2005, pp. 185–192
14. T.-C. Chen, T.-C. Hsu, Z.-W. Jiang and Y.-W. Chang, "NTUplace: A Ratio Partitioning Based Placement Algorithm for Large-Scale Mixed-Size Designs," in *Proc. ACM/IEEE International Symposium on Physical Design*, 2005, pp. 236–238
15. K. Doll, F. Johannes and K. Antreich, "Iterative Placement Improvement by Network Flow Methods," *IEEE Transactions on Computer-Aided Design of Integrated Circuits and Systems*, vol. 13(10), pp. 1189–1200, 1994
16. H. Eisenmann and F. M. Johannes, "Generic Global Placement and Floorplanning," in *Proc. ACM/IEEE Design Automation Conference*, 1998, pp. 269–274
17. S. Goto, "An Efficient Algorithm for the Two-Dimensional Placement Problem in Electrical Circuit Layout," *IEEE Transactions on Circuits and Systems*, vol. 28(1), pp. 12–18, 1981
18. D. Hill, "Method and System for High Speed Detailed Placement of Cells Within an Integrated Circuit Design," *US Patent 6370673*, 2001
19. B. Hu and M. Marek-Sadowska, "Fine Granularity Clustering-Based Placement," *IEEE Transactions on Computer-Aided Design of Integrated Circuits and Systems*, vol. 23(4), pp. 527–536, 2004
20. B. Hu, Y. Zeng and M. Marek-Sadowska, "mFAR: Fixed-Points-Addtion-Based VLSI Placement Algorithm," *Proc. ACM/IEEE International Symposium on Physical Design*, 2005, pp. 239–241
21. A. B. Kahng, I. Markov and S. Reda, "On Legalization of Row-Based Placements," in *Proc. IEEE Great Lakes Symposium on VLSI*, 2004, pp. 214–219
22. A. B. Kahng and Q. Wang, "Implementation and Extensibility of an Analytic Placer," in *Proc. ACM/IEEE International Symposium on Physical Design*, 2004, pp. 18–25
23. A. B. Kahng and S. Reda, "Zero-Change Netlist Transformations: A New Technique for Placement Benchmarking," *IEEE Transactions on Computer-Aided Design of Integrated Circuits and Systems*, vol. 25(121), pp. 2806–2819, 2006
24. A. B. Kahng and S. Reda, "An Analytic Placer for Mixed-Size Placement and Timing-Driven Placement," in *Proc. IEEE International Conference on Computer-Aided Design*, 2004, pp. 565–572
25. A. B. Kahng, P. Tucker and A. Zelikovsky, "Optimization of Linear Placements for Wirelength Minimization with Free Sites," in *Proc. IEEE Asia and South Pacific Design Automation Conference*, 1999, pp. 241–244
26. A. B. Kahng and Q. Wang, "Implementation and Extensibility of an Analytic Placer," *IEEE Transactions on Computer-Aided Design* 24(5) (2005), pp. 734–747
27. A. B. Kahng, S. Reda, and Q. Wang, "APlace: A General Analytic Placement Framework," in *Proc. ACM/IEEE International Symposium on Physical Design*, 2005, pp. 233–235
28. A. B. Kahng, S. Reda, and Q. Wang, "Architecture and Details of a High Quality, Large-Scale Analytical Placer," in *Proc. International Conference Computer-Aided Design*, 2005, pp. 891–898

29. A. B. Kahng and X. Xu, “Accurate Pseudo-Constructive Wirelength and Congestion Estimation,” in *Proc. ACM International Workshop on System-Level Interconnect Prediction*, 2003, pp. 61–68
30. A. A. Kennings and I. L. Markov, “Analytical Minimization of Half-Perimeter Wirelength”, *Proc. IEEE/ACM Asia and South Pacific Design Automation Conference*, Jan. 2000, pp. 179–184
31. G. Karypis, R. Aggarwal, V. Kumar and S. Shekhar, “Multilevel hypergraph partitioning: Application in VLSI domain,” in *Proc. ACM/IEEE Design Automation Conference*, 1997, pp. 526–529
32. G. Karypis and V. Kumar, “Multilevel k-way hypergraph partitioning,” in *Proc. ACM/IEEE Design Automation Conference*, 1999, pp. 343–348
33. A. Khatkhate, C. Li, A. R. Agnihotri, M. C. Yildiz, S. Ono, C.-K. Koh and P. H. Madden, “Recursive Bisection Based Mixed Block Placement,” in *Proc. ACM/IEEE International Symposium on Physical Design*, 2004, pp. 84–89
34. C. Li, M. Xie, C.-K. Koh, J. Cong and P. H. Madden, “Routability-Driven Placement and White Space Allocation,” in *Proc. IEEE International Conference on Computer-Aided Design*, 2004, pp. 394–401
35. G.-J. Nam, “ISPD 2006 Placement Contest: Benchmark Suite and Results,” in *Proc. ACM/IEEE International Symposium on Physical Design*, 2006, pp. 167–167
36. G.-J. Nam, C. Alpert, P. Villarrubia, B. Winter and M. Yildiz, “The ISPD2005 Placement Contest and Benchmark Suite,” in *Proc. ACM/IEEE International Symposium on Physical Design*, 2005, pp. 216–219
37. W. Naylor, “Non-Linear Optimization System and Method for Wire Length and Delay Optimization for an Automatic Electric Circuit Placer,” *US Patent 6301693*, 2001
38. B. Obermeier, H. Ranke and F. M. Johannes, “Kraftwerk – A Versatile Placement Approach,” in *Proc. ACM/IEEE International Symposium on Physical Design*, 2005, pp. 242–244
39. J. A. Roy, D. A. Papa, S. N. Adya, H. H. Chan A. N. Ng, J. F. Lu and I. L. Markov, “Capo: Robust and Scalable Open-Source Min-Cut Floorplacer,” in *Proc. ACM/IEEE International Symposium on Physical Design*, 2005, pp. 224–226
40. C. Sechen and K. W. Lee, “An Improved Simulated Annealing Algorithm for Row-Based Placement,” in *Proc. IEEE International Conference on Computer-Aided Design*, 1987, pp. 478–481
41. G. Sigl, K. Doll and F. M. Johannes, “Analytical Placement: A Linear or a Quadratic Objective Function?,” in *Proc. ACM/IEEE Design Automation Conference*, 1991, pp. 427–431
42. W.-J. Sun and C. Sechen, “Efficient and Effective Placement for Very Large Circuits,” *IEEE Transactions on Computer-Aided Design of Integrated Circuits and Systems*, vol. 14(5), pp. 349–359, 1995
43. T. Taghavi, X. Yang, B. K. Choi, M. Wang and M. Sarrafzadeh, “DRAGON2005: Large-Scale Mixed-Size Placement Tool,” in *Proc. ACM/IEEE International Symposium on Physical Design*, 2001, pp. 245–247
44. N. Viswanathan and C. Chu, “FastPlace: Efficient Analytical Placement Using Cell Shifting, Iterative Local Refinement and a Hybrid Net Model,” in *Proc. ACM/IEEE International Symposium on Physical Design*, 2004, pp. 26–33
45. N. Viswanathan and C. Chu, “FastPlace: An Analytical Placer for Mixed-Mode Designs,” in *Proc. ACM/IEEE International Symposium on Physical Design*, 2005, pp. 221–223
46. J. Vygen, “Algorithms for Large-Scale Flat Placement,” in *Proc. ACM/IEEE Design Automation Conference*, 1997, pp. 746–751

47. J. Vygen, "Algorithms for Detailed Placement of Standard Cells," in *Design, Automation and Test in Europe*, 1998, pp. 321–324
48. M. Wang, X. Yang and M. Sarrafzadeh, "DRAGON2000: Standard-Cell Placement Tool for Large Industry Circuits," in *Proc. IEEE International Conference on Computer-Aided Design*, 2001, pp. 260–263
49. X. Yang, B.-K. Choi and M. Sarrafzadeh, "Routability Driven White Space Allocation for Fixed-Die Standard-Cell Placement," in *Proc. ACM/IEEE International Symposium on Physical Design*, 2002, pp. 42–47

8

FastPlace: An Efficient Multilevel Force-Directed Placement Algorithm

Natarajan Viswanathan, Min Pan and Chris Chu
Department of Electrical and Computer Engineering, Iowa State University, Ames, IA 50011-3060
{nataraj, panmin, cnchu}@iastate.edu

8.1 Introduction

Placement is a critical component in the physical synthesis design flow of large-scale integrated circuits and is a major contributor to timing closure results. It is often run multiple times during various stages of the physical synthesis flow. In addition, circuit sizes that need to be handled by placement algorithms are steadily increasing to over tens of millions of modules. Hence, it is necessary to have efficient and scalable placement algorithms that can produce high-quality solutions satisfying a variety of design objectives.

An important constraint that needs to be handled by current placers is that of placement congestion. Placement is typically run in an iterative manner along with timing optimization transforms like buffer insertion and gate sizing. Additionally, it has a major impact on the subsequent routing stage. Hence, placement algorithms should be congestion aware so as to provide space for the subsequent timing optimization and routing stages.

In this chapter, we describe *FastPlace* an efficient congestion aware multilevel force-directed placement algorithm for large-scale mixed-size designs. The key features of *FastPlace* are:

- A *multilevel framework* [24] within the global placement stage to handle large-scale placement circuits. This is achieved by employing a two-level clustering scheme – an initial netlist based fine-grain clustering, followed by a netlist and physical based coarse-grain clustering that uses information from an initial placement of the fine-grain clusters.

- A *Hybrid net model* [22] to speed up the quadratic program solver. The Hybrid net model is a combination of the traditional clique and star net models. It results in a substantial reduction in the number of nonzero entries in the connectivity matrix as compared to the clique model thereby resulting in a significant speed-up of the quadratic program solver.
- An efficient *Cell Shifting* technique [22, 23] to spread the modules during the early stages of the placement flow. This technique roughly maintains the relative order of the modules obtained by solving the quadratic program in both the horizontal and vertical directions.
- An *Iterative Local Refinement* technique [22, 24] to reduce the wirelength based on the half-perimeter measure. This technique is applied on a coarse global placement and is highly effective in simultaneously reducing the wirelength while spreading the modules. It can also effectively handle placement blockages and placement congestion constraints.
- A *macro-block legalization* technique [23] to resolve overlaps among the macro-blocks present in the circuit. For any representation specifying the relative positions of the macros, the legalizer uses an optimal *Iterative Clustering Algorithm* to place the macros with minimum perturbation from their global placement positions.
- An efficient and robust *standard-cell legalization* technique [23] that operates on the segments created in the placement region due to the presence of placement blockages. This technique satisfies segment capacities and legalizes the standard-cells within the segments.
- A fast and effective *detailed placement algorithm* [19] that can work on both row-based standard-cell placement and placement in the presence of fixed macros. This algorithm consists of four techniques – *Global Swap*, *Vertical Swap*, *Local Cell Re-ordering* and *Single-segment Clustering* and is highly effective in further reducing the placement wirelength.

The rest of this chapter is organized as follows: In Sect. 8.2 we give an overview of our algorithm with emphasis on the global placement flow. Section 8.3 describes the generic quadratic placement methodology. Sections 8.4–8.9 describe the individual components of *FastPlace* in detail. Experimental results are presented in Sect. 8.10 followed by the conclusions in Sect. 8.11

8.2 Overview of the Algorithm

The entire flow of our placement algorithm is summarized in Figure 8.1. It is divided into three stages (1) congestion aware global placement using a multilevel framework, (2) legalization of the macro-blocks followed by legalization of the standard-cells and (3) detailed placement. In this section, we give an overview of the multilevel global placement framework and describe the individual components of the flow in more detail in the subsequent sections.

The multilevel global placement framework used within *FastPlace* is summarized in Figure 8.2. It follows the classical hierarchical flow that has been used in

Stage 1: Global Placement

Level 1: Initial Placement

1. Construct fine-grain clusters using netlist based clustering
2. Solve initial quadratic program
3. Repeat
   a. Perform *regular Iterative Local Refinement* on fine-grain clusters
4. Until the placement is roughly even

Level 2: Coarse Global Placement

5. Construct coarse-grain clusters using netlist and physical based clustering
6. Repeat
   a. Solve the convex quadratic program
   b. Perform *cell-shifting* on coarse-grain clusters and add spreading forces
7. Until the placement is roughly even
8. Repeat
   a. Perform *density-based Iterative Local Refinement* on coarse-grain clusters
   b. Perform *regular Iterative Local Refinement* on coarse-grain clusters
   c. Perform *cell-shifting* on coarse-grain clusters
9. Until the placement is quite even

Level 3: Refinement of fine-grain clusters

10. Un-cluster coarse-grain clusters
11. Perform *density-based Iterative Local Refinement* on fine-grain clusters
12. Perform *regular Iterative Local Refinement* on fine-grain clusters

Level 4: Refinement of flat netlist

13. Un-cluster fine-grain clusters
14. Perform *density-based Iterative Local Refinement* on flat netlist
15. Perform *regular Iterative Local Refinement* on flat netlist

Stage 2: Legalization

16. Legalize and fix movable macro-blocks using *Iterative Clustering Algorithm*
17. Move standard-cells among segments to satisfy segment capacities
18. Legalize standard-cells within segments

Stage 3: Detailed Placement

Fig. 8.1. Outline of the placement algorithm (*Source*: [24] © 2007 IEEE).

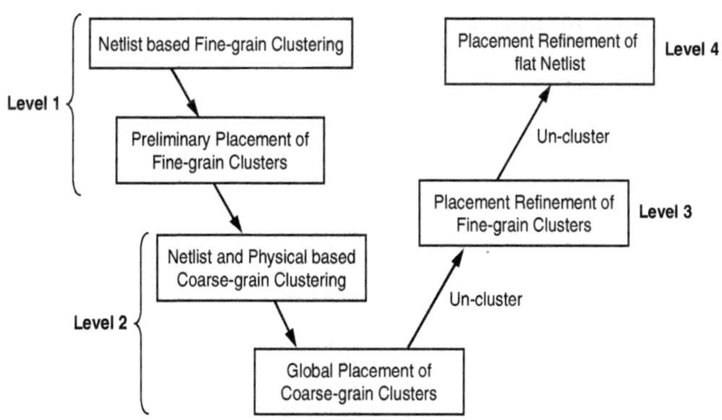

Fig. 8.2. Multilevel global placement framework (*Source*: [24] © 2007 IEEE).

many existing placement algorithms [2, 3, 5, 10, 12, 18]. In Level 1 of the multilevel flow, we create fine-grain clusters using a netlist-based connectivity score and perform a fast initial placement of the fine-grain clusters. In Level 2 we perform a second level of clustering in which we use a netlist- and physical-based clustering score to generate coarse-grain clusters. We then perform global placement on the coarse-grain clustered netlist until the clusters are evenly distributed over the placement region. Since the number of modules at this level is significantly less as compared to the original flat netlist, this step is quite fast and greatly contributes to the overall efficiency of the placement algorithm. After the placement of the coarse-grain clusters, we perform a series of un-clustering and placement refinements in Levels 3 and 4, finally yielding a global placement solution of the original flat netlist.

8.3 Quadratic Placement Methodology

The quadratic placement approach uses the analogy of springs to model the connectivity between the modules of a circuit. During quadratic placement, the total potential energy of the springs, which is a quadratic function of their length, is minimized to produce a placement solution. This is equivalent to a force equilibrium state in the spring system.

The circuit netlist that describes the connectivity between the modules is a weighted hypergraph $G = (V, E)$, where $V = \{v_1, v_2, \ldots, v_m\}$ is the set of vertices representing the modules to be placed and $E = \{e_1, e_2, \ldots, e_n\}$ is the set of hyperedges representing the connections or nets between the modules. Each net $e \in E$ has a weight w_e that reflects the criticality of this net.

In order to model the circuit by a spring system, the hypergraph needs to be transformed into a graph by using a suitable net model. This is equivalent to saying that each net in the hypergraph needs to be transformed into a set of two-pin nets. To perform this transformation we use the Hybrid net model which is described in Sect. 8.4. For the following discussion on quadratic placement we assume that this transformation has been applied.

Let m be the total number of movable modules in the circuit and (x_i, y_i) the coordinates of the center of module i. A placement of the circuit is given by the two m-dimensional vectors $\mathbf{x} = (x_1, x_2, \ldots, x_m)$ and $\mathbf{y} = (y_1, y_2, \ldots, y_m)$.

Consider a net between two movable modules i and j in the circuit. Let W_{ij} be its weight. Then the cost of the net between the two modules is

$$\frac{1}{2} W_{ij}[(x_i - x_j)^2 + (y_i - y_j)^2] \tag{8.1}$$

If a movable module i is connected to a fixed module f with coordinates (x_f, y_f), then the cost of the net is given by

$$\frac{1}{2} W_{if}[(x_i - x_f)^2 + (y_i - y_f)^2] \tag{8.2}$$

The objective function that sums up the cost of all the nets can be written in matrix notation as [8]

$$\Phi(x, y) = \frac{1}{2}x^{\mathrm{T}}Cx + d_x^{\mathrm{T}}x + \frac{1}{2}y^{\mathrm{T}}Cy + d_y^{\mathrm{T}}y + \text{constant} \tag{8.3}$$

where C is an $m \times m$ symmetric positive definite matrix and d_x, d_y are m-dimensional vectors. Since (8.3) is separable into $\Phi(x, y) = \Phi(x) + \Phi(y)$, only the x-dimension is considered for subsequent discussion, which is

$$\Phi(x) = \frac{1}{2}x^{\mathrm{T}}Cx + d_x^{\mathrm{T}}x + \text{constant} \tag{8.4}$$

From (8.1), the cost in the x-direction between two movable modules i and j is

$$\frac{1}{2}W_{ij}(x_i^2 + x_j^2 - 2x_i x_j) \tag{8.5}$$

If c_{ij} is the entry in row i and column j of matrix C, then the first and second terms in expression (8.5) contribute W_{ij} to c_{ii} and c_{jj}, respectively. The third term contributes $-W_{ij}$ to c_{ij} and c_{ji}. From expression (8.2), the cost in the x-direction between a movable module i and a fixed module f is

$$\frac{1}{2}W_{if}(x_i^2 + x_f^2 - 2x_i x_f) \tag{8.6}$$

The first term in expression (8.6) contributes W_{if} to c_{ii}. The third term contributes $-W_{if}x_f$ to the vector d_x at row i and the second term contributes to the constant part of (8.4).

The objective function (8.4) is then minimized by solving the system of linear equations represented by

$$Cx + d_x = 0 \tag{8.7}$$

Equation (8.7) thus gives the solution to the unconstrained problem of minimizing the quadratic objective function in (8.4).

8.4 Hybrid Net Model

Since matrix C is sparse, symmetric and positive definite, we solve (8.7) by the pre-conditioned Conjugate Gradient method using the Incomplete Cholesky Factorization of matrix C as the preconditioner [1, 15]. It is well-known that the runtime of the solver is directly proportional to the number of non-zero entries in matrix C. This in turn, is equal to the number of two-pin nets in the circuit. Hence, a suitable net model is required to transform the netlist hypergraph into a graph (or a set of two-pin nets) so as to have minimal number of non-zero entries in matrix C.

In this respect, we propose a Hybrid net model that is a combination of the clique and star net models. In the subsequent discussion we prove the equivalence of the clique and star models, and hence the validity of using the Hybrid net model in quadratic placement.

8.4.1 Clique and Star Net Models

The clique model is the traditional model used in analytical placement algorithms. In the clique model, a k-pin net is replaced by $k(k-1)/2$ two-pin nets forming a clique. If the weight of the k-pin net is W, then some commonly used values for the two-pin nets of the clique are $W/(k-1)$, $2W/k$, etc. The clique model for a 5-pin net is illustrated in Figure 8.3(a).

In the star model, each net has a star node to which all the pins of the net are connected. Hence, a k-pin net will yield k two-pin nets. The star model for a 5-pin net is illustrated in Figure 8.3(b).

In [16] Mo et al. use the star net model within a macro-cell placer on the MCNC92 macro block benchmarks. They report an average reduction of 30% in the number of two-pin nets as compared to using the clique model. Vygen [26] also switches to a star model for very large nets to reduce the number of terms in the objective function, but has not shown the validity of mixing the clique and star models in quadratic placement. In addition, neither paper has discussed the method to set the weight of the nets introduced by the star model.

Equivalence of the Clique and Star Net Models

We now prove that for a k-pin net of weight W, if we set the weight of the two-pin nets introduced, to γW in the clique model and $k\gamma W$ in the star model for any γ, the clique model is equivalent to the star model in quadratic placement. Therefore, the two models can be used interchangeably.

Lemma 8.1. *For any net in the star model, the star node under force equilibrium is at the center of gravity of all the pins of the net.*

Proof. Consider a k-pin net. Let x_s be the x-coordinate of the star node and let W_s be the weight of the two-pin nets introduced. Then the total force on the star node by all the pins is given by

$$F = \sum_{j=1}^{k} W_s(x_j - x_s)$$

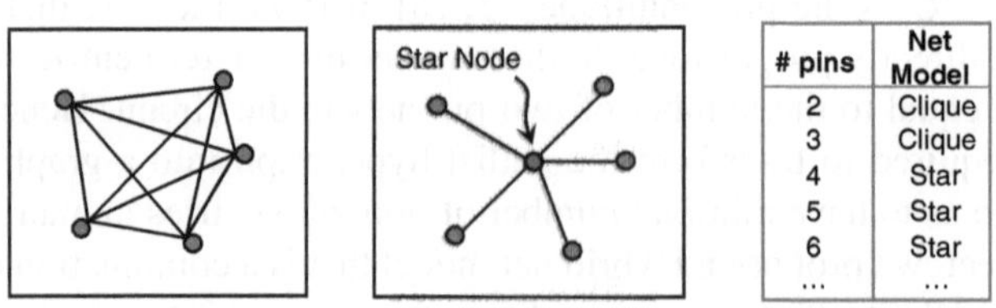

(a) Clique Model (b) Star Model (c) Hybrid Net Model

Fig. 8.3. Net models (*Source*: [22] © 2005 IEEE).

Under force equilibrium, the total force $F = 0$. Therefore,

$$x_s = \frac{\sum_{j=1}^{k} x_j}{k} \tag{8.8}$$

Hence the lemma follows. □

Theorem 8.2. *For a k-pin net, if the weight of the two-pin nets introduced is set to W_c in the clique model and kW_c in the star model, the clique model is equivalent to the star model in quadratic placement.*

Proof. For the clique model, the total force on a pin i by all the other pins is given by

$$F_i^{clique} = W_c \sum_{j=1, j\neq i}^{k} (x_j - x_i) \tag{8.9}$$

For the star model, all the pins of the net are connected to the star node. The force on a pin i due to the star node is given by

$$\begin{aligned} F_i^{star} &= kW_c\,(x_s - x_i) \\ &= kW_c\left(\frac{\sum_{j=1}^{k} x_j}{k} - x_i\right) \quad \text{by Lemma 1} \\ &= W_c\left(\sum_{j=1}^{k} x_j - kx_i\right) \\ &= W_c \sum_{j=1, j\neq i}^{k} (x_j - x_i) \\ &= F_i^{clique} \end{aligned}$$

As the forces are same in the two models, they are equivalent. □

8.4.2 Hybrid Net Model

The Hybrid net model that we propose uses a clique model for two-pin and three-pin nets, and a star model for nets with four or more pins. Within *FastPlace* we set $\gamma = 1/(k-1)$ as it works well experimentally.

By using the star model for nets with four or more pins, we will generate significantly less two-pin nets and consequently fewer non-zero entries in the matrix C as compared to the clique model. We use the clique model for two-pin nets so as to not introduce one extra net and one extra variable (corresponding to the star node) per two-pin net as in [16]. We choose to use the clique model as opposed to the star for three-pin nets because (a) if two modules are connected by more than one two-pin or three-pin net in the original netlist, then the resulting two-pin nets generated by the clique model between them can be combined and will only introduce a single non-zero entry in the matrix C; (b) it will not introduce an extra variable corresponding to the star node.

Comparing the Hybrid net model with the clique model for a k-pin net, we see that the clique model introduces $k(k-1)/2$ non-zero entries in matrix C. This is quadratic to the number of pins in the net. Whereas, the Hybrid net model will only introduce k non-zero entries in the matrix. This is linear to the number of pins in the net. This reduction in the non-zero entries in matrix C not only results in a significant speed-up of the conjugate gradient solver, but also results in a significantly lower memory usage to store the matrix.

In Table 8.1 we compare the clique and Hybrid net models in terms of the number of non-zero entries introduced in the connectivity matrix C and the runtime of the conjugate gradient solver. We use the ISPD 2005 placement contest benchmarks for our comparison. From Table 8.1 it can be seen that on average, the Hybrid net model leads to 10.26× fewer non-zero entries in the matrix and results in a 5.45× speed-up in the conjugate gradient solver.

In this section, we have described the method to set the weights of the two-pin nets introduced by the clique and star models. Consequently, we have proven the equivalence of the two models and hence the validity of mixing them in quadratic placement. Based on the proof, the main novelty of the Hybrid net model is that we can use the star model even for nets with just four or more pins. We no longer have to restrict its usage to only high-degree nets. If a combination of the clique and star models are used within quadratic placement, the Hybrid net model will give the minimum possible non-zero entries in matrix C. To the best of our knowledge, the aforementioned proof and treatment of the star model has not been reported in prior literature.

Table 8.1. Clique net model vs. Hybrid net model.

Circuit	# Non-zero Entries (Clique)	(Hybrid)	Non-zero Entries *(Clique/ Hybrid)*	Solver Runtime *(Clique/ Hybrid)*
adaptec1	7189306	1851871	3.88	2.78
adaptec2	10728119	2040590	5.26	4.05
adaptec3	22241459	3624646	6.14	3.12
adaptec4	25252138	3630561	6.96	4.25
bigblue1	10279652	2247013	4.57	3.40
bigblue2	124232100	3848452	32.28	10.69
bigblue3	68052887	7187232	9.47	5.15
bigblue4	225504129	16666571	13.53	10.16
Average			**10.26**	**5.45**

8.5 Cell Shifting

Solving (8.7) essentially minimizes the quadratic objective function. However, it does not consider the overlap among modules. Therefore, the resulting placement has a lot of overlap and is not spread over the placement region. To resolve overlaps among the modules and spread them over the placement region we employ an efficient Cell Shifting technique.

During Cell Shifting, the placement region is divided into equal sized bins, such that on average, each bin can accommodate about four modules. We call this the regular bin structure. Based on the current placement, the utilization of each bin (U_i) is then computed. U_i is defined as the ratio of the total area of all the modules overlapping with bin i to the bin area. The modules are then shifted around the placement region based on their respective bins and its current utilization. Since Cell Shifting is independent and similar in the x and y dimensions, we only describe the case where the modules are shifted in the x-dimension.

8.5.1 Shifting of Standard-cells

Shifting of cells is a two step process. In the first step, based on the utilization of all the bins in a particular row of the regular bin structure, an irregular bin structure reflecting the current bin utilization is constructed. As an example, let the utilization of all the bins in one of the rows be as depicted in Figure 8.4(a). Then the irregular bin structure constructed from the regular bin structure is as shown in Figure 8.4(b). To get the equation for the irregular bin structure, from Figure 8.4 let:

- OB_i: Coordinate of the boundary of bin i in the regular bin structure.
- NB_i: Coordinate of the boundary of bin i in the irregular bin structure.

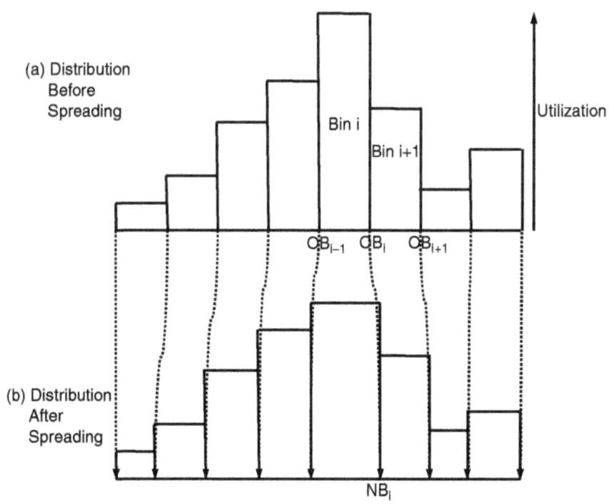

Fig. 8.4. (**a**) Regular bin structure (**b**) irregular bin Structure and utilization after shifting (*Source*: [22] © 2005 IEEE).

Then,

$$NB_i = \frac{OB_{i-1}(U_{i+1} + \delta) + OB_{i+1}(U_i + \delta)}{U_i + U_{i+1} + 2\delta} \tag{8.10}$$

The idea behind Cell Shifting is to even out the utilization among adjacent bins. Hence, (8.10) constructs the new bin such that it averages the utilization of bin i and bin $i + 1$. The reason for having the parameter δ is as follows: Let, $\delta = 0$ and $U_{i+1} = 0$, then from equation (8.10) it can be seen that, $NB_i = OB_{i+1}$ and $NB_{i+1} = OB_i$. This results in a cross-over of bin boundaries in the irregular bin structure which results in an improper mapping of the cells. To avoid this problem, we experimentally set the parameter δ to a value of 1.5.

In the second step, every cell present in a particular bin in the regular bin structure is then linearly mapped to the corresponding bin in the irregular bin structure. As a result of this mapping, cells in bins with a high utilization will shift in a manner so as to reduce the utilization of the bin and the overlap among themselves. For performing the linear mapping of cells, If

- x_j: x-coordinate of cell j in bin i before mapping.
- $x\prime_j$: x-coordinate of cell j in bin i after mapping.

Then,

$$x\prime_j = \frac{NB_i(x_j - OB_{i-1}) + NB_{i-1}(OB_i - x_j)}{OB_i - OB_{i-1}} \tag{8.11}$$

To control the actual distance moved by any cell during shifting, we introduce two *movement control parameters* α_x and α_y (<1) for the x and y dimensions. Hence, once the coordinates of cell j after mapping have been obtained from (8.11), the actual distance moved by the cell is $\alpha_x|x\prime_j - x_j|$. The *movement control parameters* are increasing functions that are inversely proportional to the maximum bin utilization and have a very small value during the early stages of placement. As a result, cells will move by a very small distance during the initial placement iterations. When the placement is spread out, the cells will not have a tendency to shift over large distances. α_x and α_y can then take a larger value to accelerate convergence.

8.5.2 Shifting of Macro-Blocks

For standard-cells, the width of the bins in the regular bin structure is greater than the average width of the cells. Hence, the movement of any cell has an influence on the utilization of only the adjacent bins. On the other hand, the movement of a macro will influence the utilization of all the bins spanned by it. Therefore, to move a macro during Cell Shifting we consider a larger region that is proportional to the size of the macro.

Shifting of the macros follows the same two-step process as the standard-cells. The only difference being the construction of the irregular bin structure. Figure 8.5 illustrates the construction of the irregular bin structure for horizontal shifting. From Figure 8.5(a), for the regular bin structure, let

- N: Total number of bins spanned by the macro.
- x_span: Total number of columns spanned by the macro.

- OB_L: x-coordinate of the left boundary of the leftmost bins spanned by the macro.
- OB_R: x-coordinate of the right boundary of the rightmost bins spanned by the macro.
- U_C: Sum of the utilizations of the N bins spanned by the macro (shaded region with lines to the bottom right).
- U_L: Sum of the utilizations of N bins to the left of the macro. (shaded region with lines to the bottom left).
- U_R: Sum of the utilizations of N bins to the right of macro. (shaded region with lines to the bottom left).

From Figure 8.5(b), for the unequal bin structure, let:

- NB_L: x-coordinate of the left boundary of the leftmost bins spanned by the macro.
- NB_R: x-coordinate of the right boundary of the rightmost bins spanned by the macro.

Then,

$$NB_L = \frac{(OB_L - x\_span)(U_C + \delta) + OB_R(U_L + \delta)}{U_L + U_C + 2\delta} \tag{8.12}$$

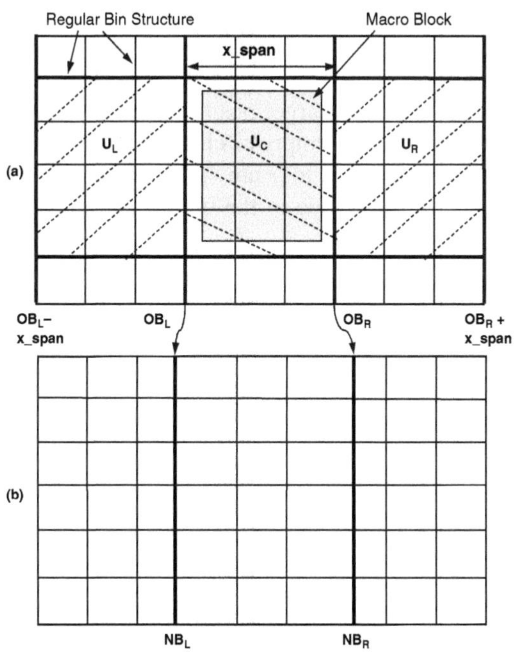

Fig. 8.5. (**a**) Regular bin structure; (b) irregular bin structure for macro shifting (*Source*: [23] © 2006 IEEE).

$$NB_R = \frac{OB_L(U_R + \delta) + (OB_R + x\_span)(U_C + \delta)}{U_R + U_C + 2\delta} \tag{8.13}$$

For performing the linear mapping, if:

- x: x-coordinate of the macro before mapping.
- $x^{'}$: x-coordinate of the macro after mapping.

Then,

$$x^{'} = \frac{NB_R(x - OB_L) + NB_L(OB_R - x)}{OB_R - OB_L} \tag{8.14}$$

8.5.3 Addition of Spreading Forces

Once the modules have been shifted, additional forces need to be added to them so that they do not collapse back to their previous positions during the next quadratic program step. This is achieved by connecting each module to a corresponding pseudo-pin added at the boundary of the placement region. The pseudo-pin and pseudo-net addition is illustrated in Figure 8.6.

Let (x_j^f, y_j^f) be the target position of module j after Cell Shifting. At the target position, the module will experience a force due to its connections with the other modules in the netlist. This force can also be viewed as the force required to move the module from its original position (before Cell Shifting) to the target position. The spreading force added to the module corresponds to this force experienced by the module in its target position.

To illustrate the addition of the spreading force, consider Figure 8.6. When module j (solid circle) is moved to its target position, it will experience a force due to the other modules connected to it (empty circles). When determining this force, we assume that the modules connected to j are still in their original positions prior to Cell Shifting. The resultant force due to the modules connected to j is then given

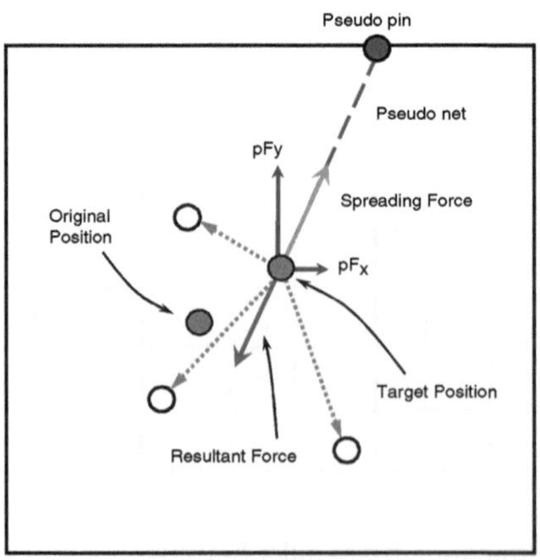

Fig. 8.6. Pseudo-pin and pseudo-net addition (*Source*: [22] © 2005 IEEE).

by the "Resultant Force" vector. The spreading force has the same magnitude as the "Resultant Force" but is in the opposite direction.

To determine the position of the pseudo-pin and the spring constant of the pseudo-net, If:

- pF_x: x-component of the spreading force.
- pF_y: y-component of the spreading force.
- pD: Distance between the pseudo-pin and the target position of module j.

Then, the position of the pseudo-pin is determined by the intersection of the "Spreading Force" vector with the chip boundary. A pseudo-net for module j is one that connects it from its target position to its pseudo-pin. The spring constant for the pseudo-net is then given by $\beta = \sqrt{pF_x^2 + pF_y^2}/pD$. During each iteration of placement, the spreading force and pseudo-pin of the previous iteration are discarded and a new spreading force and corresponding pseudo-pin position is determined for each module.

Since the pseudo-pin is a fixed pin present at the boundary, we know from expression (8.2) and the subsequent analysis in Sect. 8.3, that only the diagonal of matrix C and the d_x and d_y vectors need to be updated for every module. Hence, it takes only a single pass of $O(m)$ time, where m is the total number of movable modules in the circuit, to regenerate the connectivity matrix for the next quadratic program step. Thus we have incorporated an extremely fast Cell Shifting technique to spread the cells over the placement region.

8.6 Iterative Local Refinement

Since the quadratic objective function is only an indirect measure of the linear wirelength, it does not yield the best possible result in terms of the linear wirelength. To offset this disadvantage, some form of linearization needs to be introduced within the quadratic placement methodology. We achieve this by incorporating an Iterative Local Refinement (ILR) technique within *FastPlace*.

The ILR technique is a key component of our placement flow. This technique uses the actual position of a module and the half-perimeter bounding rectangle measure of all the nets connected to the module to move it around the placement region. It acts on a coarse global placement and is highly effective in minimizing the wirelength while simultaneously spreading the modules over the placement region. In addition, it can also seamlessly handle placement blockages and placement congestion constraints.

To handle placement congestion constraints, we separate the ILR technique into two components: (a) a density-bin-based ILR (*d-ILR*) and (b) the regular ILR (*r-ILR*). The core algorithm to move the modules, within both the components is the same and hence we only describe it in the context of the *r-ILR*.

8.6.1 Bin Structure for *r-ILR*

The *r-ILR* also employs a regular bin structure to estimate the utilization of a placement region and move the modules. During the first step of the *r-ILR* the width and height of each bin is set to 5× that of the bin used during Cell Shifting. Such large bins are constructed to have a global view of the current placement and enable modules to move over long distances. This is done to minimize the wirelength of long nets that might span a large part of the placement region. During subsequent steps, the width and height of the bins are gradually brought down to the values used in the Cell Shifting step. As a result, the movement of the modules gets progressively localized.

8.6.2 ILR for Simultaneous Spreading and Wirelength Minimization

During any iteration of the ILR, once the placement region has been binned, we traverse through all the modules and determine their respective *source* bins. For every module present in a bin, we compute eight scores that correspond to moving the module to its nearest eight neighbouring bins. For calculating the score, we assume that a module is moving from its current position in a *source* bin to the same relative position in the *target* bin. The score for each move is a weighted sum of two components: the first being the change in the wirelength for the move and the second being a function of the change in the bin utilization.

The wirelength component is computed as the sum of the half-perimeter of the bounding rectangle (HPWL) of all the nets connected to the module. Since it directly takes the HPWL into account, it is more accurate than the quadratic objective function. For the utilization component to accurately reflect the placement distribution, we define a utilization weight for each bin in the placement region. This weight is a function of the bin utilization and is constantly updated based on the current placement distribution. Hence, a sparse bin will have a low utilization weight so that more modules can be moved into it, whereas a dense bin will have a higher utilization weight so that modules can be moved out of the bin. As the weights are a function of the bin utilization, they are constantly updated and prevent oscillations in terms of the movement of the modules.

If all eight scores are negative, the module will remain in the current bin. Otherwise it is moved to the *target* bin with the highest score for the move. During one iteration of the ILR, we go through all the modules in the placement region and follow the above steps for moving the module. Subsequently, this iteration is repeated until there is no significant improvement in the wirelength.

8.6.3 ILR for Handling Placement Blockages

Most circuits contain a number of placement blockages in the form of fixed macros. Analytical placement techniques often place a lot of movable modules on top of the fixed macros. These modules have to be moved out of the fixed macros in an effective manner with minimal increase in the wirelength.

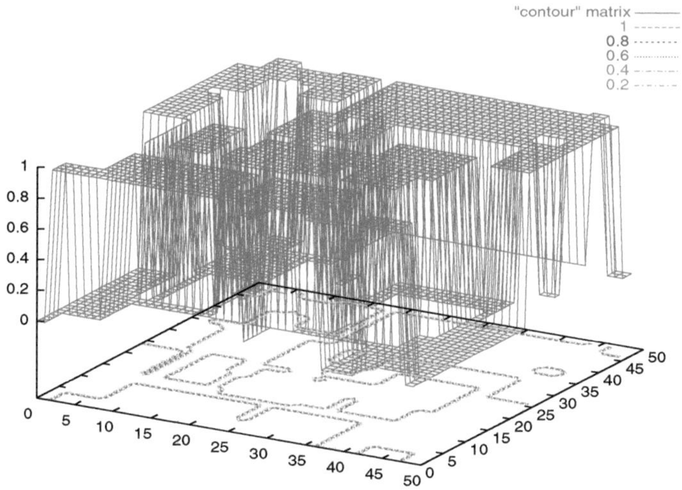

Fig. 8.7. Initial contour map depicting placement blockages (*Source*: [24] © 2007 IEEE).

To handle fixed macros during placement, using the ILR bin structure, we construct a contour map of the placement region. Based on the fixed macros, each bin in the contour map has a value of either 1 in case it overlaps with a fixed macro or 0 otherwise. The initial contour map for one of the placement benchmarks is shown in Figure 8.7. We then use a 3×3 Laplacian as a smoothing filter to transform the entire contour map. This transformation smoothes the sharp edges in the original contour map creating the modified map as shown in Figure 8.8. During the initial steps of ILR, the smoothing transform is run for a large number of iterations so that modules can easily move over and cross a fixed macro if required. During the final steps of ILR, the smoothing transform is run for lesser number of iterations so that the edges are quite steep. This will enable the modules to slide down the slope and be moved out of the fixed macro. As a result of the smoothing each bin will now have a contour height associated with it.

More precisely, for a cell i currently in bin m, we use the following notations:

- α: Weight for the wirelength component.
- $wl_i(m)$: Half-perimeter wirelength when i is in bin m
- $wl_i(n)$: Half-perimeter wirelength when i is in bin n
- $\beta(m)$: Weight of the utilization component for bin m.
- $\beta(n)$: Weight of the utilization component for bin n.
- $U(m)$: Utilization function for bin m
- $U(n)$: Utilization function for bin n
- γ: Weight for the contour component.
- $C(m)$: Contour height of bin m
- $C(n)$: Contour height of bin n

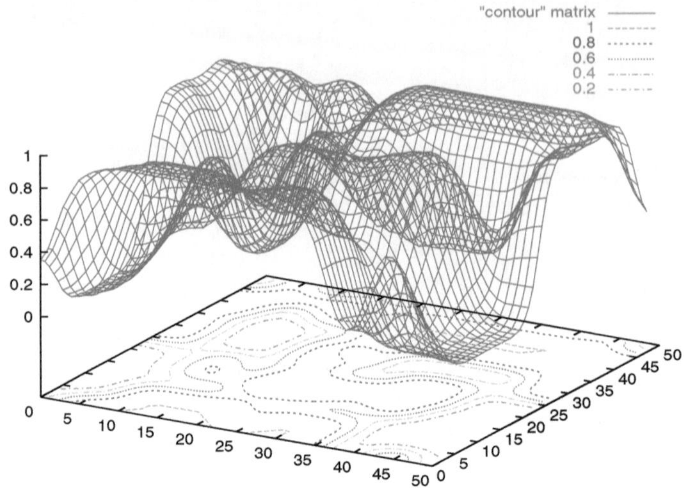

Fig. 8.8. Contour map after smoothing transform (*Source*: [24] © 2007 IEEE).

Then the consolidated score for the move from bin m to bin n accounting for wirelength, bin utilization and placement blockage is given by

$$s_i(m,n) = \alpha[wl_i(m) - wl_i(n)] + [\beta(m)U(m) - \beta(n)U(n)] + \gamma\,[C(m) - C(n)]$$

8.6.4 ILR for Placement Congestion Control

To reduce placement congestion, designers often run placement algorithms with specific *placement_target_density* values. To determine the placement density, the placement region is binned using a pre-defined grid. Usually, the grid is square with the dimensions being a multiple of the standard-cell row height. The *density* of a bin is then defined as the ratio of the total area of movable objects within the bin to the total available free space within the bin. The *placement_target_density* basically specifies the maximum allowed occupation for any bin in the placement region. Satisfying the *placement_target_density* constraint means that the *density* of all the bins should be less than or equal to the *placement_target_density* value.

To handle the *placement_target_density* constraint, we use the *d-ILR* or the density-bin-based ILR along with the *r-ILR*. The *d-ILR* uses the global pre-defined grid structure used for placement *density* computation to calculate the score and move a module from its *source* to *target* bin. Once the *d-ILR* is performed, we then run the *r-ILR* as before in which the bin sizes are set and changed as described in Sect. 8.6.1. The interaction between the *d-ILR* and the *r-ILR* can be seen in Figure 8.9 which shows the decrease in the size of the bins from the *d-ILR* stage to the end of the *r-ILR* stage.

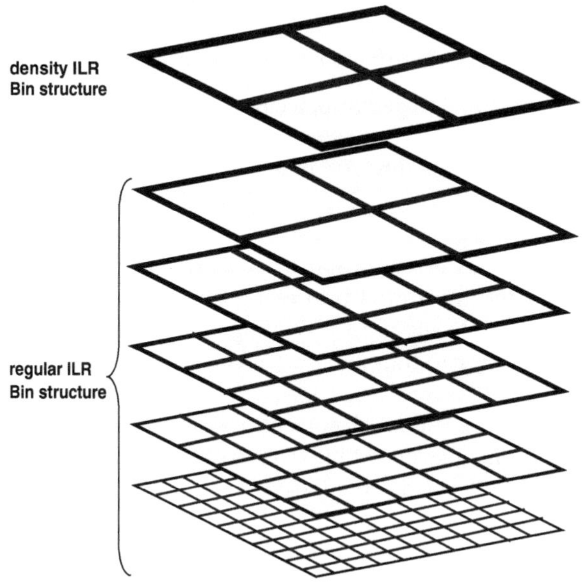

Fig. 8.9. Bin Structure for Iterative Local Refinement (*Source*: [24] © 2007 IEEE).

8.7 Clustering for Placement

The circuit sizes that need to be handled by current placement algorithms are steadily increasing towards tens of millions of modules. In such a scenario, a flat placement methodology may not be effective in producing a good quality solution within a reasonable amount of runtime. Hence, for efficient and scalable placement algorithm design, a hierarchical approach is beneficial. To this effect many placers follow a hierarchical or multilevel approach for placement [2, 3, 10, 12, 18, 21]. Circuit clustering is an attractive method to reduce the placement problem size for large- scale VLSI designs. If clustering is performed in a careful manner, it can also yield better wirelength along with faster runtime as compared to flat placement approaches.

In our multilevel global placement framework we use clustering in a *persistent* context as defined by [18]. In other words, we use clustering at the beginning of the placement flow to pre-process the input netlist so as to reduce the placement problem size.

8.7.1 Two-Level Clustering Scheme

To facilitate the description of the clustering, we define:

- M: The total number of modules before clustering.
- α: The clustering ratio which is defined as the ratio of the number of modules before and after clustering.
- *target_number_of_modules*: M/α.

- a_i: The area of module i.
- *average_cluster_area*: $\Sigma_{i=1}^{M} a_i / target\_number\_of\_modules$

To reduce the circuit size for global placement, we follow a two-level clustering scheme as shown in Figure 8.2. For each level we set $\alpha = 2$ resulting in a $4\times$ reduction in the number of movable modules in the final coarse-grain clustered netlist.

During the first level of clustering we create fine-grain clusters of about 2–3 modules per cluster. This clustering is solely based on the connectivity information between the modules in the original flat netlist. Since it is performed at the beginning of placement, we restrict it to fine-grain clustering to minimize any loss in placement quality due to incorrect clustering. In fact, it was demonstrated in [9] that fine-grain clustering can improve placement efficiency with negligible loss in placement quality.

We then perform a fast, initial placement of the fine-grain clusters. The purpose of this step is to get some placement information for the next clustering level. Since each cluster in the first level has only around 2–3 modules, the initial placement of the clusters closely resembles an initial placement of the original flat netlist. We then create coarse-grain clusters by performing a second level of clustering. In this level, we consider both, the connectivity information between the clusters and their physical locations as obtained from the initial placement. We believe that generating coarse-grain clusters based on actual placement information, is better than generating them by a solely netlist-based approach; and such an approach would further minimize any loss in (or even improve) the final placement wirelength.

The key difference between our clustering scheme and the ones followed in [2, 4, 12, 18] is that we use actual placement information while forming coarse-grain clusters, whereas the other approaches generate coarse-grain clusters solely based on the netlist information[1]. Our approach closely resembles that of [10]. The difference being that [10] uses two-levels of netlist-based clustering followed by physical clustering, whereas we only use one level of fine-grain netlist-based clustering.

After experimenting with a variety of clustering techniques, we chose to use the *Best-Choice* clustering technique described in [18] for both the levels of clustering. In Figure 8.10, we summarize the modified version of the *Best-Choice* clustering algorithm using Lazy-Update speed-up technique that is employed within our two-level clustering scheme.

From Figure 8.10 there are three key parameters within our clustering scheme:

- $s(j,k)$: The netlist-based clustering score between two modules j and k.
- *max_cluster_area*: The upper-bound on the cluster area.
- *distance_threshold*: The distance threshold used for the physical clustering.

Within our clustering scheme, the netlist-based clustering score between two modules j and k is given by

$$s(j,k) = \frac{\Sigma_{v \in N} w_v}{a_j + a_k}$$

[1] Note that in [3, 4], placement information is used for clustering during the second V-cycle.

```
Algorithm Clustering

Phase 1: Construct Initial Priority-queue (PQ)
        For each module j
          1. Find closest module k and clustering score s(j, k)
          2. Insert triple (j, k, s) into PQ with s as the key

Phase 2: Form Clusters
        while (number_of_modules > target_number_of_modules)
          1. Pick top triple (j, k, s) from PQ
          2. if j is marked invalid
             3. Re-calculate closest module k' and clustering score s'(j, k')
             4. Insert triple (j, k', s') into PQ
          5. else
             6. if fine-grain clustering
                7. if (a(j) + a(k)< max_cluster_size) cluster j and k into new module j'
             8. if netlist+physical clustering
                9. Calculate d(j, k) the distance between j and k
                10. if (d(j, k)< distance_threshold and a(j)+a(k)< max_cluster_size)
                       cluster j and k into new module j'
             11. Update netlist based on the clustering
             12. For module j' find closest module k' and clustering score s'(j', k')
             13. Insert triple (j', k', s') into PQ with s' as the key
             14. Mark neighbours of j' as invalid
```

Fig. 8.10. Best-choice clustering algorithm with placement information (*Source*: [24] © 2007 IEEE).

where N is the set of nets connecting the two modules and $w_\nu = 1/|k|$ where k is the degree of net ν.

Controlling the area of the clusters is highly imperative. Otherwise, a cluster can get progressively larger by absorbing smaller clusters around it. This is often detrimental and leads to bad solution quality. Having an area term in the denominator of the clustering score biases the clustering technique to pick modules that will not result in forming huge clusters. In addition, we also impose an upper-bound on the cluster area using the *max_cluster_area* parameter. Within our clustering scheme the *max_cluster_area* is set to $5\times$ *average_cluster_area*. This results in the formation of balanced clusters.

It is quite possible that two modules that have a very high connectivity score do not actually end up being close to each other in the final placement. This can happen because of the influence of the other nets/modules connected to them. Hence, during the second level of clustering, even though we rank and pick the modules based on their connectivity score we cluster the modules only if the distance between them, as obtained from the initial global placement is within a certain threshold. In our clustering scheme we experimentally set the *distance_threshold* to 10% of the maximum chip dimension.

8.8 Legalization

The global placement solution of force-directed placers typically has overlaps among the modules that need to be resolved. Our legalization stage is divided into two steps. In the first step we ignore the standard-cells and resolve overlaps among the macro-blocks and assign them to legal positions. In the next step we fix all the macros and legalize the standard-cells.

8.8.1 Legalization of Macro-Blocks

Since the movement of the macro-blocks has a significant impact on the wirelength, the aim of the macro-block legalization is to maintain the global placement positions of the macros as much as possible. If we denote the global placement position of a macro as its *target position*, then the macro block legalization problem is to resolve overlaps among all the macros while minimizing the total perturbation of the macros from their *target* positions.

We formulate this problem as a minimum-perturbation fixed-outline floorplanning problem. We use the sequence-pair [17] to represent the floorplan and enforce the non-overlapping constraints among the macros. Any other floorplanning representation can also be easily incorporated within our approach. Formally, the minimum perturbation placement problem can be described as:

Minimum Perturbation Floorplan Realization (MPFR) Problem:

Given: n macros with target coordinates (x_i^*, y_i^*) for $i = 1, \ldots, n$ and a sequence pair (p, q).

Determine: Legalized coordinates (x_i, y_i) s.t. $\sum_{i=1}^{n} |x_i - x_i^*| + |y_i - y_i^*|$ is minimized.

In the following sections, we first describe the *Iterative Clustering Algorithm* that is used to generate a placement of the macros for a given sequence-pair. We then describe the top-level flow for macro-block legalization using simulated annealing. Since the horizontal and vertical non-overlapping constraints can be handled independently, we only discuss the horizontal problem.

Iterative Clustering Algorithm

The basic idea behind the *Iterative Clustering Algorithm* is that if we know which macros abut with each other to form a cluster in the optimal solution, then the position of the cluster is easy to find. In Figure 8.11 we give the pseudo-code of the Iterative Clustering Algorithm.

For horizontal placement, we first find the immediate left and right neighbours of the macros. These neighbours are associated with the non-transitive edges in the horizontal constraint graph and can be found in $O(n^2)$ time. We then place the macros one at a time from left to right according to the sequence p. In case a macro overlaps

Iterative Clustering Algorithm:
1. Find the immediate left and right neighbours of all macros
2. **for** $i = 1$ **to** n
3. Place macro p_i in its target position
4. Let C be a new cluster consisting of p_i
5. **while** C overlaps with other clusters **do**
6. Merge C with the closest cluster on its left
7. Let C be the new cluster formed
8. Shift C to its optimal position
9. **if** macro m in C is at its target position **do**
10. Detach m from C if necessary and goto step 8
11. **endwhile**
12. **endfor**

Fig. 8.11. Iterative clustering algorithm (*Source*: [23] © 2006 IEEE).

with an existing cluster then the clustering is updated according to steps 5–11. The condition in step 5 and the closest cluster in step 6 can be determined by considering the constraints of the immediate left neighbours of modules in C. The shifting in step 8 is easy according to the following lemma.

Lemma 8.3. *For a cluster C, its position is optimal if the number of macros perturbed to the left from their target positions is equal to the number perturbed to the right.*

Since we add macros from left to right, macros will always be added to the right of a stationary cluster. So the clusters will always shift left. Therefore, it is very easy to find the correct shift amount of the newly formed clusters. In step 9, after shifting a cluster C, a macro $m \in C$ may potentially reach its target position. If m does not have any right neighbours belonging to cluster C, then it should be detached from the cluster. Otherwise, it will move with the cluster during subsequent steps and will not be in its optimal position. The condition to detach m can be checked by looking at its immediate right neighbours.

Although the while loop in steps 5–11 looks complicated, we can show with careful implementation and analysis that the runtime complexity of the Iterative Clustering Algorithm is $O(n^2)$.

Macro-block Legalization by Simulated Annealing

The aim of the top-level simulated annealing framework is to obtain a sequence pair such that the corresponding placement obtained from the *Iterative Clustering Algorithm* will resolve overlaps among the macros with minimum perturbation from their global placement positions. Another factor to be considered during placement is that the macros have to be placed in legal positions within the core region. Hence, the cost function for simulated annealing is a weighted sum of the total perturbation along with a penalty for being out of bound.

Let (p, q) represents the sequence pair. Then, the initial sequence for p/q is generated by sorting the macros in ascending order according to the Manhattan distance from the upper left/lower left corner to their target positions. This sequence pair closely corresponds to the original placement and is usually quite good. Hence, a low-temperature annealing is sufficient to generate a good result. Besides, we restrict each annealing move to randomly exchange two adjacent macros in one of the two sequences so as to not disturb the current solution significantly.

In Figures 8.12 and 8.13 we plot the placement of the macros before and after legalization for the circuit ibm06-HB. From the two figures, we can see that the macros have moved by a very small amount from the global placement solution.

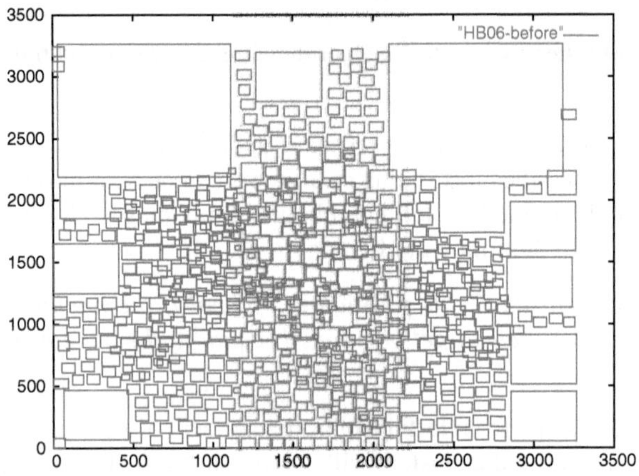

Fig. 8.12. Circuit ibm06-HB before legalization of macro-blocks.

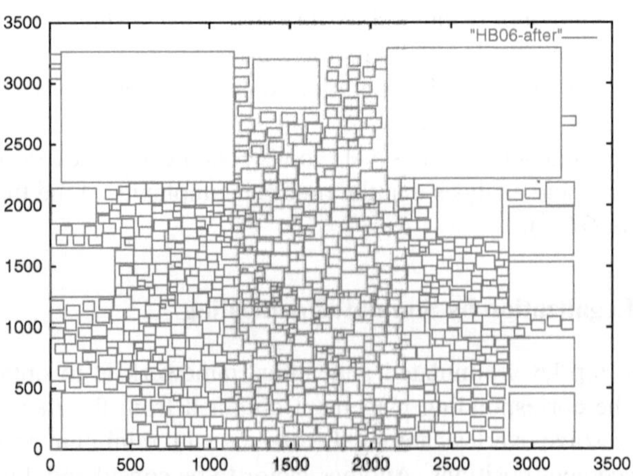

Fig. 8.13. Circuit ibm06-HB after legalization of macro-blocks.

8.8.2 Legalization of Standard-Cells

Once the overlaps among the movable macros have been resolved, we fix their positions for all subsequent steps and treat them as placement blockages. We then divide each row in the core region into placement segments based on the overlap of the blockages with the row. A placement segment is defined as the maximal part of a row that is not covered by a placement blockage. We then move the standard-cells among the segments to satisfy their respective capacities. Finally, we legalize the standard cells within the segments.

To move the cells among the segments, we use a greedy heuristic similar to the Iterative Local Refinement technique. For every cell present in a segment, we compute eight *scores* based on moving the cell to its nearest eight neighbouring segments. For calculating the score, we assume that a cell is moving from its current position in a *source* segment to the nearest possible position in the *target* segment. Each score is a weighted sum of two components: The first being the half-perimeter wirelength reduction for the move. The second being a function of the utilization of the source and target segments. Since the legalization technique is mainly used to even out the placement and bring all the segments within capacity, a higher weight is assigned to the second component. If all the scores are negative, the cell will remain in the current segment. Otherwise, it will move to the target segment with the highest score for the move. During one iteration, we traverse through all the segments that are above capacity and follow the above steps for cell movement. Subsequently, this iteration is repeated until all the segments are within their respective capacities. We then assign the cells to legal positions within each segment.

8.9 FastDP: Efficient and Effective Detailed Placement

To further reduce the wirelength of a legalized global placement, we use an efficient and effective detailed placer called *FastDP*. Our detailed placer works only on the standard-cells in a legalized row-based standard-cell placement or a placement in which the macro-blocks have been fixed.

FastDP consists of four key techniques: *Global Swap*, *Vertical Swap*, *Local Re-ordering*, and *Single-Segment Clustering*. The flow of our detailed placement algorithm is summarized in Figure 8.14. We first apply the Single-Segment Clustering technique to obtain a relatively good starting solution for the main loop of the algorithm. In the main loop, Global Swap, Vertical Swap, and Local Re-ordering are employed to reduce the wirelength until there is no significant improvement. Finally, we re-apply the Single-Segment clustering to get better positions for the cells within the segments without changing their order.

8.9.1 Global Swap

Given that all the other cells in the circuit are fixed, the "optimal region" for a cell i is defined as the region to place i where the wirelength is optimal. Accordingly, the

Detailed Placement Algorithm

Perform Single-Segment Clustering

Repeat

  Perform Global Swap

  Perform Vertical Swap

  Perform Local Re-ordering

Until no significant improvement in wirelength

Repeat

  Perform Single-Segment Clustering

Until no significant improvement in wirelength

Fig. 8.14. Detailed placement flow (*Source*: [19] © 2005 IEEE).

idea behind Global Swap is to find the "optimal region" for a cell i in the placement region and swap i with a cell j or a space s in its "optimal region". In the following sections, we first describe the method to find the "optimal region" for any cell. We then discuss the method to consider the effect of overlap during a swap operation. Finally, we describe the Global Swap technique based on the "optimal region" and the penalty method to deal with overlap. We then provide experimental results to show the effectiveness of the Global Swap technique.

Optimal Region

The "optimal region" for a cell is determined based on the median idea of [7] as follows.

For a cell i, we traverse all the nets connecting to it (noted as N_i) and find their bounding boxes. Here, cell i is excluded from the nets when computing their bounding boxes. For each net $p \in N_i$, we find its bounding box ($x_l[p]$, $x_r[p]$, $y_l[p]$, $y_u[p]$ – the left, right, lower, and upper boundaries). From [7], the optimal position for i is given by (x_{opt}, y_{opt}), where x_{opt} and y_{opt} are the medians of the x series ($x_l[1], x_r[1], x_l[2], x_r[2], \ldots,$) and y series ($y_l[1], y_u[1], y_l[2], y_u[2], \ldots,$) of bounding boxes. In general, the optimal position is a rectangular region rather than a point as the total number of elements in the x and y series are even. This region is defined as the "optimal region" for cell i. In some cases, the "optimal region" can degrade to a point/line when the two medians of the x and/or the y series carry the same value.

Figure 8.15 shows the optimal region for cell 1. There are three nets connecting to cell 1 ($Net1$, $Net2$ and $Net3$). The nets are denoted by closed dashed lines. The bold rectangles are the bounding boxes for the nets excluding cell 1. The light lines are the grids constructed by the x series ($x_l[1], x_r[1], x_l[2], x_r[2], x_l[3], x_r[3]$) and y series ($y_l[1], y_u[1], y_l[2], y_u[2], y_l[3], y_u[3]$). The shadowed region is the optimal region for cell 1.

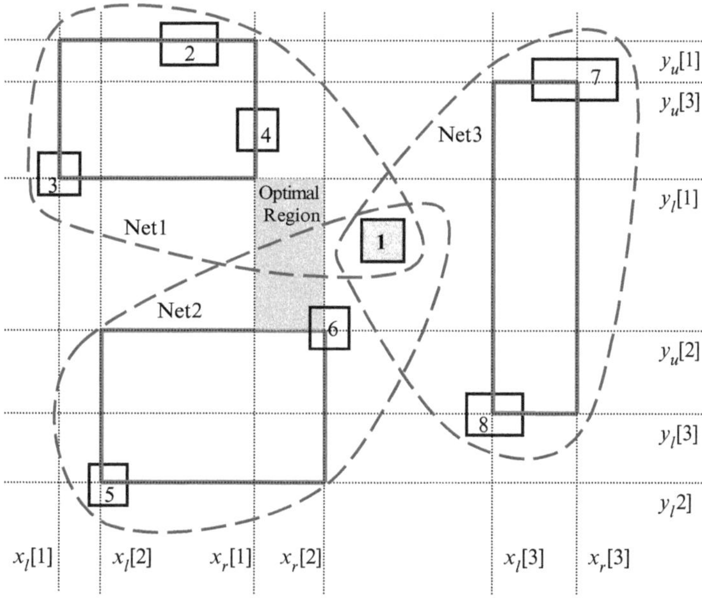

Fig. 8.15. Optimal region for cell 1 (*Source*: [19] © 2005 IEEE).

Modelling the Effect for Overlap

After finding the optimal region for a cell i, we want to move i into the region. However, overlaps may be created if we swap i with a cell/space in the optimal region. This may happen because of the difference in the sizes of i and the cell/space that is picked for swap. Since no overlap is allowed in the placement solution, a consequent legalization has to be done. Therefore, before performing Global Swap, we need to consider the effect of any possible overlap on the wirelength. Usually, overlaps are resolved by shifting the neighbouring cells in the segment with the overlap. In case we need to shift the neighbouring cells, we introduce a penalty for this shifting effect.

In order to characterize the penalty more accurately, we have two types of penalties: $P1$ and $P2$. $P1$ is the penalty on shifting the closest two cells to resolve overlap. $P2$ is the penalty on shifting cells other than the closest two cells. Figure 8.16 illustrates two examples to compute $P1$ and $P2$. The bold boxes are the cells and the light boxes are segments. The dashed boxes show the positions of the swapped cells after swap. Consider the case where we swap cell i (width w_i) in segment seg_i with cell j (width w_j) in segment seg_j that is in the optimal region of i. Assume $w_i > w_j$. The two cells left and right to j are j_1 and j_2. The widths of the two closest spaces left to j are s_1 and s_2, and the widths of two closest spaces right to j are s_3 and s_4. The total width of spaces s_1, s_2, s_3, s_4 is S1.

$P1$ is the wirelength penalty caused by shifting j_1 and j_2. If $S1 \geq (w_i - w_j)$, the total shift of j_1 and j_2 to resolve overlap is $(w_i - w_j) - (s_2 + s_3)$. We make $P1$

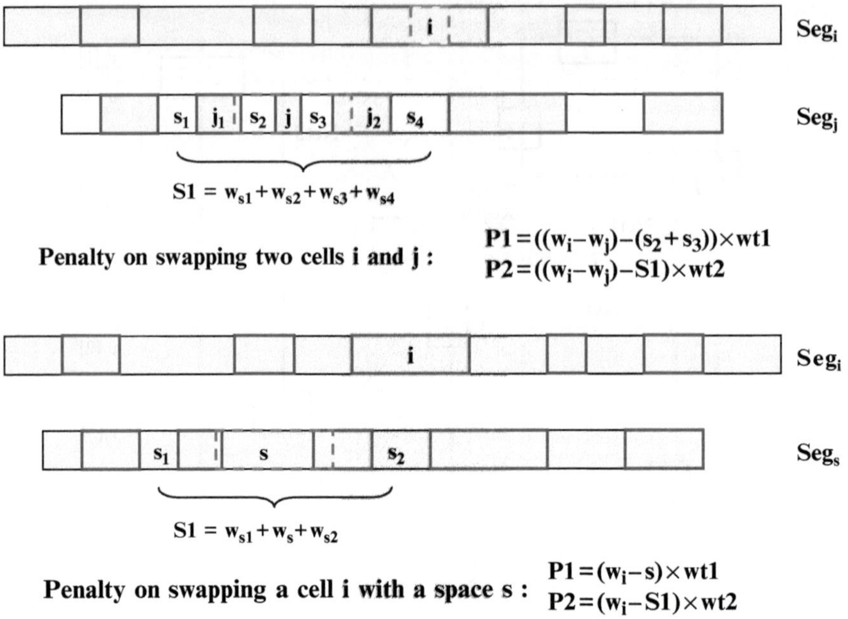

Fig. 8.16. Penalty on swapping two cells and swapping a cell with a space (*Source*: [19] © 2005 IEEE).

proportional to this shift. If $S1 \le (w_i - w_j)$, only shifting j_1 and j_2 cannot resolve the overlap and we need to shift more cells in seg_j. $P2$ is the penalty of shifting cells other than j_1 and j_2 in seg_j. In this case $P2$ is proportional to the shift on cells other than j_1 and j_2, which is $(w_i - w_j) - S1$. Hence, we set $P1$ and $P2$ as follows:

$$\begin{aligned} P1 &= ((w_i - w_j) - (s_2 + s_3)) \times wt1 \\ P2 &= ((w_i - w_j) - S1) \times wt2 \end{aligned} \tag{8.15}$$

where $wt1$ and $wt2$ are weights. For the case where we swap i with a space s, the method to get the penalty is similar to that for swapping two cells. The only difference is that the width difference is $w_i - 0 = w_i$ and $S1$ is the sum of the widths of s, the closest space left to s and the closest space right to s.

Since the shifts in $P1$ and $P2$ have the dimension of length, the two weights $wt1$ and $wt2$ are just constants with no dimension. Because we do not want to disturb the original placement too much by legalization, large overlap is discouraged by setting $wt2$ higher than $wt1$.

Global Swap Based on Optimal Region

Based on the optimal region and the penalty on overlap, we develop a Global Swap technique. Since there could be several cells and spaces in the optimal region, we have many choices. We use a term "benefit" B as a measurement for selecting the cell or space in the optimal region. The "benefit" for a swap has two components: one is the difference between the total wirelength before and after the swap, the other is the penalty charged on the created overlap. If the wirelength before and after the swap are W_1 and W_2, respectively, the "benefit" can be obtained by (8.16).

$$B = (W_1 - W_2) - P1 - P2 \tag{8.16}$$

If $B > 0$, it means that we can benefit from the swap. Otherwise, the resulting placement could be worse than original. Of course, the "benefit" we compute is not accurate because the real wirelength change due to resolving the overlap is hard to measure. Based on this "benefit", we perform the swap as follows. For each standard-cell i, we find its optimal region and try to swap it with every cell j and space s in the optimal region of i. We measure the "benefit" for each tentative swap and pick the j or s with the best "benefit" to perform the swap. If the best "benefit" has a value less than zero, we do not perform a swap on i.

8.9.2 Vertical Swap

During Global Swap, it is possible to not find a good candidate cell or space in the optimal region of cell i to perform a swap. This could be because of two reasons. First, the size of i is large and the optimal region of i is congested. Hence, the segments that span the optimal region do not have enough space to hold i. Second, in order to hold i, many cells have to be shifted to remove the overlap created by swapping, which introduces a high penalty.

To increase the possibility for a good swap and reduce the vertical wirelength locally, we have a Vertical Swap technique similar to the Global Swap. The idea of Vertical Swap is to move a cell vertically towards its optimal region. This technique is not as greedy as Global Swap and only moves a cell up or down by one row. For a cell i, if the optimal region is above/below its current position, then a few cells in the segment above/below i are considered to be the candidates. We use the same penalty as in Global Swap to estimate the effect of overlap and pick the best candidate for the swap. We observed that interleaving the Vertical Swap with Global Swap results in a much faster decrease in the wirelength as compared to only applying the Global Swap. We believe the reason for this is because the Vertical Swap is not very greedy and has more flexibility in moving the cells. At the same time, it may aid the Global Swap. In addition, this technique is much faster than Global Swap as the number of candidate cells considered for the swap are much less than in Global Swap.

8.9.3 Local Re-Ordering

With Vertical Swap fixing local vertical errors, we need a technique to fix local horizontal errors. Therefore, we propose a very fast Local Re-ordering technique to handle this problem. For any n consecutive cells within a segment, we try all possible left-right ordering of the cells and pick the order giving the best wirelength. To determine the positions of the cells, we consider them as a group and make the left(right) boundary of the group as the left(right) boundary of the first(last) cell in the original order. Then for each order, we distribute the cells evenly within the left and right boundaries. Hence, in each order, all the cell positions are fixed and the wirelength evaluation is much faster than considering multiple choices for the cell positions.

In our implementation, we set $n = 3$. The reason being that $n = 2$ which is pair-wise swapping is too constrained whereas, $n = 4$, will be four times slower and we also do not observe any significant improvement as compared to $n = 3$. Compared to the conventional window-based technique, Local Re-ordering has a 3-cell window in one row/segment and is very local.

8.9.4 Single-Segment Clustering

After the main loop of *FastDP*, we fix the cells and their ordering within each segment. We now further reduce the wirelength by moving the cells within the segments. For a legalized placement, if we fix the order of the cells in one segment and the positions of the cells in all other segments, the problem becomes a fixed-order single segment problem described later.

Fixed-Order Single Segment Placement Problem

Given a segment S in the placement region with n standard-cells $C_1, C_2, \ldots, C_n$, whose left-to-right order is fixed (C_i is left to C_j if $i < j$), with all cells not in S being fixed. Find a non-overlapping placement for the segment S so that the total half-perimeter wirelength is minimized.

This problem is basically the same as the Single-Row Problem in [13], where an optimal dynamic programming algorithm is proposed to solve the problem. In this section, we describe a more efficient algorithm that can also find the optimal solution.

First, we define some terms used in our algorithm. A *cluster* is a cell or a group of cells abutted together (retaining the original order of cells). *Clustering* is the operation to abut two clusters to form a new cluster (the width of the new cluster is the sum of the widths of the original clusters). The wirelength function of the x-coordinate of a cluster is a convex piecewise linear function $W(x)$ when all other objects are fixed. The slopes for the linear pieces are $\ldots, -3, -2, -1, 0, 1, 2, 3, \ldots$ The slope 0 part is the optimal region in x-direction for the cluster. The points where the function changes slope are called *bounds*. These bounds are the left and right boundaries of the bounding boxes for the nets connecting to the cluster but excluding the cluster itself. *Optimal Region Center* of a cluster is the middle point of the optimal region in x-direction when all the objects (standard-cells and macro-blocks) not part of the cluster are fixed.

In order to find the optimal region for a cluster C in segment S, we need to fix the positions of all the other objects except the cells in S. Therefore, if C has connections to any cells in S, the bounds for C cannot be determined. However, since we fix the order of the cells in S, we know the left-right orders between the cells. We use this information to get the bounds so that the optimality of the solution will not be affected. The method to get the bounds is as follows. When computing the bounding box for any net N connecting to C, if N is connecting to a cell C' in S, we will assume C' at the end of the segment S, i.e., if C' is left to C, we assume C' is at the left end of segment S; otherwise, C' is at the right end of segment S. Although we are not using the real position for C', we will not affect the optimality of the position of C because the left–right order of C and C' has to be maintained.

The main idea of the algorithm is to put every cluster at its *Optimal Region Center*. If there is overlap between two clusters, we perform clustering and form a new cluster. The new cluster will not be broken at any later stage. Then we put the new cluster at its *Optimal Region Center*. We iteratively perform clustering until all the clusters are put at their *Optimal Region Center* without any overlap. If any optimal region boundary is out of the segment range, we will assign it to the closest segment boundary. In this way, no cell will be put out of the segment.

Theorem 8.4. *The Single-Segment Clustering Algorithm finds the optimal solution for the Fixed-Order Single Segment Placement Problem.*

Proof. First, we prove that if the clusters are formed correctly, the solution obtained by our algorithm is optimal. Since all the clusters are located at their optimal positions in the end, if there is no overlap between any clusters, the solution is optimal. However, if some clusters overlap with each other, they should be clustered together, a contradiction to our assumption that the clusters are formed correctly.

Then, we prove that the clusters are formed correctly by our algorithm. To show this, assume on the contrary that the clustering in the optimal solution is different from our solution. Consider a gap in the optimal solution surrounded by a pair of cells a and b that are in the same cluster in our solution. Suppose a and b are clustered together when we merge clusters A and B in some step of our algorithm. See Figure 8.17 for an illustration. Without loss of generality, we can assume there is no gap within cluster A and within cluster B in the optimal solution. Otherwise, we can consider the gap within cluster A or cluster B instead. Since we merge cluster A and cluster B together at some point, A and B cannot be at the optimal region at the same

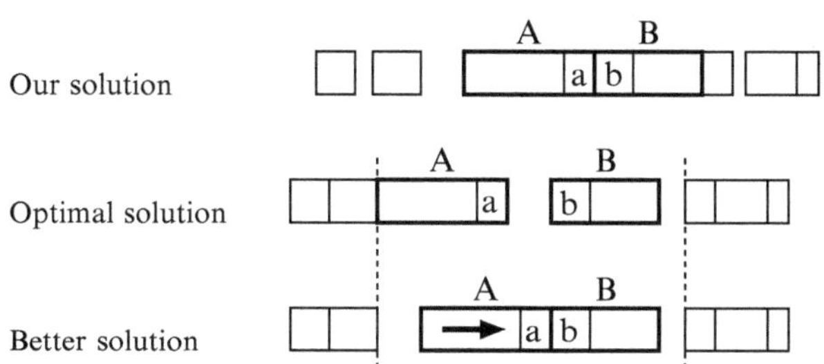

Fig. 8.17. Proof of optimality of Single-segment Clustering Algorithm (*Source*: [19] © 2005 IEEE).

time if their order is not changed. For any solution, either A wants to move right or B wants to move left (or both) to reduce the wirelength. We can always generate a better wirelength than the optimal solution by moving either A or B towards the gap without creating any overlap. A contradiction. Thus, our solution is optimal. □

We now analyze the complexity of the algorithm. There are n cells in total, and the maximum number of clustering operation is $n - 1$. In the clustering operation, every step needs constant time except merging the two bounds lists. The merge takes linear time to the number of bounds m. The complexity of the algorithm is $O(nm)$. However, in practice, it can be much better. In our implementation, we are not keeping all the bounds for the clusters. Instead, we only keep a small constant number of bounds for every cluster. Therefore, the merge also takes constant time. The total complexity of the algorithm is $O(n)$. Of course, it will compromise the optimality, but experiments show that even using a small constant will not degrade the solution appreciably. In our implementation, the constant we choose is 16.

Although this algorithm can find optimal solution, we still need to run it iteratively as it is only optimal when all cells not in the current segment are fixed. Since we are changing the cell positions segment by segment, we need to run several iterations to find good positions for all the cells.

8.10 Experimental Results and Analysis

FastPlace is implemented in C and the executables of the latest release are available for download from [25]. The current version of our algorithm – *FastPlace3.0* was tested on the ISPD-2005, ISPD-2006, PEKO-MS ISPD-2005 and PEKO-MS ISPD-2006 benchmark suites. All the experiments were run using a 64-bit binary on a 2.5 GHZ AMD Opteron 252 machine with 8 GB RAM.

8.10.1 Runtime Analysis of the Algorithm

In Table 8.2 we give a break-up of the total runtime of the algorithm on the ISPD-2005 and ISPD-2006 placement contest benchmarks. It can be seen that even for circuits with over two million movable modules (bigblue4, newblue7), *FastPlace3.0* takes less than 1 1/2 h to generate a placement solution. This demonstrates the efficiency and scalability of the placement algorithm.

8.10.2 ISPD-2005 Placement Contest Benchmarks

In Table 8.3, we compare the half-perimeter wirelength of our placer with *mPL6*, *APlace 2.0* and *Capo10.2*. *APlace 2.0* [12, 14] is a faster version of the placer [11] used in the ISPD-2005 placement contest, *mPL6* [4] comprises of enhanced versions of the placers described in [3, 6] and *Capo10.2* is the latest version of the placer described in [20]. It can be seen that on average, we are 4% better in terms of half-perimeter wirelength as compared to *APlace 2.0*, 1% higher as compared to *mPL6* and 11% better as compared to *Capo10.2*, respectively.

Table 8.2. Break-up of total runtime (all values in seconds).

Circuit	Global Placement					Leg'n	Detailed Placement	Total
	Clustering	QP Solver	Cell Shifting	ILR	Post-Process			
adaptec1	10	4	3	154	9	11	89	282
adaptec2	13	5	6	220	14	11	139	411
adaptec3	24	12	29	611	31	35	494	1248
adaptec4	22	10	17	494	26	24	219	823
bigblue1	13	7	4	230	16	10	162	445
bigblue2	44	11	9	451	24	38	508	1092
bigblue3	48	28	88	1731	63	101	919	2993
bigblue4	144	62	135	2245	112	163	1877	4757
adaptec5	49	25	55	1339	43	189	145	1848
newblue1	27	20	16	263	0	44	804	1175
newblue2	22	11	7	459	22	67	338	928
newblue3	56	6	28	381	32	20	1249	1783
newblue4	38	21	38	762	29	108	115	1112
newblue5	92	45	94	1601	62	386	270	2555
newblue6	145	43	56	1706	81	87	683	2805
newblue7	130	78	234	2494	126	257	430	3760

Table 8.3. HPWL Comparison of *FastPlace3.0* with *mPL6*, *APlace2.0* and *Capo10.2* on the ISPD-2005 benchmark suite.

Circuit	Half-Perimeter Wirelength			
	FastPlace3.0	$\frac{mPL6}{FP3.0}$	$\frac{APlace2.0}{FP3.0}$	$\frac{Capo10.2}{FP3.0}$
adaptec1	77536552	1.00	1.01	1.18
adaptec2	93331216	0.99	1.03	1.08
adaptec3	212891104	1.01	1.03	1.07
adaptec4	197050832	0.99	1.06	1.06
bigblue1	95618032	1.01	1.05	1.14
bigblue2	153419088	0.99	1.00	1.06
bigblue3	362728672	0.95	1.13	1.10
bigblue4	831285312	1.00	1.05	1.16
Average		**0.99**	**1.04**	**1.11**

In Table 8.4 we compare the wirelength results of *FastPlace3.0* in *default mode* with that of other placers reported during the ISPD 2005 placement contest. For the contest, all the placers were given the benchmark circuits in advance. There was no limit on the CPU time and the placers were allowed to have *separate parameters* for each individual circuit to obtain the best possible wirelength. From Table 8.4, the contest version of *APlace* is on average 3.3% better than our placer in terms of

Table 8.4. HPWL Comparison of *FastPlace3.0* with other academic placers on the ISPD-2005 benchmark suite.

Placer	Circuit						Average
	adaptec2	adaptec4	bigblue1	bigblue2	bigblue3	bigblue4	
APlace	0.94	0.95	0.99	0.94	0.99	1.00	0.967
FastPlace 3.0	**1.00**	**1.00**	**1.00**	**1.00**	**1.00**	**1.00**	**1.000**
mFAR	0.98	0.97	1.02	1.10	1.05	1.05	1.029
Dragon	1.01	1.02	1.07	1.04	1.05	1.09	1.047
mPL	1.04	1.02	1.03	1.13	1.02	1.09	1.054
Capo	1.07	1.07	1.13	1.12	1.05	1.32	1.129
NTUplace	1.07	1.05	1.11	1.24	1.14	1.39	1.167
Fengshui	1.32	1.71	1.20	1.86	1.30	1.25	1.440
Kraftwerk	1.69	1.79	1.56	2.10	1.81	1.69	1.773

Table 8.5. Comparison of *FastPlace3.0* with other academic placers on the ISPD-2006 benchmark suite using the ISPD-2006 placement contest scoring function.

Placer	Circuit								Avg
	ad5	nb1	nb2	nb3	nb4	nb5	nb6	nb7	
Kraftwerk	1.01	1.19	1.00	1.00	1.01	1.04	1.00	1.00	1.03
mPL6	1.00	1.06	1.07	1.17	1.00	1.02	1.00	1.00	1.04
FastPlace3.0	**1.07**	**1.18**	**0.99**	**1.09**	**0.99**	**1.12**	**0.97**	**0.90**	**1.04**
NTUPlace2	1.02	1.00	1.07	1.16	1.03	1.00	1.04	1.07	1.05
mFAR	1.09	1.23	1.09	1.16	1.09	1.13	1.03	1.04	1.11
APlace3	1.26	1.20	1.05	1.13	1.35	1.21	1.06	1.05	1.16
Dragon	1.08	1.21	1.29	1.90	1.05	1.13	1.03	1.23	1.24
DPlace	1.26	1.55	1.77	1.36	1.14	1.35	1.23	1.25	1.36
Capo	1.16	1.57	1.64	1.44	1.22	1.28	1.32	1.46	1.39

half-perimeter wirelength. Whereas, our results are better than the reported results of all the other placers.

8.10.3 ISPD-2006 Placement Contest Benchmarks

In Table 8.5 we compare our placement results with that of other placers reported during the ISPD 2006 placement contest. We use the same scoring function as the contest which is a weighted function of the wirelength, placement congestion and runtime. On average, our score is only 1% higher than the best reported results during the placement contest. Looking at individual results, on 4 of the 8 circuits we obtain the best results among all the placers.

In Table 8.6 we report the half-perimeter wirelength, scaled-overflow, and scaled half-perimeter wirelength of *FastPlace3.0* on the ISPD-2006 placement contest benchmarks. To determine the scaled half-perimeter (S_HPWL) values on these circuits, we use the same function as the ISPD-2006 placement contest, which is

$$S\_HPWL = HPWL \times (1 + 0.01 \times Scaled\_overflow\_per\_bin) \tag{8.17}$$

Table 8.6. HPWL, scaled-overflow per bin, and scaled HPWL on the ISPD-2006 benchmark suite.

Circuit	HPWL	SO/Bin	S_HPWL
adaptec5	440638720	13.6210	500658247
newblue1	77234368	0.8070	77857680
newblue2	207180096	0.5333	208284948
newblue3	291394656	0.5289	292935881
newblue4	297285504	9.2112	324669094
newblue5	581120768	11.7824	649590908
newblue6	541387840	1.3056	548456045
newblue7	1087625344	1.2011	1100689254

Table 8.7 gives the runtime comparison of our placer with other placers reported during the ISPD 2006 placement contest. This is a direct comparison of the runtime as the machine specifications for the contest were the same as the one on which we ran our experiments. On average, the runtime of our placer is the least among all the placers.

Table 8.7. Runtime comparison of *FastPlace3.0* with other academic placers on the ISPD-2006 benchmark suite.

Placer	Circuit								Avg
	ad5	nb1	nb2	nb3	nb4	nb5	nb6	nb7	
FastPlace 3.0	**1.00**	**1.00**	**1.00**	**1.00**	**1.00**	**1.00**	**1.00**	**1.00**	**1.00**
Kraftwerk	1.78	0.97	1.09	0.51	2.49	2.91	1.91	1.99	1.70
mPL6	4.47	1.92	6.56	5.44	5.23	4.83	4.29	7.55	5.04
NTUplace2	5.68	1.84	4.77	3.73	6.72	8.00	4.94	5.71	5.17
mFAR	3.72	2.16	3.12	1.66	5.72	4.47	4.33	5.18	3.80
APlace3	10.97	3.66	5.96	7.01	13.48	12.84	10.38	14.59	9.86
Dragon	1.22	0.84	1.76	0.66	1.34	1.38	1.38	2.63	1.40
DPlace	1.56	0.87	6.89	0.58	1.48	1.78	1.44	2.53	2.14
Capo	5.26	2.18	6.08	3.41	6.23	8.16	6.59	14.62	6.57

8.10.4 PEKO-MS Benchmarks

In Table 8.8 we report the wirelength and runtime results of *FastPlace3.0* on the PEKO-MS ISPD-2005 benchmarks. From column 3, it can be seen that on average, we are 1.73× higher than the optimal wirelength reported on these benchmarks.

Finally, in Table 8.9, we report the wirelength, scaled-overflow, and runtime results of our placer on the PEKO-MS ISPD-2006 benchmarks. From column 3, it can be seen that on average, we are 2.11× higher than the optimal half-perimeter wirelength reported on these benchmarks. Looking at the scaled half-perimeter wirelength, from column 6, we are on average, 3.25× higher than the scaled half-perimeter wirelength of the optimal placement.

Table 8.8. HPWL and Runtime on the PEKO-MS ISPD-2005 benchmark suite.

Circuit	HPWL		Runtime (s)			
	FastPlace3.0 (×10e6)	$\frac{FP3.0}{opt\_wl}$	Global Placement	Legalization	Detailed Placement	Total
adaptec1	31.62	1.58	91	8	77	176
adaptec2	40.13	1.61	117	3	88	207
adaptec3	73.25	1.79	265	19	178	463
adaptec4	67.53	1.71	304	22	148	474
bigblue1	34.06	1.63	193	2	100	295
bigblue2	69.92	1.65	351	26	167	543
bigblue3	173.36	1.84	655	57	424	1136
bigblue4	343.14	2.00	2042	134	947	3123
Average		**1.73**				

Table 8.9. HPWL, Scaled-overflow per bin and runtime on the PEKO-MS ISPD-2006 benchmark suite.

Ckt	HPWL		SO/Bin		S_HPWL	Runtime (s)			
	FP3.0 ($\times 10e7$)	$\frac{FP3.0}{opt\_wl}$	FP3.0	optimal	$\frac{FP3.0}{opt}$	GP	Leg'n	DP	Total
ad5	17.09	2.79	100.291	9.988	5.09	413	176	51	640
nb1	4.47	2.18	13.267	1.734	2.43	193	131	46	371
nb2	5.28	1.89	45.400	10.294	2.50	271	135	77	483
nb3	8.15	2.61	81.334	9.552	4.33	269	105	71	444
nb4	7.57	1.72	98.478	9.260	3.12	278	128	67	473
nb5	20.95	2.43	84.804	9.584	4.10	493	11	66	571
nb6	12.59	1.53	45.613	8.361	2.06	820	298	354	1472
nb7	27.22	1.75	45.055	7.065	2.36	1586	922	527	3035
Average		**2.11**			**3.25**				

8.11 Conclusions

In this chapter, we describe *FastPlace* an efficient and scalable force-directed placement algorithm for large-scale mixed-size circuits. The global placement stage of *FastPlace* relies on quadratic wirelength optimization and uses a multilevel framework to improve the scalability of the algorithm. It uses a *Hybrid Net Model* to speed-up the quadratic program solver, an efficient *Cell-Shifting* technique to spread the modules and an efficient linearization technique – *Iterative Local Refinement* to minimize the half-perimeter wirelength. The legalization stage uses an optimal *Iterative Clustering Algorithm* for macro-block legalization and a robust segment-based standard-cell legalization technique. The detailed placement stage uses a variety of transforms like *Global Swap*, *Vertical Swap*, *Local Cell Re-ordering*, and *Single-segment Clustering* for further wirelength improvement.

FastPlace can effectively handle placement blockages and placement congestion constraints. It produces competitive results as compared to other state-of-the-art academic placers but at a much lesser runtime. Such an ultra-fast placer is very much needed in present day iterative physical synthesis flows to achieve timing closure without a significant runtime overhead.

References

1. R. Barrett et al. *Templates for the Solution of Linear Systems: Building Blocks for Iterative Methods*. SIAM, 2nd edition, 1994
2. T. Chan, J. Cong, T. Kong, and J. Shinnerl. Multilevel optimization for large-scale circuit placement. In *Proc. IEEE/ACM Int. Conf. Comput.-Aided Design*, pages 171–176, 2000
3. T. Chan, J. Cong, and K. Sze. Multilevel generalized force-directed method for circuit placement. In *Proc. Int. Symp. Phys. Design*, pages 185–192, 2005
4. T. F. Chan, J. Cong, J. R. Shinnerl, K. Sze, and M. Xie. mPL6: Enhanced multilevel mixed-size placement. In *Proc. Int. Symp. Phys. Design*, pages 212–214, 2006
5. C. C. Chang, J. Cong, and X. Yuan. Multi-level placement for large-scale mixed-size IC designs. In *Proc. Asia and South Pacific Design Automat. Conf.*, pages 325–330, 2003
6. J. Cong and M. Xie. A robust detailed placement for mixed-size ic designs. In *Proc. Asia and South Pacific Design Automat. Conf.*, pages 188–194, 2006
7. S. Goto. An efficient algorithm for the two-dimensional placement problem in electrical circuit layout. *IEEE Trans. Circuits and Systems*, CAS-28(1):12–18, 1981
8. K. M. Hall. An r-dimensional quadratic placement algorithm. *Manage. Sci.*, 17:219–229, 1970
9. B. Hu and M. Marek-Sadowska. Fine granularity clustering for large scale placement problems. In *Proc. Int. Symp. Phys. Design*, pages 67–74, 2003
10. B. Hu and M. Marek-Sadowska. Multilevel fixed-point-addition-based VLSI placement. *IEEE Trans. Comput.-Aided Design*, 24(8):1188–1203, August 2005
11. A. B. Kahng, S. Reda, and Q. Wang. APlace: A general analytic placement framework. In *Proc. Int. Symp. Phys. Design*, pages 233–235, 2005
12. A. B. Kahng, S. Reda, and Q. Wang. Architecture and details of a high quality, large-scale analytical placer. In *Proc. IEEE/ACM Int. Conf. on Comput.-Aided Design*, pages 890–897, 2005
13. A. B. Kahng, P. Tucker, and A. Zelikovsky. Optimization of linear placements for wirelength minimization with free sites. In *Proc. Asia and South Pacific Design Automat. Conf.*, pages 241–244, 1999
14. A. B. Kahng and Q. Wang. Implementation and extensibility of an analytic placer. *IEEE Trans. Comput.-Aided Design*, 24(5):734–747, May 2005
15. D. S. Kershaw. The Incomplete Cholesky-Conjugate Gradient method for the iterative solution of systems of linear equations. *J. Comp. Phys.*, 26:43–65, 1978.
16. F. Mo, A. Tabbara, and R. Brayton. A force-directed macro-cell placer. In *Proc. IEEE/ACM Intl. Conf. on Computer-Aided Design*, pages 177–180, 2000
17. H. Murata, K. Fujiyoshi, S. Nakatake, and Y. Kajitani. VLSI module placement based on rectangle-packing by the sequence pair. *IEEE Trans. Comput.-Aided Design*, 15(12):1518–1524, December 1996
18. G.-J. Nam, S. Reda, C. J. Alpert, P. G. Villarrubia, and A. B. Kahng. A fast hierarchical quadratic placement algorithm. *IEEE Trans. Comput.-Aided Design*, 25(4):678–691, April 2006

19. M. Pan, N. Viswanathan, and C. Chu. An efficient and effective detailed placement algorithm. In *Proc. IEEE/ACM Intl. Conf. on Comput.-Aided Design*, pages 48–55, 2005
20. J. A. Roy, S. N. Adya, D. A. Papa, and I. L. Markov. Min-cut floorplacement. *IEEE Trans. Comput.-Aided Design*, 25(7):1313–1326, July 2006
21. T. Taghavi, X. Yang, B.-K. Choi, M. Wang, and M. Sarrafzadeh. Dragon2005: Large-scale mixed-size placement tool. In *Proc. Intl. Symp. Phys. Design*, pages 245–247, 2005
22. N. Viswanathan and C. C.-N. Chu. FastPlace: Efficient analytical placement using cell shifting, iterative local refinement and a hybrid net model. *IEEE Trans. Comput.-Aided Design*, 24(5):722–733, May 2005
23. N. Viswanathan, M. Pan, and C. Chu. Fastplace 2.0: An efficient analytical placer for mixed-mode designs. In *Proc. Asia and South Pacific Design Automat. Conf.*, pages 195–200, 2006
24. N. Viswanathan, M. Pan, and C. Chu. Fastplace 3.0: A fast multilevel quadratic placement algorithm with placement congestion control. In *Proc. Asia and South Pacific Design Automat. Conf.*, pages 135–140, 2007
25. N. Viswanathan, M. Pan, and C. Chu. FastPlace: An Analytical Placer for Large-scale VLSI Circuits.
url=http://www.public.iastate.edu/~nataraj/FastPlace.html.
26. J. Vygen. Algorithms for large-scale flat placement. In *Proc. ACM/IEEE Design Automat. Conf.*, pages 746–751, 1997

9

mFAR: Multilevel Fixed-Points Addition-Based VLSI Placement

Bo Hu[1] **and Malgorzata Marek-Sadowska**[2]
[1]Velogix Inc.
[2]Department of Electrical and Computer Engineering, University of California, Santa Barbara
hu@velogix.com, mms@ece.ucsb.edu

9.1 Introduction

The rapid advance of VLSI technology has created an increasing demand for high-quality placement tools. A placer has to deliver solutions that meet all the design requirements in a rapid fashion without wasting any computational resources. The nanometer technology makes it possible to integrate billions of transistors in a single chip. Such a design complexity, combined with the increasingly stringent market pressure, requires a very efficient implementation of the placement algorithms. A modern design scenario usually involves several iterations between the logic synthesis and physical design before timing closure can be achieved. From a design iteration point of view, an efficient placement algorithm is essential. Moreover, shrinking feature sizes introduce a full spectrum of deep submicron effects, such as interconnect dominance, crosstalk, IR drop, etc., which challenge the chip designers more than ever before. A placer needs to address explicitly timing, congestion, signal integrity, etc., so that the design can be signed off in a timely manner to meet the shrinking market window.

For more than three decades, quadratic programming has been attracting interest from researchers in academia and industry. A direct application of a nonconstrained quadratic programming formulation to placement often results in excessive cell overlapping. To solve this problem, we introduce fixed points into the unconstrained quadratic-programming formulation. Fixed points act as pseudo cells located at fixed positions. They can be used to pull out cells from the dense regions to reduce the cell-overlapping. We present an in-depth study of the placement technique based on fixed-points addition and theoretically prove that fixed points are a generalization of the constant additional forces in [4]. We developed a multilevel placer based upon the fixed-point technique, and we demonstrate that it produces competitive placement results compared to the existing state-of-the-art placers.

The rest of the chapter is organized as follows. In Sect. 9.2, we present background knowledge on quadratic placement. In Sect. 9.3, we present the concept of fixed points and fixed-points addition. In Sect. 9.4, we discuss the application of fixed-points addition in the quadratic placement flow. In Sect. 9.5, we introduce a multilevel placement flow based on the fixed-points technique. In Sect. 9.6, we present the experimental results on various academic benchmarks. Then this chapter is summarized in Sect. 9.7.

9.2 Background

Let N denote the number of movable cells in a circuit and denote the coordinates of the center point of a cell i. A net is modeled as a clique whose edges are assigned weights of $\frac{1}{k-1}$, where k is the degree of the net. The weight of an edge is also referred to as an *edge-strength*.

A *connection* $c(i, j)$ is a pair of cells that share at least one net. The *weight* on a connection is the summation of all edge-weights introduced by the nets incident to a pair of cells. We use $w_{i,j}$ to denote the weight on a connection $c(i, j)$. In this work, we use the terms *nets*, *interconnects*, and *wires* interchangeably.

A nonconstrained quadratic placement problem minimizes the cost function, which is a summation of the squares of connection lengths as stated in (9.1):

$$cf = \sum_{i,j} w_{i,j} \left((x_i - x_j)^2 + (y_i - y_j)^2\right) \tag{9.1}$$

Each connection $c(i, j)$ in the circuit contributes two weighted quadratic terms in (9.1), $w_{i,j}(x_i - x_j)^2$ in the x direction, and $w_{i,j}(y_j - y_j)^2$ in the y direction.

Let $\bar{p}$ denote the coordinate vector $(x_1, x_2, \ldots, x_n, y_1, y_2, \ldots, y_N)$, (9.1) can be rewritten in a matrix form:

$$cf = \frac{1}{2}\bar{p}^T C \bar{p} + \bar{d}^T \bar{p} + \text{const} \tag{9.2}$$

where $\bar{d}$ is introduced by the connections between movable cells and fixed IO pads.

To minimize *cf*, we must solve a system of linear equations:

$$C\bar{p} + \bar{d} = 0 \tag{9.3}$$

Traditionally, quadratic formulation as shown above is also referred to as a force-directed approach. If we model the movable cells and fixed input/output (IO) pads as objects and nets as springs, we can consider the netlist as a system of objects connected by springs with different strengths (weights). Minimizing (9.1) is equivalent to putting the system in a force-equilibrium state. In this state, the force applied on each movable cell by all the connected springs is equal to zero.

Placements achieved by minimizing (9.1) result in excessive overlapping among cells. This issue was addressed by Proud [11] and Gordian [9] by adopting a partitioner in the quadratic placement flow. In [4], an additional constant force related to cell density was introduced to eliminate the overlaps and to evenly distribute the cells. With that formulation, placement can be obtained by solving a sequence of modified linear equation systems:

$$C\bar{p} + \bar{d} + \bar{e} = 0 \tag{9.4}$$

In (9.4), $\bar{e}$ is the vector of forces that remain constant while solving (9.4). The force-equilibrium state is now maintained not only by the strings connecting the movable cells but also by the constant forces $\bar{e}$. In the attractor-and-repeller approach [5], the authors present a formulation that includes a target spreading distance between cells and the fixed dummy cells used as attractors. In FAR [6], we generalized the idea of additional forces in [4] and removed the target distances that might result in excessive net stretching in [5]. FastPlace [12] enhances the fixed-points addition-based approach with a new cell-spreading strategy and a local refinement procedure. We should mention that the quadratic formulation does not exactly match the placement objective of total linear wirelength-minimization. As a result, it may produce a suboptimal placement even if the optimal quadratic solution is found. There have been attempts to incorporate linear wirelength into quadratic formulation. In [10], the authors proposed to weigh the quadratic wirelength in such a way that they approximate the linear wirelengths.

9.3 Fixed Points

In this section, we begin with definitions of concepts related to fixed-points and force-equilibrium state, followed by discussion of fixed-points addition in the context of quadratic placement. In the following, $\bar{i}$ denotes the coordinate vector of a cell i and $\bar{f}$ denotes a fixed point f. In addition, (x_i, y_i) denotes a point or a location, and $\overline{(x_i, y_i)}$ is a vector directed from the coordinate origin (0, 0) to (x_i, y_i).

9.3.1 Fixed-Points and Force-Equilibrium State

Definition 1 *Fixed point f is a dimensionless pseudo cell on a chip plane positioned at (x_f, y_f). $H(f)$ denotes the cell connected to f. We assume that each fixed point has one and only one host.*

Definition 2 *The connection $c(f, H(f))$ between a fixed point f and its host cell $H(f)$ is a pseudo connection, since it does not exist in the original design. Each pseudo connection is assigned a weight $w_{H(f),f}$. The length of a pseudo connection, $l_{H(f),f}$, is a distance between $H(f)$ and f. In vector algebra, $l_{H(f),f} = \left|\overline{H(f)} - \bar{f}\right|$.*

Definition 3 *The force introduced by a fixed point f acts as an attracting force $\overline{F(H(f), f)}$ on $H(f)$ · $\overline{F(H(f), f)}$ is defined as*

$$\overline{F(H(f), f)} = w_{H(f),f} \cdot \left(\bar{f} - \overline{H(f)}\right). \tag{9.5}$$

The force $\overline{F(H(f), f)}$ points from $H(f)$ to f and its magnitude is equal to the distance between $H(f)$ and f, weighted by $w_{H(f),f}$. If more than one fixed point is connected to a cell i, $\overline{FF(i)}$ denotes the total force introduced from all fixed points.

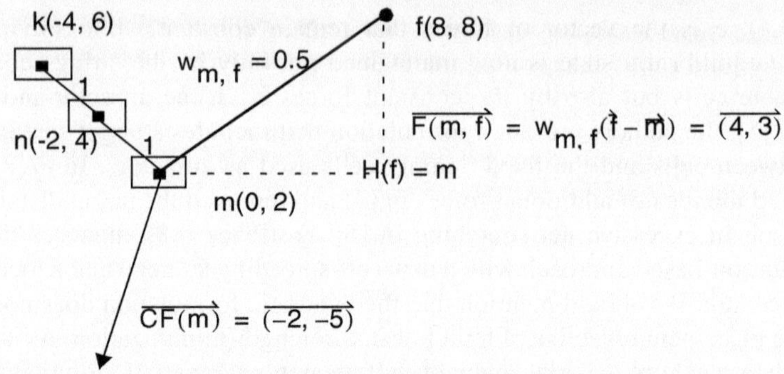

Fig. 9.1. An example of a fixed point, constant force, and force-equilibrium state.

Figure 9.1 gives an example. The fixed point f at (8,8) is connected to its host cell m at (0, 2). Suppose the weight on $c(m, f)$ is 0.5, the attracting force $\overline{F(m, f)}$ is equal to $0.5 \times (\overline{(8, 8)} - \overline{(0, 2)}) = \overline{(4, 3)}$. Since f is the only fixed point connected to m, $FF(m)$ is also equal to $\overline{(4, 3)}$.

Definition 4 *A real connection is a connection between two real cells.*

Definition 5 *An intrinsic force $\overline{I(i, j)}$ on a cell i is the attracting force that cell j exerts on i. $\overline{IF(i)}$ denotes all the intrinsic forces applied on the cell i by real connections incident to i. The intrinsic force $\overline{I(i, j)}$ is defined as*

$$\overline{I(i, j)} = w_{i,j} \cdot (\bar{j} - \bar{i}). \tag{9.6}$$

For example, in Figure 9.1, m has only one real connection $c(m, n)$ ($w_{m,n} = 1$) with the cell n at $(-2, 4)$. $\overline{IF(m)} = \overline{I(m, n)} = 1 \times (\overline{(-2, 4)} - \overline{(0, 2)}) = \overline{(-2, 2)}$.

Definition 6 *A constant force, $\overline{CF(i)}$, is the force externally applied on a real cell i. $\overline{CF(i)}$ is defined as*

$$\overline{CF(i)} = \bar{c} \tag{9.7}$$

In (9.7), $\bar{c}$ is a constant vector. By definition, $\overline{CF(i)}$ does not depend on the location of the cell i. For example, in Figure 9.1, a constant force $\bar{c} = \overline{(-2, -5)}$ is applied on the cell m. Even if m moves to another location, $\overline{CF(m)}$ remains the same. In contrast, $\overline{F(m, f)}$ and $\overline{I(m, n)}$ changes as m is relocated.

Definition 7 *A cell i is in a force-equilibrium state if $\overline{FF(i)} + \overline{IF(i)} + \overline{CF(i)} = 0$, otherwise, it is in disequilibrium. A placement is in a force-equilibrium state if and only if all the movable cells are in their equilibrium states.*

In Figure 9.1, the cell n at $(-2, 4)$ is in force-equilibrium state because $\overline{FF(n)} + \overline{IF(n)} + \overline{CF(n)} = \overline{(0, 0)} + \overline{IF(n, k)} + \overline{IF(n, m)} + \overline{(0, 0)} = \overline{(0, 0)}$. The cell m is also in force-equilibrium state since $\overline{FF(m)} + \overline{IF(m)} + \overline{CF(m)} = \overline{(4, 3)} + \overline{(-2, 2)} + \overline{(-2, -5)} = \overline{(0, 0)}$. The vector $\overline{(0, 0)}$ above indicates that the total forces applied in both x and y directions are zero. If we change the constant force applied on a cell m from $\overline{(-2, -5)}$ to $\overline{(-2, -15)}$, we compute that $\overline{FF(m)}+\overline{IF(m)}+\overline{CF(m)} = \overline{(0, -10)}$. Vector $\overline{(0, -10)}$ indicates that there is no force acting on the cell m in the x direction, but a force of magnitude 10 is applied in the y direction. Consequently, cell m is in disequilibrium state.

9.3.2 Fixed-Points Addition

Fixed points are added to the existing placement to:

1. Achieve the force-equilibrium state
2. Perturb the placement toward a specific direction

We present the following two theorems related to the point (1) above.

Theorem 9.1. *Any given initial placement with fixed IO pads and movable standard cells can be transformed into a force-equilibrium state by adding one fixed point to each movable cell.*

Proof. For any cell in disequilibrium state, we can always add one fixed point to cancel the total intrinsic forces. Specifically, for each movable cell i, we first compute the $\overline{IF(i)}$. Then we introduce a fixed-point f connected to the cell i such that $\overline{F(i, f)} = -\overline{IF(i)}$. Obviously, $\overline{F(i, f)}$ and $\overline{IF(i)}$ cancel each other and so the cell i is in force-equilibrium state. □

Theorem 9.2. *Any initial placement can be transformed to a force-equilibrium state in an infinite number of ways.*

The proof of Theorem 9.2 follows from the fact that there are infinite number of combinations of $\bar{f} - \overline{H(f)}$ and $w_{H(f),f}$ that produce exactly the same $\overline{F(H(f), f)}$ required to achieve the force-equilibrium state. For example, suppose that $\overline{F(i, f)} = \overline{(5, 5)}$ is required to cancel an intrinsic force $\overline{IF(i)} = \overline{(-5, -5)}$. We may either make $\bar{f} - \overline{H(f)} = \overline{(5, 5)}$ and $w_{i,f} = 1$, or $\bar{f} - \overline{H(f)} = \overline{(10, 10)}$ and $w_{i,f} = 0.5$.

Fixed points that keep the current placement in a force-equilibrium state are called the controlling fixed points. Theorem 9.1 tells us that we need is only to add one controlling fixed point per cell, if we want to transform a placement into force-equilibrium state. Theorem 9.2 says that for each fixed point, we have a flexibility of choosing different combinations of lengths and strengths, as long as the attracting force and the intrinsic forces cancel each other out.

Once all the cells are in the force-equilibrium state, more fixed points can be added to perturb the current placement. These are the perturbing fixed points that disrupt the balance and cause redistribution of cells toward a new equilibrium state.

For example, in Figure 9.2(a), we show two movable cells, i and j, and a controlling fixed point f. The forces acting on cell i are $\overline{I(i, f)} + \overline{F(i, f)} = \overline{(0, 4)} + \overline{(0, -4)} = \overline{(0, 0)}$. The controlling fixed point f at (2, −4) holds the cell

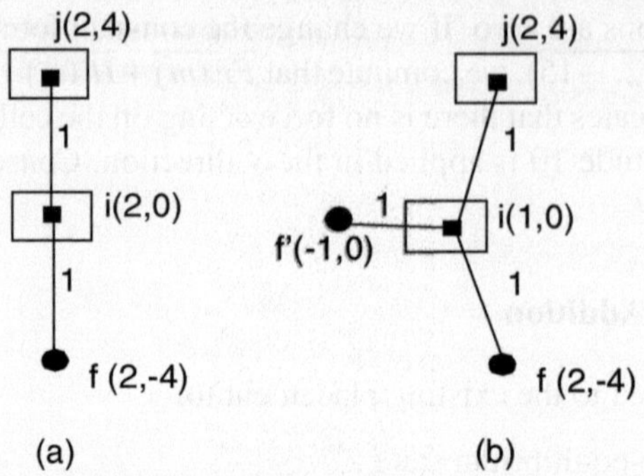

Fig. 9.2. Controlling fixed-point and perturbing fixed-point.

i in a force-equilibrium state. In Figure 9.2(b), we add a perturbing fixed point f' at $(-1, 0)$. If the cells i and j stayed at their original locations, i would be in disequilibrium because $\overline{I(i, f)} + \overline{F(i, f)} + \overline{F(i, f')} = \overline{(-3, 0)}$. But the force acting on i moves it to the left to a new force-equilibrium state, as illustrated in Figure 9.2(b).

Similarly to fixed points, the constant additional forces $\overline{CF(i)}$ introduced in [4] can be also used to maintain a force-equilibrium state and to perturb the current placement. In fact, the constant forces can be used in exactly the same way as the fixed points. Suppose that an intrinsic force $\overline{IF(i)}$ is acting on i. A constant force $\overline{CF(i)} = -\overline{IF(i)}$ can be added to i. As $\overline{CF(i)}$ and $\overline{IF(i)}$ cancel each other, the cell i is in force-equilibrium state. To perturb the cell location, another constant force is applied on the cell in the same way as demonstrated in Figure 9.2(b).

With the addition of fixed points, the quadratic formulation in (9.9) is modified as follows:

$$f = \sum_{i,j} \left((x_i - x_j)^2 + (y_i - y_j)^2\right) + \sum_{f} w_{H(f),f} \left((x_{H(f)} - x_f)^2 + (y_{H(f)} - y_f)^2\right) \tag{9.8}$$

where each fixed point f introduces two weighted quadratic terms, namely $w_{H(f),f}$ $(x_{H(f)} - x_f)^2$ and $w_{H(f),f}(y_{H(f)} - y_f)^2$, into the original formulation. Since f is fixed at (x_f, y_f), it behaves exactly the same as the fixed IO pads. Consequently, only diagonal elements of matrix C and vector $\bar{d}$ in (9.3) need to be updated to reflect the changes in (9.8).

However, because of the similarity between fixed points and IO pads, the pseudoconnections introduced by fixed points cannot be distinguished from real connections by a quadratic solver. It is possible that the solver may favor optimization of the pseudoconnections instead of the real ones. Applying fixed points into the placement should be carefully studied.

9.4 Fixed-Points Addition-Based Placement

In this section, we compare the fixed-points and constant forces in the context of quadratic placement and prove that the fixed points are a generalization of the constant forces. Fixed-point-based placement will be discussed next.

9.4.1 Fixed Points vs. Constant Forces

We will compare fixed points and constant forces in the following categories:

Flexibility. The following theorem characterizes the flexibility of the fixed points as related to the constant forces.

Theorem 9.3. *Fixed point is a generalization of a constant force.*

Proof. A fixed point f is able to mimic the constant force $\overline{CF(i)}$ applied on a cell i by using the combination of an infinitely large length $|\bar{f} - \bar{i}|$ and infinitely small weight $w_{i,f}$ while making $\overline{F(i,f)} = w_{f,i} \cdot (\bar{f} - \bar{i}) = \overline{CF(i)}$. Since $|\bar{f} - \bar{i}|$ is infinitely large, that is, f is infinitely far away from i, the movement of a cell i within the chip's boundary has no effect on $\overline{F(i,f)}$, thus $\overline{F(i,f)} = w_{f,i} \cdot (\bar{f} - \bar{i}) = \overline{CF(i)}$ does not depend on the position of a cell i and remains constant. □

Controllability. Fixed points guarantee that in the force-equilibrium state all movable cells stay within a predefined boundary (for example, the chip boundary).

Theorem 9.4. *Using only fixed points guarantees that in the force-equilibrium state all the movable cells will be located in the area bounded by a box containing the fixed points and IO pads.*

Proof. Let the bounding box formed by the fixed points and IO pads be denoted by (x_l, y_b) (the left-bottom) and (x_r, y_t) (the right-top) coordinates. In the following, we prove that in the force-equilibrium state no movable cell is placed outside the x-direction bounds $[x_l, x_r]$. Suppose that this is not true and that in a force-equilibrium state, there are cells i on the right side of the right bound x_r. For those cells $x_i > x_r$, let j correspond to a cell with the largest x_j. This cell is connected to some cells to its left. Since j is the right-most cell, it is impossible for it to be in a force-equilibrium state because there are no right-bound forces required to cancel the attaching forces induced by the cells on its left side. This contradicts the condition that the current placement is in a force-equilibrium state. So the assumption that some cells are located on the right side of the right bound x_r is invalid. Similarly, we can prove that the cells cannot be located on the left side of the left bound x_l in the force-equilibrium state. Therefore all the cells are located within the bound of $[x_l, x_r]$. With the same reasoning, we can prove that cells will not be placed outside of the y-direction bounds $[y_b, y_t]$. □

By Theorem 9.3, constant forces are such fixed points whose lengths are infinitely large. So the bounding box in Theorem 9.4 is the whole 2D plane. As a result, with constant forces acting on cells, there is no guarantee that in the equilibrium state the cells will be placed in a finite-bounded region.

9.4.2 Fixed Points in Global Placement

We will consider global placement as a sequence of transformations, each composed of the following three stages.

First Stage – Adding Controlling Fixed Points

The first stage is to transform the present placement, which either was obtained as a result of previous transformation or was an initial placement, to a force-equilibrium state by introducing one controlling fixed point per cell. For a cell i, we add the controlling fixed point f to balance the intrinsic force, $\overline{IF(i)}$, therefore $\overline{F(i, f)} = -\overline{IF(i)}$. After deciding the weight $w_{i,f}$, we obtain the corresponding $\bar{f} - \bar{i} = \overline{F(i, f)}/w_{i,f}$. The location of f is $\bar{f} = \bar{i} + \overline{F(i, f)}/w_{i,f}$, where $\bar{i}$ is the coordinate vector of the cell i. Obviously, different selections of $w_{i,f}$ result in different $\bar{f} - \bar{i}$ and different locations of f.

Second Stage – Adding Perturbing Fixed Points

Now we add perturbing fixed points to eliminate overlapping. We adopt a cell-spreading strategy similar to that in [12] but with a more global view of the present cell distribution.

We impose a $H_g \times W_g$ global bin structure on the placement area as shown in Figure 9.3. H_g is the number of global bins in vertical direction and W_g is the number of global bins in horizontal direction. A global bin is referred to as $b_{r,c}$, where r is its row index and c the column index. A bin may hold approximately 4–5 standard cells of average size. We denote by $C(b)$ the capacity of a global bin b. $A(b)$ denotes the total area of the cells inside b. A bin b has a cell overflow if $A(b) > C(b)$. An evenly distributed global placement implies that no significant cell overflow (or none) occurs. To achieve this, we should move cells from those bins with overflow to those without. We should add perturbing fixed points in such a way that after the transformation, cells can be moved out of the dense bins. We describe below the procedure of adding the perturbing fixed points based on virtual boundary shift.

Let us consider a vertical boundary between the global bins $b_{r,c}$ and $b_{r,c+1}$ and compute the minimum number of cells that have to be transferred across this boundary to achieve balanced cell distribution in the row r. Let $C_{r,(0,c)}$ and $C_{r,(c+1,W_g)}$

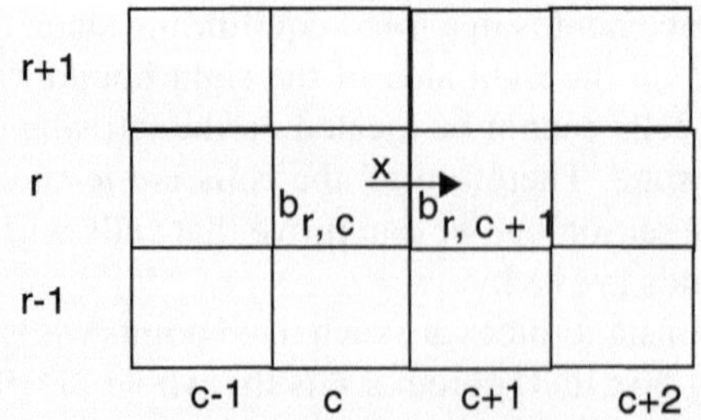

Fig. 9.3. Global bin structure for placement transformation.

be the total row capacity on the left and right sides of the boundary; let $A_{r,(0,c)}$ and $A_{r,(c+1,W_g)}$ be the total actual cell area in the row r on the left and right sides of the boundary. The minimum cell flow $x_{r,(c,c+1)}$ across the boundary between $b_{r,c}$ and $b_{r,c+1}$ can be computed as follows:

$$\frac{A_{r,(0,c)} - x_{r,(c,c+1)}}{C_{r,(0,c)}} = \frac{A_{r,(c+1,W_g)} + x_{r,(c,c+1)}}{C_{r,(c+1,W_g)}} \tag{9.9}$$

Note that when computing the minimum flow $x_{r,(c,c+1)}$ using (9.9), we consider the balance of the entire row r.

A positive $x_{r,(c,c+1)}$ indicates that cells should be moved from left to right across the boundary, while negative value indicates movement in the reverse direction. If the boundary is virtually movable, positive $x_{r,(c,c+1)}$ suggests that the boundary shifts to the right to introduce more capacity and decrease the cell density. The distance a boundary virtually shifts is decided by the magnitude of $x_{r,(c,c+1)}$ as follows:

$$s_{r,(c,c+1)} = \frac{x_{r,(c,c+1)}}{h(b_{r,c})} \tag{9.10}$$

where $h(b_{r,c})$ is the height of the global bin $b_{r,c}$. We refer to the boundary shift as virtual because those shifts are not executed, but instead are used to introduce perturbing fixed points.

The virtual boundary shift $s_{r,(c,c+1)}$ is proportional to $x_{r,(c,c+1)}$. In the case when $x_{r,(c,c+1)}$ is large, (this happens in the early stages of quadratic placement when a large number of cells overlap in a few global bins), the shift could be dramatic. In practice, we impose a limit on the maximum boundary shift to prevent the placement from drastic changes in one transformation. The limits on shifts in either direction is set to a half of the global bin's dimensions. After determining the amounts of virtual shifts for all the boundaries, we compute the new target cell locations (x', y') by mapping the present locations of cells according to the shifted bin boundary as follows:

$$x' = x'_{\min} + \frac{x - x_{\min}}{x_{\max} - x_{\min}} \left(x'_{\max} - x'_{\min}\right) \tag{9.11}$$

$$y' = y'_{\min} + \frac{y - y_{\min}}{y_{\max} - y_{\min}} \left(y'_{\max} - y'_{\min}\right) \tag{9.12}$$

The $(x_{\min}, y_{\min})$ and $(x_{\max}, y_{\max})$ are left-bottom and right-top coordinates of a global bin; $(x'_{\min}, y'_{\min})$ and $(x'_{\min}, y'_{\min})$ are the new coordinates after the boundary shift. It can be seen that if a global bin is expanding (the boundaries shift away from the bin center), the cell density (or equivalent cell overlapping after the mapping) will decrease because the mapping will increase the distance between cells.

Moving the cells to their target locations (x', y') determined by the boundary shifts and the mappings will decrease the cell overlaps. The perturbing fixed points are added in such a way as to attract the cells to their corresponding target locations. Specifically, for a cell i we introduce the perturbing fixed point f with $\overline{F(i, f)} = \alpha\bar{d}$ where $\bar{d}$ is a vector $\overline{(x' - x, y' - y)}$ and α is a design-specific parameter used to control the overall strength of the placement perturbation. In general, α is a function of the interconnect complexity of the design. High interconnect complexity (e.g., a large Rent's exponent [3]) suggests more global and semiglobal connections among cells and stronger attraction in quadratic formulation. As a result, the quadratic approaches (constant force or fixed-point) find it relatively more difficult to spread the cells for the designs with higher interconnect complexity than for those with low complexity. As the Rent's exponent of the design increases, α tends to be larger. For the IBM benchmark suites (which have a relatively large Rent's exponent) used in our experiments, α is set to 2.5, and for PEKO benchmarks [2] (which have a relatively small Rent's exponent), α is set to 0.5. Since similar designs can be treated by the same or similar α (we use the same α for all IBM benchmarks), in practice, designers may decide a good α based on their previous knowledge of similar designs.

Another approach to determine a good α is to start the placement with a small α and gradually modify/increase it based on the overlapping situation during placement. Since a quadratic solver with fixed-point addition can be very fast as indicated in [12], a smaller starting α does not necessarily compromise the placement efficiency. Furthermore, if we maintain different α for every cell/module in the circuit, we can deal with designs of nonuniform interconnect complexity. Note that at this point, we are only determining the force vector $\overline{F(i, f)}$. The location of f is not yet decided.

From the discussion above, we know how to determine the force induced by a controlling or perturbing fixed point. In the controlling fixed-points case, $\overline{F(i, f)} = -\overline{IF(i)}$; in the perturbing fixed-points case, $\overline{F(i, f)} = \alpha\bar{d}$. There is a flexibility of selecting proper combinations of $\bar{f} - \bar{i}$ and $w_{i,f}$ such that $\overline{F(i, f)} = w_{i,f}(\bar{f} - \bar{i})$.

We classify fixed points into two categories: on-chip and off-chip. On-chip fixed points are those located inside the chip boundary; off-chip fixed points are those outside the chip boundary. Compared to off-chip fixed-points, on-chip fixed-points have an obvious advantage of controllability, which ensures that all the cells in the force-equilibrium state are inside the chip's boundary according to Theorem 9.4. But the disadvantage of the on-chip fixed-points is that with the same force magnitude $\overline{F(i, f)}$, the weights $w_{i,f}$ of on-chip fixed-points are usually larger than those of off-chip fixed-points. If $w_{i,f}$ is large enough to be comparable to the weight of real connections, placement quality may be compromised. This is because the pseudo connections behave exactly the same as real connections and so the optimizer may favor pseudo connections instead of real ones. In the following, we use one example to illustrate this analysis.

In Figure 9.4, we show three standard cells A, B, and C with horizontal coordinates. Since x direction is independent of y direction in our force formulation, we only consider horizontal direction in the following discussion. An on-chip fixed-point f is attached to cell A. In the figure, the strengths are labeled above every connection. Note that $w_{A,f} = 1$ is 5 times larger than $w_{A,B} (= 0.2)$. As can be easily computed, cell A and B are both in force-equilibrium state. Now suppose, because of some reason, cell C is moved from $x = 7$ to $x = 8$. The movement breaks the force-equilibrium state of cell B and thus causes B to move in order to enter the new equilibrium state. In addition, the movement of B will also cause the displacement of cell A. By simple computation, we obtain the new force-equilibrium state where

A is located at $x = 1.14$ while B is at $x = 6.86$. If we calculate the wirelength of real connections, we get $(C - B) + (B - A) = 6.86$.

Now let us replace the on-chip fixed-point f with an off-chip fixed point that has the same force magnitude but much smaller weight (Now $w_{A,f}$ is 0.1 instead of 1 in Figure 9.4, and repeat the experiment above. As can be seen in Figure 9.5, f is placed at $x = -9$ due to the decreased $w_{A,f}$. If we calculate the new force-equilibrium state after C's movement, we obtain the new location for A and B, which are 1.6 and 6.94, respectively. The wirelength at new equilibrium state is $(C - B) + (B - A) = 8 - 1.6 = 6.4$, which is less than that in Figure 9.4. So from a total-wirelength-optimization point of view, the fixed-point in Figure 9.5 is a better choice than that in Figure 9.4.

The discussion above suggests that with the same force magnitude, the weight on the pseudo connection introduced by a fixed-point should be as small as possible to minimize interference with the real connections. But as the weights decrease, all the fixed points will eventually evolve into off-chip fixed-points. As Theorem 9.4 states, off-chip fixed-points cannot guarantee that all cells in the equilibrium state stay inside the chip boundary. When cells are pulled outside the chip boundary because of the introduced off-chip fixed-points, we say that cell explosion has occurred. Even if the forces introduced by particular fixed points are small, a large number of such fixed-points pointing in the same direction may still cause cell explosion (this happens often in the early placement stages when a large number of overlapping cells are

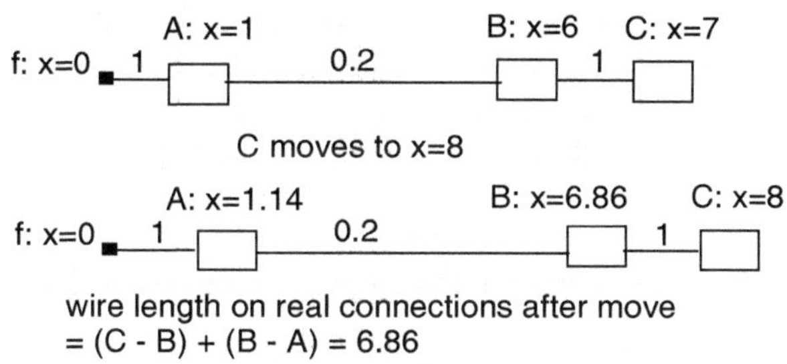

Fig. 9.4. The impact of on-chip fixed-point.

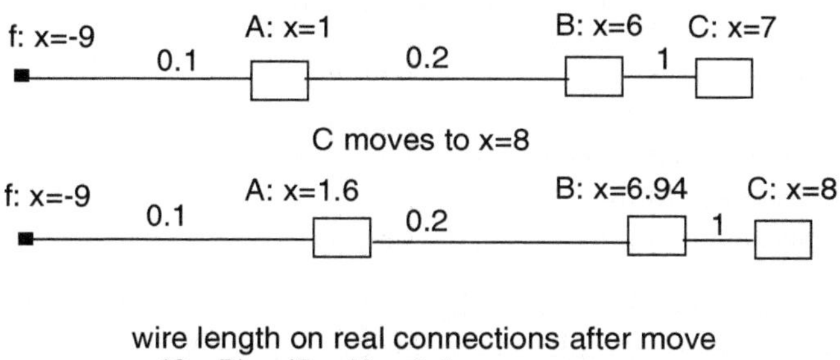

Fig. 9.5. Impact of off-chip fixed-point.

heading toward the same sparse region). If cell explosion happens, fitting the cells back into the chip boundary distorts the optimum solution produced by the quadratic solver, which leads to bad placement results.

After introducing the controlling and perturbing fixed points, we update matrix C and vector $\bar{d}$ accordingly and use a quadratic solver to find the optimum placement solutions – that is, the force-equilibrium states – for all the movable cells.

Third Stage – Refinement

The third stage is an optional local cost optimization procedure similar to the procedure described in [12]. Each cell is selected to move around its neighborhood to improve placement quality, as measured by different objectives such as total wirelength, timing, etc. After the cell movement, the placement transits from the force-equilibrium state achieved after stage 2 to a new disequilibrium state.

The placement is conducted by iterating the three-stage transformation discussed above until the cells are evenly distributed over the chip area. This is measured by the standard deviation of cell distributions over the global bin structure.

9.4.3 Detailed Placement

Detailed placement includes two phases. The first phase takes the evenly distributed global placement result, which may have a certain degree of cell overlapping, and find the legal placement for all movable macros and standard cells. In our current implementation, we first find the legal locations for all movable large macros, and then fit the cells into standard-cell-rows not occupied by either fixed or movable macros. The second phase is an optimization procedure that moves cell or swaps cells in the neighborhood if the cost decreases until no improvement is possible.

9.5 mFAR: Multilevel Fixed-Point Addition-Based Placement

Multilevel implementation of fixed-point addition-based placement (mFAR) is based on the classical V-cycle structure, which has been successfully demonstrated in various multilevel placers such as [1] and multilevel partitioners like hMetis [8].

mFAR consists of the following five major steps:

1. Fine-granularity clustering
2. Multilevel general clustering
3. Global placement and refinement of general clusters
4. Detailed placement of fine clusters
5. Detailed placement of original standard cells

mFAR starts with the fine-granularity clustering [7] whose purpose is to improve the overall efficiency of the placement flow. Fine-granularity clustering generates very small clusters with no more than three standard cells. In [7], we demonstrated that these small clusters can be determined based on the local connectivity information, and they are very useful in boosting placement efficiency. The clustering algorithm is a mutual contraction-based greedy algorithm described in [7]. The mutual contraction is a metric used to measure how strongly two circuit elements are connected. It is defined as follows:

$$c_p(u, v) = \frac{w_{u,v}}{\sum_x w_{u,x}} \frac{w_{v,u}}{\sum_y w_{v,y}} \tag{9.13}$$

In (9.13), $\sum_x w_{u,x}$ is the sum of the weights on all connections incident to cell u and $\sum_y w_{v,y}$ denotes the same for the cell v. The larger the mutual contraction is, the more likely it is that corresponding cells will be placed together. So the algorithm in [7] always looks for a clustering connection with the largest contraction and does it until the target upper limit for the cluster size is reached. The aspect ratio of a fine cluster is determined as follows: the height of a fine cluster is the same as normal standard cell height, while the cluster width is the width-summation of all the cells inside the cluster. With this aspect ratio, a legal cell placement can be easily achieved once a legal placement of fine clusters is obtained.

The step 2 creates multiple levels of clusters (general clusters), from the fine clusters. We use the same clustering algorithm as in step 1, but determine the aspect ratio of a general cluster differently. Specifically, the aspect ratio of a general cluster is set to 1; the cluster area is equal to the total cell area inside the cluster.

In both fine-granularity clustering and general clustering, the resulting number of clusters is roughly half of the original number of cells (fine-granularity clustering) or general clusters (general clustering). mFAR usually constructs 4 to 5 levels of clusters and so the number of top-level clusters is roughly 1/20 of original number of movable cells.

In *mFAR*, the step 3 performs placement transformations for general clusters as described in Sect. 9.5, and declustering alternatively until it reaches the fine-cluster level. During declustering, all the cells/clusters belonging to a higher-level cluster are placed at the center of the higher-level cluster. After each level of declustering, we iterate the same transformations as those used in the global placement, and distribute the cells/clusters.

After the global placement of fine clusters is obtained, step 4 executes the detailed placement on fine clusters. To make the flow flexible, we also embed an optional simulated annealing-based placement optimization in this step. The user can turn it on if a placement with even better quality is desired. The initial temperature is adjustable depending on how much time the user wants to allocate for placement improvement. The higher the temperature is, the better the placement quality is. Higher temperature always comes at the cost of more CPU time.

The final step is to decluster fine clusters and perform final optimization. Since the placement of fine clusters is legal after the previous step, declustering fine-clusters does not cause any illegal placement.

9.6 Experimental Results

In this section, we present experimental results on four sets of academic benchmarks: ISPD05, ISPD06, PEKO05, and PEKO06.

9.6.1 ISPD05 Placement Contest Benchmarks

Each ISPD05 Placement benchmark contains a number of fixed large macros that occupy a significant portion of placement region. As a result, the placer has to consider the location of macros and optimize the placement of movable cells accordingly. In mFAR, fixed macros are handled naturally by computing their density contributions but not applying perturbing fixed points on them. As they are involved in density computation, the perturbing fixed points induced from density distribution are consequently affected by the locations of fixed macros. As a result, perturbing fixed points are pulling movable cells away from the area occupied by macros.

Table 9.1 shows the mFAR results on all ISPD05 benchmarks. The results are obtained on the machine with Dual Xeon 2.8 GHz processor, 8G memory, and Redhat Enterprise Linux 3.0 OS. In the table, HPWL denotes half-perimeter wirelength and is given in meters. CPU times are reported in seconds. It can be noted that the results differ slightly from ISPD05 Placement contest results because the contest results are obtained by tuning the placer for each individual benchmark while the results here came from the same running settings.

9.6.2 ISPD06 Placement Contest Benchmarks

ISPD06 benchmarks share many common features with ISPD05: large number of movable instances and fixed large macros. In addition, ISPD06 also imposes a new requirement for the placer: target cell density. That is, the cell density at feasible cannot exceed a predefined number. In other words, the placer has to intentionally leave some white space for routability and/or buffer insertion in the later design stage.

To handle target cell density constraint, mFAR employs a straightforward cell-inflation approach. After the fine-granularity clustering, mFAR inflates each fine cluster horizontally by $1/d$. As a result, once the legal placement for fine cluster

Table 9.1. mFAR results on ISPD05 benchmarks.

	HPWL	CPU
Adaptec1	82.50	2,081
Adaptec2	92.79	2,761
Adaptec3	217.56	4,725
Adaptec4	197.9	4,216
Bigblue1	98.8	2,543
Bigblue2	160.4	5,654
Bigblue3	368.7	10,549
Bigblue4	865.4	18,551

is obtained. Deflation of fine-clusters automatically reserve the required white space at all places.

Table 9.2 lists mFAR results on ISPD06 benchmarks. The results are obtained on the machine with two 2.6 GHz AMD Opteron™ Processor, 8G memory, and Linux OS 2.4.21-37.ELsmp. As can be seen that newblue4 and newblue5 exhibit large overflow numbers (5.42 and 5.92 respectively). This is due to the fact that mFAR does not explicitly enforce overflow control during greedy wirelength optimization at step 5.

9.6.3 PEKO 2005

Table 9.3 lists mFAR results on PEKO 2005 benchmarks. The results are obtained on the machine with Dual Xeon 2.8 GHz processor, 8G memory, and Redhat Enterprise Linux 3.0 OS. In the table, HPWL denotes half-perimeter wirelength and is given in meters. CPU times are reported in seconds.

9.6.4 PEKO 2006

Table 9.4 lists mFAR results on PEKO 2006 benchmarks. The results are obtained on the machine with Dual Xeon 2.8 GHz processor, 8G memory, and Redhat Enterprise Linux 3.0 OS. In the table, HPWL denotes half-perimeter wirelength and is given in meters.

Table 9.2. mFAR results on ISPD06 benchmarks.

	HPWL	Overflow	CPU
Adaptec5	448.43	6.21	6,875
Newblue1	77.36	0.22	2,538
Newblue2	211.65	0.59	2,891
Newblue3	303.58	0.11	2,957
Newblue4	307.72	5.42	6,362
Newblue5	567.65	5.92	11,426
Newblue6	527.36	1.63	12,154
Newblue7	1135.80	1.58	19,483

Table 9.3. mFAR results on PEKO 2005 benchmarks.

	HPWL	CPU
Adaptec1	48.8	1,346
Adaptec2	65.1	1,623
Adaptec3	99.3	2,975
Adaptec4	87.3	3,251
Bigblue1	55.3	1,747
Bigblue2	*	*
Bigblue3	287.4	7,245
Bigblue4	618.4	17,840

*: mFAR crashed because of floating nodes

Table 9.4. mFAR results on PEKO 2005 benchmarks.

	HPWL	Overflow penalty
Adaptec5	275	209
Newblue1	136	555
Newblue2	122	183
Newblue3	161	164
Newblue4	153	208
Newblue5	323	196
Newblue6	269	176
Newblue7	566	177

9.7 Conclusions

In this chapter, we have presented an in-depth study of the fixed-point addition placement technique. We demonstrate both theoretically and empirically that fixed-point yield a useful generalization of the constant force, and a placer based on this technique produces good results on netlists of standard cells. We present *mFAR,* a multilevel, fixed-points-based placer and demonstrate its ability to produce competitive placement results.

References

1. Chang C, Cong J, Pan Z, Yuan X (2002) Physical hierarchy generation with routing congestion control. Proc. International symposium on physical design, pp 36–41
2. Chang C, Cong J, Romesis M, Xie M (2004) Optimality and scalability study of existing placement algorithms. J IEEE Transactions on Computer-Aided Design of Integrated Circuits and Systems, vol. 23, pp 537–549
3. Donath WE (1979) Placement and average interconnection lengths of computer logic. J IEEE Transactions on Circuits and Systems, vol. CAS-26, pp 272–277
4. Eisenmann H, Johannes FM (1998) Generic global placement and floor planning. Proc. Design Automation Conference, pp 269–274
5. Etawil H, Areibi S, Vannelli A (1999) Attractor-repeller approach for global placement. Proc. International Conference on Computer-Aided Design, pp 20–24
6. Hu B, Marek-Sadowska M (2002) FAR: fixed point addition and relaxation based placement. Proc. International Symposium on Physical Design, pp 161–166
7. Hu B, Marek-Sadowska M (2003) Wire length prediction based clustering and its application in placement. Proc. Design Automation Conference, pp 800–805
8. Karypis G, Aggarwal R, Kumar V, Shekhar S (1997) Multilevel hypergraph partitioning: application in VLSI domain. Proc. Design Automation Conference, pp 526–529
9. Kleinhans JM, Sigl G, Johannes FM, Antreich KJ (1991) GORDIAN: VLSI placement by quadratic programming and slicing optimization. J IEEE Transactions on Computer-Aided Design of Integrated Circuits and Systems, vol 10, issue 3, pp 356–365
10. Sigl G, Doll K, Johannes F (1991) Analytical placement: a linear or a quadratic objective function. Proc. Design Automation Conference, pp 427–432

11. Tsay RS, Kuh E, Hsu CP (1988) PROUD: A sea-of-gates placement algorithm. J IEEE Design & Test of Computers, vol. 5, issue 6, pp 44–56
12. Viswanathan N, Chu CN (2005) FastPlace: efficient analytical placement using cell shifting, iterative local refinement and a hybrid net model. J IEEE Transactions on Computer-Aided Design of Integrated Circuits and Systems, vol. 24, issue 5, pp 722–733

10

mPL6: Enhanced Multilevel Mixed-Size Placement with Congestion Control

Tony F. Chan[1], **Jason Cong**[2], **Joseph R. Shinnerl**[2], **Kenton Sze**[1] **and Min Xie**[2]
[1]UCLA Mathematics Department, Los Angeles, CA 90095-1555
[2]UCLA Computer Science Department, Los Angeles, CA 90095-1596
{chan, nksze}@math.ucla.edu, {cong, shinnerl, xie}@cs.ucla.edu

10.1 Introduction

mPL6 consists of three basic ingredients: global placement by multilevel nonlinear programming [21], discrete graph-based macro legalization followed by linear-time scan-based standard-cell legalization [26], and detailed placement [26]. It is designed for speed and scalability, low wirelength results, adaptability to complex constraints, and robustness under low white space. Compared to the 2005 implementation [19], the main improvements to mPL6 are:

1. Improved clustering by the "best-choice" heuristic [5]
2. $2\times$ reduction in the number of levels of clusters
3. More aggressive weighting of wirelength relative to overlap removal during optimization at each level
4. A faster single-V-cycle iteration flow
5. Gradual determination of the locations of large objects earlier in the multilevel flow
6. Density-sensitive legalization and detailed placement
7. Improved handling of unconnected filler cells supporting convergence to nonuniform module-area distributions

The organization of this chapter is as follows. In Sect. 10.2, we present some definitions and notations that are used in the chapter. In Sect. 10.3, we give the problem formulation of placement. In Sect. 10.4, we discuss the multilevel framework for placement. In Sect. 10.5, we present the core optimization engine – generalized force-directed algorithm, for our global placement. Following the global placement algorithm, we discuss our legalization and detailed placement algorithms in Sect. 10.6. Finally, numerical results are presented in Sect. 10.7.

10.2 Definitions and Notations

In this section, definitions and notations are presented which are used in the later sections.

Definition 10.1. *Given a real-valued $m \times n$ matrix A with a_{ij} as its ij-th entry, the transpose of A, denoted as A^{T}, is an $n \times m$ matrix defined as:*

$$A^{\mathrm{T}} \equiv \begin{pmatrix} a_{11} & a_{21} & \cdots & a_{m1} \\ \vdots & & & \vdots \\ a_{1n} & a_{2n} & \cdots & a_{mn} \end{pmatrix}$$

Definition 10.2. *$A[n]$ is denoted for a real $n \times n$ square matrix A with each ij-th entry denoted by $A[n]_{ij}$.*

Definition 10.3. *The ij-th entry, $C[n]_{ij}$, of a discrete cosine matrix $C[n]$ is defined by:*

$$C[n]_{ij} \equiv \sqrt{\frac{2 - \delta_{i1}}{n}} \cos\left(\frac{(i-1)(2j-1)\pi}{2n}\right)$$

where $1 \leq i, j \leq n$, $\delta_{i1} = 1$ if $i = 1$ and $\delta_{i1} = 0$ otherwise.

Definition 10.4. *The tensor product of two matrices, $A[m]$ and $B[n]$ with a_{ij} and b_{ij} as the ijth entry respectively, is defined as*

$$A[m] \otimes B[n] \equiv \begin{pmatrix} a_{11}B[n] & a_{12}B[n] & \cdots & a_{1m}B[n] \\ \vdots & & & \vdots \\ a_{m1}B[n] & a_{m2}B[n] & \cdots & a_{mm}B[n] \end{pmatrix}$$

which is an $mn \times mn$ matrix.

10.3 Problem Formulation

Circuit placement can be characterized as an optimization problem on a hypergraph $H = (V, E)$. Let $V = \{v_1, v_2, \ldots, v_N, v_{N+1}, \ldots, v_{N+P}\}$ represents the set of cells and $E = \{e_1, e_2, \ldots, e_m\}$ represents the set of nets. A net with degree k is called a k-pin net. The set $\{v_{N+1}, \ldots, v_{N+P}\}$ represents fixed objects or terminals, e.g., *pads* along the perimeter of the rectangular placement region R in which cells $v \in V$ are to be positioned. Each e_i is a subset of V specifying a connection called a *net* or *hyperedge* among the cells it contains.

Let (x_k, y_k) be the center coordinate of the cell v_k. The pin-to-pin half-perimeter wirelength (HPWL) of a net e, given by:

$$l(e) = \max_{v_i,v_j \in e, i<j} |(x_i + p^e_{x_i}) - (x_j + p^e_{x_j})| + \max_{v_i,v_j \in e, i<j} |(y_i + p^e_{y_i}) - (y_j + p^e_{y_j})| \quad (10.1)$$

where $(p^e_{x_i}, p^e_{y_i})$ is the relative pin position on cell v_i connecting by net e. The total HPWL

$$\sum_{e \in E} l(e)$$

is a standard estimate of wirelength used to measure the quality of the results (QoR) obtained from the placement algorithm. Throughout the chapter, we use HPWL or WL to denote the half-perimeter wirelength of a net or a circuit. For simplicity, many placement algorithms consider the center-to-center HPWL

$$\max_{v_i,v_j \in e, i<j} |x_i - x_j| + \max_{v_i,v_j \in e, i<j} |y_i - y_j| \quad (10.2)$$

for each net e, in which all the aforementioned pin offsets $(p^e_{x_i}, p^e_{y_i})$ are set to zero.

Our objective is to place the cells subject to constraints such that the total wirelength

$$\sum_{e \in E} l(e)$$

is minimized. The constraints include the challenging requirement that no two cells in the placement overlap. We consider both the standard-cell and mixed-size placement problems, in both of which all movable standard cells must be placed in given rows of R without overlapping.

Typically, placement is divided into three distinct stages: global placement, legalization, and detailed placement. During global placement, cells are spread out in the placement region R. Many global placers, including mPL6, require that the total cell area in each subregion of a given partition of R does not exceed preset limits. These *bin-density* constraints are a weakened approximation of the pairwise nonoverlap constraints. The result of global placement is then *legalized* by a discrete scheme [26, 42]; i.e., all cells are assigned discrete locations in R in such a way that no two cells overlap. The legalization scheme starts from the result of the global placement and generally attempts to perturb it as little as possible in order to obtain its strictly feasible, i.e., overlap-free, result. After legalization, the wirelength is further reduced by feasible iterative improvement, e.g., local cell swapping, in which no move is allowed to incur any constraint violation.

10.4 Multilevel Framework

Multigrid and algebraic multigrid have been successfully applied to solve partial differential equations and linear system of equations, respectively, [10, 13, 14, 52, 55]. They are highly effective and scalable. Similar ideas have been applied to large-scale optimization problems [12]. Many studies [3, 15–17, 22, 35] show that the multilevel metaheuristic is a promising approach to large-scale global optimization in the VLSI

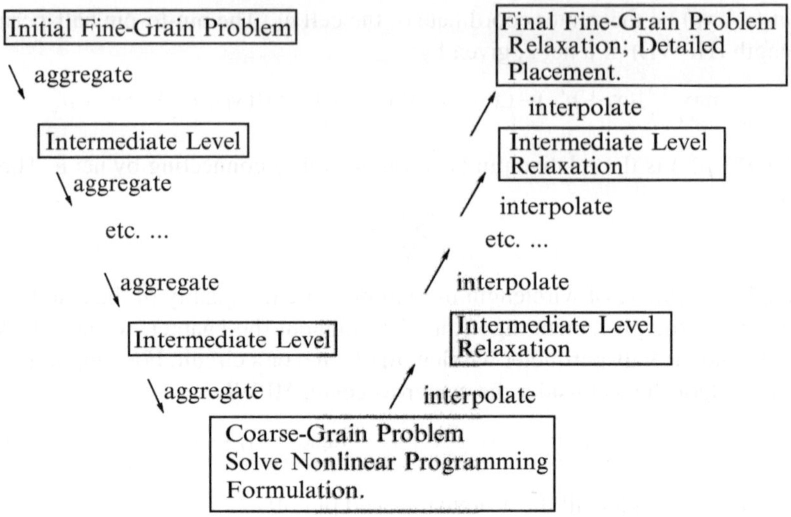

Fig. 10.1. Multilevel optimization V-cycle. © ACM, 2006. This is the author's version of the work. It is posted here by permission of ACM for your personal use. Not for redistribution. The definitive version was published in Proceedings of the 2005 International Symposium on Physical Design, pp. 185–192, April 2005 (ISPD'05) http://doi.acm.org/10.1145/1055137.1055177.

domain. It is not only used for speed-up, but also for improved solution quality. In this section, the general multilevel framework on which mPL6 is built is presented.

Figure 10.1 shows an example of the V-cycle variant of multilevel optimization for placement, as used in [15, 16, 21, 58]. The main idea of multilevel algorithms is to build a sequence of coarser-level or simplified problems to approximate the original problem (the finest level problem). Each coarse problem is an approximated and simplified problem of the finer level problem such that a good solution obtained from optimizing the coarse-level problem serves as a good starting point for optimizing the adjacent finer-level problem. The main components of multilevel algorithms are (a) coarsening (clustering), to build the coarse level problems; (b) interpolation, to transfer variables, solutions, objectives, and constraints between levels; (c) relaxation, improvement of a given solution at a given level; and (d) multilevel iteration flow, the order in which subproblem levels are visited and relaxation to them is applied. Each of these components is discussed in the following sections.

10.4.1 Coarsening

The purpose of coarsening or clustering is to build a hierarchy of approximate problem formulations. Due to the physical meaning of placement, a natural way to reduce the problem size is by recursively grouping cells into clusters and clusters into larger clusters until the number of clusters is small enough. Each level of the recursion defines a level in the cluster hierarchy to which iterative improvement is applied. Nets connecting cells within each cluster are ignored, and at each level we consider placement only of the clusters defined at that level. In this way, a hierarchy of coarsened, i.e., simplified problems of the original problem is constructed.

Whether clustered netlists are good approximations of the original netlist depends on the way the cells are clustered. Hence, there are many studies on clustering scheme [5, 23, 34, 35]. In general, most clustering schemes for placement consist of two steps. The first step is to define the affinity between two connected cells, the larger the affinity the higher the chance to cluster the cells. The next step is to decide the sequence in which cells are selected and clustered with other cells or clusters. The coarsened hypergraph is thus constructed as follows. An affinity graph is constructed by joining each vertex to exactly one of its neighbors for which it has maximal affinity. Each group of joined vertices is called a cluster and becomes a coarser level vertex. The cluster size equals the sum of the sizes of the cells it contains. Hyperedges are defined on the clusters in the obvious way: each hyperedge on the finer level becomes a hyperedge at the coarser level defined simply as the set of all coarser-level clusters containing finer-level vertices in the original finer-level hyperedge. Because singleton hyperedges are simply ignored, a smaller hypergraph is obtained at the coarser level.

A clustering scheme for hypergraph coarsening, called First Choice, is proposed in [35]. It first transforms the hypergraph into a clique model weighted graph (see Figure 10.2). Given a hypergraph $H = (V, E)$, the weight or affinity between any two vertices v and w in the clique model weighted graph is defined as:

$$r_{vw} = \sum_{e \in E | v, w \in e} \frac{1}{(|e| - 1)} \tag{10.3}$$

First Choice traverses the list of vertices in an arbitrary order. Each visited vertex is then clustered with a neighbor for which it has largest affinity (10.3). The clustering algorithm stops when the number of clusters has reached the target.

From (10.3), affinity increases with connectivity between the cells. Also, smaller net degree $|e|$ contributes increased affinity between the cells. The intuition is that cells with high affinity should stay close together in a good placement solution. In the following sections, two clustering schemes, modified First Choice and Best Choice, for placement problem are presented.

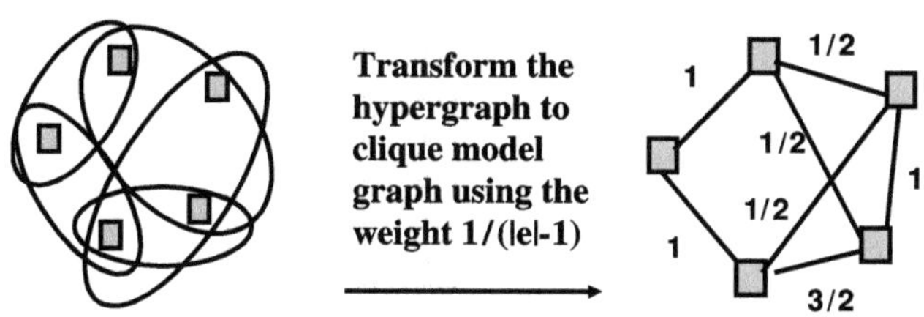

Fig. 10.2. Transformation from hypergraph to clique model weighted graph. © ACM, 2006. This is the author's version of the work. It is posted here by permission of ACM for your personal use. Not for redistribution. The definitive version was published in Proceedings of the 2005 International Symposium on Physical Design, pp. 185–192, April 2005 (ISPD'05) http://doi.acm.org/10.1145/1055137.1055177.

Modified First Choice Clustering

For placement, the affinity between vertices or cells v and w is defined as:

$$r_{vw} = \sum_{\{e \in E \mid v, w \in e\}} \frac{1}{(|e| - 1)\text{area}(e)} \tag{10.4}$$

where area(e) denotes the sum of the areas of the cells in e. The additional term area(e) is an indirect way to control the cluster size. Unlike the affinity defined in (10.3), which is proposed mainly for hypergraph partitioning, the modified affinity (10.4) targets placement. By controlling the cluster size, the coarsened hypergraph has less variation in cluster size. Unlike First Choice, cells are visited in ascending order of cell area (with preference to smaller cell degree in order to break ties). This ordering is observed to balance clusters' areas better, because (i) smaller clusters are merged first, making them larger, and (ii) merging two large clusters is less likely.

If a good initial placement is provided, the spatial information of the cells' locations can be incorporated into the affinity between cells as follows,

$$r_{vw} = \sum_{e \in E | v, w \in e} \frac{1}{(|e| - 1)\text{area}(e)\text{dist}(v, w)} \tag{10.5}$$

where dist(v, w) is the Euclidean distance between v and w. That is, vertices in close proximity have increased affinity for each other.

Best Choice Clustering

Best Choice, as proposed in [5], is a greedy clustering scheme. The affinity between any two cells v and w used in Best Choice is defined as:

$$r_{vw} = \sum_{e \in E | v, w \in e} \frac{1}{(|e|)(\text{area}(v) + \text{area}(w))} \tag{10.6}$$

where area(v) and area(w) denote the areas of cell v and cell w, respectively. In addition to the indirect control of the cluster size, Best Choice imposes a hard upper limit for cluster size. Clusters larger in area than this predefined upper bound are not formed.

Best Choice first computes the maximum affinity or score (10.6) each cell has for any of its neighbors. Symmetry of the score function (10.6) implies that the score can be associated with pairs of vertices rather than individual vertices. The pair with the largest score is clustered before the others, and, in principle, the netlist is immediately updated before the next cluster is formed. Hence, Best Choice is a greedier heuristic than First Choice, which makes a clustering pass through the entire vertex list before updating the affinity graph.

In principle, Best Choice uses the immediately updated coarsened hypergraph to compute affinities for subsequent candidate clusters after every cluster is formed.

However, as the immediate updating of all affected vertices in the hypergraph after each clustering operation is time consuming, lazy updating is instead used to reduce runtime. Instead of explicitly recomputing affinities of vertex pairs affected by a given vertex merge, the algorithm marks the affected vertex pairs as invalid and updates them only after they have been selected for merging based on their invalid score. Thus, each time a vertex pair is selected based on its invalid score, its correct updated score is computed before deciding whether to merge the vertices or not. Although the decision to merge them occasionally incurs errors, because updated scores are sometimes compared to other still invalid scores, the overall clustering error is demonstrated to be small and well worth the dramatic reduction in run time.

As with the modified affinity (10.5), a given good initial placement of the cells can be incorporated into the Best Choice affinity between cells:

$$r_{vw} = \sum_{e \in E | v, w \in e} \frac{1}{(|e|)(\text{area}(v) + \text{area}(w))\text{dist}(v, w)} \tag{10.7}$$

where $\text{dist}(v, w)$ is the Euclidean distance between v and w. That is, closer neighbors share increased affinity.

10.4.2 Relaxation

The optimization process for a given problem is referred to as relaxation. In the multilevel framework, the ideal complexity for relaxation at each level is linear time in terms of the number of variables at that level, so that the overall complexity for a V-cycle (cf. Figure 10.1) optimization is also linear.[1]

At the coarsest level where the problem size is small enough, the relaxation should be good enough to find a good solution or nearly optimal solution. At the other levels, an initial solution is given or transferred from the coarser level and serves as a starting point for the relaxation scheme. The initial solution at each level is iteratively improved by the relaxation scheme. The generalized force-directed formulation of relaxation used by mPL6 is described below in Sect. 10.5.

10.4.3 Interpolation

Interpolation is used to transfer solutions from level to level. For example, given a candidate placement at a coarse level, we compute from it a corresponding candidate placement at the adjacent finer level by interpolation.

A simple interpolation scheme, called constant interpolation, is to simply assign each cluster's location to its component subclusters. This interpolation gives an

---

[1] Under the mild assumption that the number of variables at each level is at most r times the number of variables at the adjacent finer level, for fixed fraction $r \in (0, 1)$ independent of level number, the total number of variables at all levels is linear in the number variables at the finest level, as $1 + r + r^2 + \cdots + r^L < 1/(1 - r)$ for any positive number of levels L. Typical values of r in practice are below 1/2.

immediate solution for the finer-level problem. However, for placement, constant interpolation puts all the subclusters on top of each other at the same location. This approach creates significant overlapping between cells which may not serve as a good initial solution for some particular placement techniques.

A more sophisticated interpolation scheme, called weighted or AMG-based interpolation, is proposed in [15]. It creates less overlap than constant interpolation. A graph model of connectivity (cf. Figure 10.2) is employed to define the interpolation; the weight of edge e_{ij} is

$$w(e_{ij}) = \sum_{\{e \in E \mid i, j \in e\}} \frac{1}{(|e| - 1)} \tag{10.8}$$

For efficiency, only edges with weights above a certain threshold (currently 1/4) are used; all other edges are ignored. Finer-level vertices v_i within each cluster with the highest vertex degree (using cell area to break tie) are designated as "C-points" (C-points are finer-level vertices serving as coarser-level representatives) and are given the positions of their parent clusters. C-point locations are fixed during interpolation. The remaining points are designated as "F-points" (finer-level representatives) and are placed at the weighted average of the positions of the C-points to which they are connected. Once an F-point has been placed, it can be treated like a C-point and used to influence the positioning of other F-points to which it has connections (cf. Figure 10.3). Moreover, since the process depends on the vertex order, iterations are used to allow all interconnected nodes to influence each others' positions. For this purpose, the nodes are ordered by decreasing connectivity $w(v_i) = \sum_j w(e_{ij})$, following (10.8).

If an initial solution is given for the original problem, it must be transferred to coarse levels so that it can be utilized and further improved through coarse-level optimization. The interpolation of a solution to a coarser level is relatively simple – the coarser level vertex or cluster position is determined by the average positions of its component subclusters.

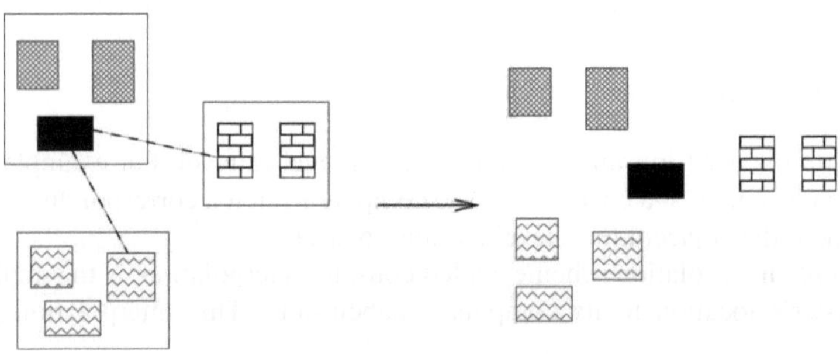

Fig. 10.3. AMG-based weighted interpolation.

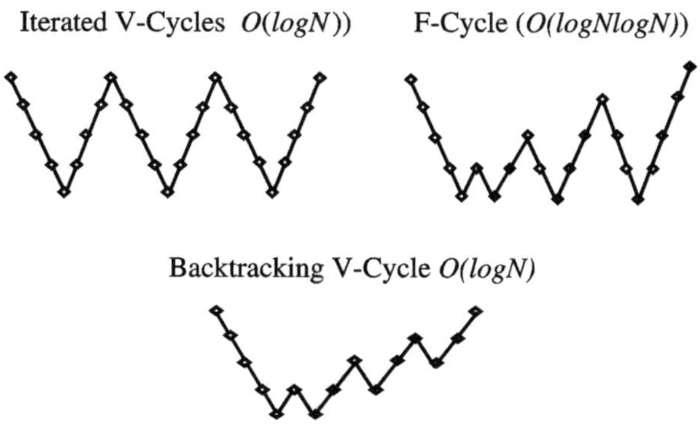

Fig. 10.4. Iterated multilevel flow alternatives. © ACM, 2006. This is the author's version of the work. It is posted here by permission of ACM for your personal use. Not for redistribution. The definitive version was published in Proceedings of the 2006 Conference on Asia South Pacific Design Automation (ASP-DAC'06), pp. 188–194 http://doi.acm.org/10.1145/1118299.1118353.

10.4.4 Multilevel Flow

Multilevel flow defines the sequence of problem levels along which solutions are computed and propagated. Figure 10.4 shows three different flow sequences. The simplest multilevel flow is V-cycle optimization. In a V-cycle, coarse level problems are constructed recursively until the problem size is small enough for the relaxation to find a good solution. The solution at the coarsest level is then recursively interpolated and relaxed back to the the original problem at the finest level.

The F-cycle (Figure 10.4) is a more extensive flow sequence based on recursive V-cycles. In an F-cycle flow, a complete V-cycle is applied to every coarse level problem, in order from the coarsest level to the finest. Hence, the solution at the coarsest level is continually refined as new improvements at finer levels are propagated back to it. F-cycles provide extensive relaxation at the expense of added runtime. A backtracking V-cycle flow (Figure 10.4) is a trade-off between speed and quality and may be viewed as a compromise between the V-cycle and F-cycle flows.

The multilevel flow can be iterated to further improve the quality of the solution. The solution from the previous multilevel flow can be incorporated into the clustering scheme affinity (10.5) and (10.7) to speed up convergence.

10.5 Generalized Force-Directed Algorithm

In this section, an effective constrained-minimization algorithm for a smooth approximation of placement is proposed. Accurate smooth approximations to both the wirelength objective and the pairwise nonoverlap constraints are described. A nonlinear-programming algorithm derived as a generalized force-directed (GFD) algorithm is presented. GFD is based on the well-known Uzawa algorithm [4] for mathematical programming, i.e., constrained numerical optimization.

As practical placement problems may have tens of millions of cells and require solutions in fast runtime (close to linear complexity), several heuristic techniques are incorporated to accelerate convergence. GFD can be viewed as a generalization of the Poisson-based force-directed method first proposed in [28]. The GFD algorithm is embedded in multilevel framework for better scalability and better global optimization.

10.5.1 Constrained Minimization Problem Formulation

In this section we present smooth approximations to the wirelength objective (10.2) and pairwise nonoverlapping constraint of the placement problem.

Smooth Wirelength Approximation

Since (10.2) is not differentiable and the constraints are highly nonconvex, a globally optimal placement is hard to locate. In order to employ advanced numerical solution methods, continuous differentiable functions are used to approximate (10.2) Many studies, for example [16,28,39], use a quadratic wirelength approximation given by:

$$\sum_{e \in E}\left(\sum_{v_i, v_j \in e, i<j} |x_i - x_j|^2 + \sum_{v_i, v_j \in e, i<j} |y_i - y_j|^2\right) \tag{10.9}$$

The advantage of using the quadratic wirelength objective is that its unique[2] unconstrained minimizer can be obtained by solving a positive definite linear system of equations. However, the quadratic model overpenalizes long nets and in so doing may produce a highly suboptimal half-perimeter wirelength placement solution.

In mPL6, we use the following more accurate half-perimeter wirelength approximation objective [9,21,40,48] given by:

$$\begin{aligned} \eta \sum_{e \in E} (\log \sum_{v_k \in e} \exp(x_k/\eta) + \log \sum_{v_k \in e} \exp(-x_k/\eta) \\ + \log \sum_{v_k \in e} \exp(y_k/\eta) + \log \sum_{v_k \in e} \exp(-y_k/\eta)) \end{aligned} \tag{10.10}$$

where the smaller η, the more accurate the approximation. It is a smooth convex function. Also, the number of terms in (10.10) is significantly less than that in (10.9). However, it is more costly to compute the exponential function, and η cannot be set too small, due to machine precision and numerical stability. In our experiments, we scale the placement problem so that all the cell locations are between 0 and 1. The value of η is then set to 0.01.

[2] Uniqueness of an optimal quadratic-wirelength solution is ensured if and only if at least one fixed terminal exists.

We have also proposed and studied another accurate approximation to (10.2) using L_p-norm:

$$\sum_{e \in E} \left(\left(\sum_{v_k \in e} x_k^p \right)^{1/p} - \left(\sum_{v_k \in e} x_k^{-p} \right)^{-1/p} + \left(\sum_{v_k \in e} y_k^p \right)^{1/p} - \left(\sum_{v_k \in e} y_k^{-p} \right)^{-1/p} \right) \tag{10.11}$$

since the first term and the second term tend to $\max\{x_k\}$ and $\min\{x_k\}$, respectively, as p tends to infinity. We set $p = 32$ in experiments so that x^p and $x^{1/p}$ can be computed efficiently. A slightly different approximation using L_p-norm is proposed in [37].

Smooth Constraints Approximation

As the pairwise cells nonoverlap constraints are highly nonconvex and difficult to satisfy during global placement, we relax the constraints by using the bin density constraints described below.

We divide the placement region R into $m \times n$ uniform nonoverlapping subregions (bins) B_{ij}, $1 \leq i \leq m$, $1 \leq j \leq n$ such that $\cup_{i,j} B_{ij} = R$. Let h_x and h_y be the bin width and bin height, respectively. Define D_{ij} to be the average density in the bin B_{ij} which is given by:

$$D_{ij}(x, y) = \sum_{k=1} a_{ij}(v_k)/(h_x h_y) \tag{10.12}$$

where $a_{ij}(v_k)$ is the fractional area of cell v_k lying inside bin B_{ij} (see Figure 10.5).

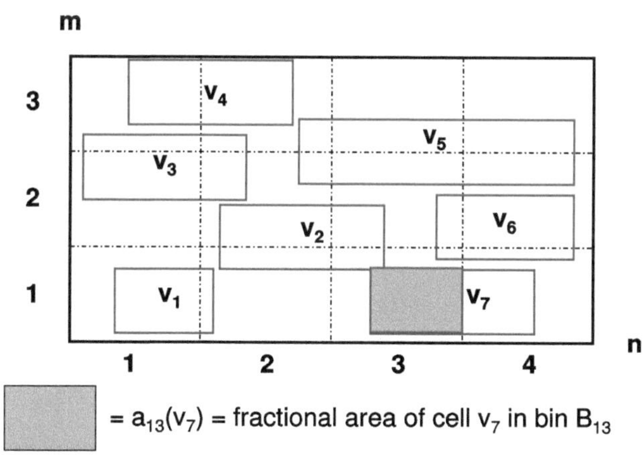

Fig. 10.5. Illustration of fractional cell area in a 3 × 4 bins region. © ACM, 2006. This is the author's version of the work. It is posted here by permission of ACM for your personal use. Not for redistribution. The definitive version was published in Proceedings of the 2006 Conference on Asia South Pacific Design Automation (ASP-DAC'06), pp. 188–194 http://doi.acm.org/10.1145/1118299.1118353.

We consider the constrained minimization problem:

$$\begin{aligned} \min \quad & W(x, y) \\ \text{s.t.} \quad & D_{ij} = K, \quad 1 \le i \le m, 1 \le j \le n \end{aligned} \tag{10.13}$$

where D_{ij} is the average density in B_{ij} defined through (10.12) and $K(\le 1)$ is the total cells area divided by the area of the placement region R. In the following discussions, we assume that $K = 1$, that is, total cells area equals the area of the placement region. This assumption is reasonable, because artificial movable "filler" cells unconnected to any real cells can always be added to the problem to increase total cell area.

In general we can have different density target $K_{ij}(\le K)$ for each bin B_{ij} to reflect uneven density requirement due to preplaced blocks etc. The nonuniform density requirements problem can be reduced to the uniform density case if dummy fixed density F_{ij} is prepositioned in each bin B_{ij} so that $K_{ij} + F_{ij} = K$. This fact is stated formally in the following theorem.

Theorem 10.5. *A global optimal solution of the following constrained minimization problem*

$$\begin{aligned} \min \quad & W(x, y) \\ \text{s.t.} \quad & D_{ij} = K_{ij}, \quad 1 \le i \le m, 1 \le j \le n \end{aligned} \tag{10.14}$$

can be obtained from a global optimal of (10.13).

Proof. *Let F_{ij} be the amount of fixed density added to each bin B_{ij} such that $F_{ij} + K_{ij} = K$. Then a global optimal solution of (10.13) is satisfying the constraints in (10.14). It is also the global optimal solution for (10.14) because any solution satisfying the constraints in (10.14) with smaller objective value contradicts the optimality of the solution we obtained in (10.13).* □

The current problem is to find a placement that minimizes the wirelength $W(x, y)$ such that cells are evenly distributed over the region. However, it is difficult to solve the above problem, because the density function is not differentiable. To make the problem easier to solve, we use the inverse Laplace transformation [29] to smooth the density function. The smoothing operator $\Delta_\epsilon^{-1}\mathrm{d}(x, y)$ is defined by solving the following Helmholtz equation:

$$\begin{cases} \Delta\psi(x, y) - \epsilon\psi(x, y) = \mathrm{d}(x, y), & (x, y) \in R \\ \frac{\partial \psi}{\partial \nu} = 0, \quad (x, y) \in \partial R \end{cases} \tag{10.15}$$

where $\epsilon > 0$, ∂R is the boundary of R, ν is the outer unit normal vector pointing outside the boundary, $\mathrm{d}(x, y)$ is the continuous density function obtained from the distribution of cell area, and Δ is a differential operator given by:

$$\Delta \equiv \frac{\partial^2}{\partial x^2} + \frac{\partial^2}{\partial y^2} \tag{10.16}$$

The inverse operator $\Delta_\epsilon^{-1}\mathrm{d}(x, y)$ is well defined, as (10.15) has a unique solution for any $\epsilon > 0$ [29]. Because the solution of (10.15) gains two more derivatives [29] than

$d(x, y)$, ψ is a smoothed version of the density function. In other words, any solution to (10.15) must be at least twice differentiable.

We use the finite difference method [45] to discretize the problem (10.15) using the bin grids we defined earlier. Neumann boundary conditions are used in the discretization scheme. Let $\psi_{i,j}$ denote the value of ψ at the center of Bin B_{ij}. The approximation equations of (10.15) are given by:

$$\frac{\psi_{i+1,j} - 2\psi_{i,j} + \psi_{i-1,j}}{h_y^2} + \frac{\psi_{i,j+1} - 2\psi_{i,j} + \psi_{i,j-1}}{h_x^2} - \epsilon\,\psi_{i,j} = D_{ij}$$
$$\forall\, 1 \le i \le m, 1 \le j \le n \qquad (10.17)$$

where

$$\begin{aligned} \psi_{0,j} &= \psi_{1,j} \quad \forall\, 1 \le j \le n \\ \psi_{m+1,j} &= \psi_{m,j} \quad \forall\, 1 \le j \le n \\ \psi_{i,0} &= \psi_{i,1} \quad \forall\, 1 \le i \le m \\ \psi_{i,n+1} &= \psi_{i,n} \quad \forall\, 1 \le i \le m \end{aligned} \qquad (10.18)$$

and D_{ij} is the average density in B_{ij}. Let $L_\epsilon[mn]$ be the matrix corresponding to the earlier linear system. Then $\Psi = (\psi_{11}, \psi_{12}, \ldots, \psi_{mn})^{\mathrm{T}}$ can be computed by solving the following linear system

$$L_\epsilon[mn]\Psi = \mathbf{D} \qquad (10.19)$$

where $\mathbf{D} = (D_{11}, D_{12}, \ldots, D_{mn})^{\mathrm{T}}$. Problem (10.19) can be solved in $O(mn \log mn)$ by fast discrete cosine transform [18]. The matrix $L_\epsilon[mn]$ can be diagonalized by discrete cosine matrix, that is,

$$L_\epsilon[mn] = (C[m] \otimes C[n])^{\mathrm{T}} \Lambda_\epsilon[mn](C[m] \otimes C[n]) \qquad (10.20)$$

where $C[n])$ is the $n \times n$ discrete cosine matrix and $\Lambda_\epsilon[mn]$ is a diagonal matrix with diagonal entries being the eigenvalues of $L_\epsilon[mn]$. The eigenvalues of $L_\epsilon[mn]$ can be computed analytically and are given by:

$$-\frac{4}{h_y^2}\sin^2\left(\frac{(i-1)\pi}{2m}\right) - \frac{4}{h_x^2}\sin^2\left(\frac{(j-1)\pi}{2n}\right) - \epsilon \quad 1 \le i \le m,\ 1 \le j \le n$$
$$(10.21)$$

Since $(C(m) \otimes C[n])^{-1} = (C[m] \otimes C[n])^{\mathrm{T}}$ and the multiplication $(C[m] \otimes C[n])\mathbf{D}$ or $(C[m] \otimes C[n])^{\mathrm{T}}\mathbf{D}$ can be computed in $O(mn \log mn)$ time [30, 53], the solution of (10.19) is given by:

$$\Psi = (C[m] \otimes C[n])^{\mathrm{T}} \Lambda_\epsilon^{-1}[mn](C[m] \otimes C[n])\mathbf{D} \qquad (10.22)$$

which can be computed in $O(mn \log mn)$ time.

Now we can reformulate the problem (10.13) as:

$$\begin{aligned} \min \quad & W(x, y) \\ \text{s.t.} \quad & \psi_{ij}(x, y) = \bar{K}_\epsilon \quad 1 \le i \le m, 1 \le j \le n \end{aligned} \qquad (10.23)$$

where $\Psi = L_\epsilon[mn]^{-1}\mathbf{D}$ and $\bar{K}_\epsilon \mathbf{1} = L_\epsilon[mn]^{-1}K\mathbf{1} = -K/\epsilon\mathbf{1}$ is a constant vector where $\mathbf{1} = (1, \ldots, 1)^{\mathrm{T}}$. We have smooth objective function and constraints, making (10.23) solvable by mathematical programming, as discussed next.

10.5.2 Problem Solver

In this section, a generalization of the force-directed method [28], called GFD algorithm, is presented to solve (10.23).

GFD Algorithm

Many nonlinear programming techniques might be used to solve (10.23). We use the Uzawa algorithm [4], because (i) it does not require a Hessian inversion to find a minimizer satisfying the KKT condition [9], and (ii) it can be viewed as a generalization of the force-directed method [28].

Applying the Uzawa algorithm to (10.23) gives the following iterative scheme:

$$\begin{cases} \nabla W(x^{k+1}, y^{k+1}) + \sum_{i,j} \lambda_{ij}^k \nabla \psi_{ij}(x^k, y^k) = 0 \\ \lambda_{ij}^{k+1} = \lambda_{ij}^k + \alpha(\psi_{ij}(x^{k+1}, y^{k+1}) - \bar{K}_\epsilon) \end{cases} \tag{10.24}$$

where $\boldsymbol{\lambda}^k$ is the vector of Lagrange multipliers estimates at the kth iteration, α is a parameter to control the rate of convergence, and x^k and y^k are the cells center locations at the kth iteration.

The gradient of ψ_{ij} with respect to the center location of cell v_k is approximated by the forward difference scheme [45]

$$\nabla_{x_k} \psi_{ij}(x, y) = \frac{\psi_{i,j+1} - \psi_{i,j}}{h_x} \quad \text{and} \quad \nabla_{y_k} \psi_{ij}(x, y) = \frac{\psi_{i+1,j} - \psi_{i,j}}{h_y} \tag{10.25}$$

if the center location of cell v_k is inside B_{ij} and zero otherwise. Using a backward difference scheme or a central difference scheme [45] instead of forward difference scheme (10.25) gives similar results.

In each step of the iterative scheme (10.24), we solve a nonlinear equation by applying a time marching scheme [2, 51] to a corresponding ordinary differential equation (ODE):

$$\begin{cases} \begin{pmatrix} \frac{\partial x(t)}{\partial_t} \\ \frac{\partial y(t)}{\partial_t} \end{pmatrix} = -(\nabla W(x(t), y(t)) + \sum_{i,j} \lambda_{ij} \nabla \psi_{ij}(x(t), y(t))) \\ (x(0), y(0)) \text{ is a given initial placement} \end{cases} \tag{10.26}$$

where $(x(t), y(t))$ denotes the placement at time t. The solution of the nonlinear equation is just the steady-state solution of the above ODE. This method can be considered a gradient descent scheme for the Lagrangian function

$$L(x(t), y(t), \boldsymbol{\lambda}) \equiv W(x(t), y(t)) + \sum_{i,j} \lambda_{ij}(\psi_{ij} - \bar{K}_\epsilon) \tag{10.27}$$

because

$$\frac{\mathrm{d}L(x(t), y(t), \boldsymbol{\lambda})}{\mathrm{d}t} = -\left\| \frac{\partial L(x(t), y(t), \boldsymbol{\lambda})}{\partial t} \right\|_2^2 < 0 \tag{10.28}$$

One can also think of this approach as minimizing the sum of the wirelength objective and violated-constraint penalty at each iteration. We solve the above ODE by the explicit Euler method [45], which gives the following iterative scheme:

$$\begin{cases} \begin{pmatrix} x^{k+1} \\ y^{k+1} \end{pmatrix} = \begin{pmatrix} x^{k} \\ y^{k} \end{pmatrix} - \tau(\nabla W(x^k, y^k) + \sum_{i,j} \lambda_{ij} \nabla \psi_{ij}(x^k, y^k)) \\ (x^0, y^0) \quad \text{is a given initial placement,} \end{cases} \tag{10.29}$$

where (x^k, y^k) are the locations of cells at the kth step, and τ is the time step. The time step τ has to be small enough to guarantee convergence. An analytical upper bound for τ depends on the Hessian of the Lagrangian function (10.27) and is hard to determine. In practice, the initial value of τ is reduced by a constant ratio and the previous solution is restored, if the iterative scheme (10.29) does not converge.

The GFD (Generalized Force-directed) algorithm used to solve (10.23) is given in Table 10.1. The algorithm takes in the number of outer iterations and the stopping

```
GFD(outer_iters, stop_percent)
if initial placement not given
   use the unconstrained minimizer of the quadratic
   wirelength objective as an initial solution.
endif
compute nnb = number of nonzero density bins.
set P = the set of pads and fixed cells.
set M = the set of movable cells.
set inner_iters = |M|.
set μ = 1.5. (Experiments show that it is a good
trade-off between runtime and wirelength)
for i = 1 to outer_iters
   set α = sqrt(max{|P|,1}) / (h_x h_y log|M|).
   κ = min{100i / outer_iters, stop_percent}.
   λ = 0.
   for j = 1 to inner_iters
      if nnb not increased
         α = μα.
      endif
      λ = λ − α(Ψ − K_ε).
      solve the ODE (10.26) by explicit Euler method (10.29).
      compute nnb.
      if more than κ% nonzero density bins
         break.
      endif
   endfor
call detailed placement.
endfor
```

Table 10.1. GFD algorithm. © ACM, 2006. This is the author's version of the work. It is posted here by permission of ACM for your personal use. Not for redistribution. The definitive version was published in Proceedings of the 2005 International Symposium on Physical Design, pp. 185–192, April 2005 (ISPD'05) http://doi.acm.org/10.1145/1055137.1055177.

criterion for inner iteration. Parameter α is used to speed up convergence. Parameter μ is the factor by which α is increased, as needed, in order to spread cells into empty bins. Parameter κ is the percentage of the number of nonzero density bins for the stopping criterion. M is the set of movable cells, and P is the set of pads and fixed cells. Since one can only get a local minimizer by solving (10.24), the initial solution is important. The outer iterations can be considered a continuation method, where the solution at each outer iteration is used as an initial solution for the next outer iteration.

We use uniform bin grids, and the number of bins is roughly equal to the number of cells. Since the global placement produced by the GFD algorithm contains overlapping cells, a discrete algorithm is used to legalize the solution which is discussed in more details in Sect. 10.6.

Comparisons with APlace and Kraftwerk

In this section, we compare the GFD algorithm with well-known analytical placers APlace [40] and Kraftwerk [28].

APlace [40] is an analytical placer based on [48]. The problem formulation considered in [40] is the same as (10.13) but it uses a bell-shaped function [48] to smooth the density constraint locally. In our case, however, the inverse Laplace transformation (10.15) smooths the density function globally and the smoothed function can be computed very efficiently. APlace uses a penalty method, in which the constrained problem (10.13) is recast as a sequence of unconstrained problems formed by adding weighted squared constraint violations to the wirelength objective. Each value of the weight defines a distinct unconstrained subproblem in the sequence. Each such unconstrained subproblem is approximately solved by nonlinear conjugate gradients. As the weight on the constraint violations is increased, the total constraint violation in computed placements is reduced, and the sequence of unconstrained solutions gradually converges to a feasible placement [31].

Kraftwerk is an analytical placement algorithm utilizing the force-directed method proposed in [28]. In [28], the divergence of the forces $\mathbf{f}(x, y) = (f_1, f_2)^{\mathrm{T}}$ is assumed to be proportional to the density; that is,

$$\frac{\partial f_1}{\partial x} + \frac{\partial f_2}{\partial y} = c \cdot \mathrm{d}(x, y) \tag{10.30}$$

where c is a constant. Assuming that the forces do not form closed loops, there exists a scalar function $\phi(x, y)$ satisfying

$$\nabla\phi(x, y) = \mathbf{f}(x, y) \tag{10.31}$$

Combining (10.30) and (10.31) gives the following equation

$$\Delta\phi(x, y) = c \cdot \mathrm{d}(x, y) \tag{10.32}$$

with boundary conditions that the magnitude of the forces $\nabla\phi(x, y)$ is zero at infinity.

Comparing (10.15) with (10.32), the main difference is the boundary condition, if we choose small ϵ. The boundary condition in our GFD formulation (10.15) says that the forces pointing outside the boundary are zero. This GFD formulation is more useful computationally than assuming the forces are zero at infinity, because we want to place the cells inside a finite region.

Moreover, the force-directed method in Kraftwerk [28] can be considered a special case of GFD (10.24). Kraftwerk uses a quadratic-wirelength objective (10.9) for $W(x, y)$ and iteratively solves

$$\begin{pmatrix} C & 0 \\ 0 & C \end{pmatrix} \begin{pmatrix} x^{k+1} \\ y^{k+1} \end{pmatrix} + \begin{pmatrix} \mathbf{p}_x \\ \mathbf{p}_y \end{pmatrix} + \tau_k \begin{pmatrix} \mathbf{f}_x^k \\ \mathbf{f}_y^k \end{pmatrix} = 0 \tag{10.33}$$

until all cells are well distributed over the chip region. C, $\mathbf{p}_x$ and $\mathbf{p}_y$ are derived from $\nabla W(x, y)$. The τ_k is a scalar to control the movement of cells in each iteration. The horizontal force $\mathbf{f}_x^k$ and the vertical force $\mathbf{f}_y^k$ acting on the cells are given by $\sum(\nabla_{x_1}\phi_{ij}, \ldots, \nabla_{x_N}\phi_{ij})^{\mathrm{T}}$ and $\sum(\nabla_{y_1}\phi_{ij}, \ldots, \nabla_{y_N}\phi_{ij})^{\mathrm{T}}$, respectively, computed based on the placement solution at the kth iteration. Clearly, this is a particular case of (10.24) by setting $\lambda_{ij}^k = \tau^k$. One can expect the above fixed point iteration requires a small enough τ_k for convergence. But we know that $\boldsymbol{\lambda}^k$ is the Lagrange multiplier for (10.23) which has to be large enough to get a well-distributed placement. Also, the $\boldsymbol{\lambda}^k$ is a vector in (10.24) and has each of its components acting as a scaling factor for the forces induced in the corresponding bins. These arguments show that (i) the GFD algorithm is more general and more robust than Kraftwerk, and (ii) GFD overcomes the well-known [59] shortcoming of ad hoc force scaling used in [28].

10.5.3 Analysis and Enhancements of the GFD Algorithm

In this section, analysis and enhancements of the GFD algorithm are presented. Due to the high complexity of the placement problem — up to millions of cells must be placed subject to pairwise nonoverlap constraints — heuristics are necessary in building a highly scalable placement algorithm. Several important heuristic enhancements to the GFD algorithm are discussed below which make it much more robust and stable.

Multilevel Implementation

It is well known that the constrained minimization problem (10.23) has myriad local minima. For a good local optimal solution to be found by the nonlinear programming GFD algorithm, a good initial solution is necessary. The multilevel optimization framework presented in Sect. 10.4 provides not only a scalable framework but also a good initial placement for the GFD algorithm at each level. The recursive coarsening ensures that the coarsest-level problem is small enough for the GFD algorithm to find a good solution, which is then recursively interpolated and iteratively improved to finer levels.

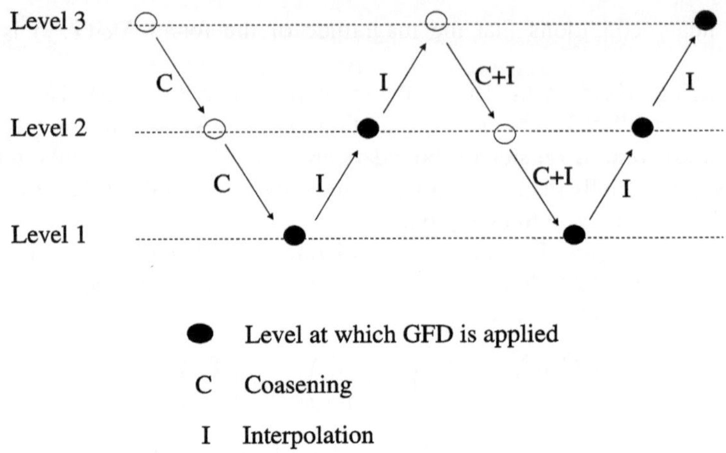

Fig. 10.6. Two V-cycle multilevel framework makes use of GFD algorithm. © ACM, 2006. This is the author's version of the work. It is posted here by permission of ACM for your personal use. Not for redistribution. The definitive version was published in Proceedings of the 2006 Conference on Asia South Pacific Design Automation (ASP-DAC'06), pp. 188–194 http://doi.acm.org/10.1145/1118299.1118353.

- Set the $stop\_percent = 95$
- Increase α in the GFD algorithm whether nnb is increased or not
- Reduce the number of bins to half of the default
- Reduce the amount of cell swapping in the detailed placement

Table 10.2. mPL5-fast overview. © ACM, 2006. This is the author's version of the work. It is posted here by permission of ACM for your personal use. Not for redistribution. The definitive version was published in Proceedings of the 2005 International Symposium on Physical Design, pp. 185–192, April 2005 (ISPD'05) http://doi.acm.org/10.1145/1055137.1055177.

In our multilevel framework, the cluster ratio ranges from 0.1 to 0.4, where cluster ratio is the number of clusters created at the current level divided by the number cells at the finer level. The coarsest-level problem size is less than around 1% of the original problem size or 500, whichever is smaller.

Figure 10.6 shows an example of multilevel optimization framework making use of the GFD algorithm. It is a two V-cycle placement in which each V-cycle consists of three levels. Level 3 in the figure represents the original problem and Level 1 represents the coarsest level problem. In practice, the relaxation at the finest level in the first V-cycle is skipped to reduce the computations, which is found to be a good trade-off between quality and speed. A more detailed algorithm is shown in Table 10.3. It is named mPL5.

A fast mode of mPL5, named mPL5-fast, has also been developed and it has the runtime reduction features described in Table 10.2.

```
use modified FC (cf. (10.4)) to coarsen the hypergraph,
with cluster ratio 0.4, until the number of cells reaches target.
set nl = number of levels.
set stop_percent = 97
% suppose level nl is the finest level corresponding
% to the original hypergraph.
for i = 1 to nl − 1
   set distri_percent = min(50 + 50 * i/nl, 90).
   call GFD(1, distri_percent) for level i.
   interpolate placement from level i to level i + 1.
endfor
% start the second V-cycle.
use modified geometric-based FC (cf. (10.5))
to coarsen the hypergraph until the number of
cells reaches target.
placement from first V-cycle is interpolated to coarse
levels during the coarsening.
set nl = number of levels.
for i = 1 to nl − 1
   set distri_percent = min(50 + 50 * i/nl, 90).
   call GFD(1, distri_percent) for level i.
   interpolate placement from level i to level i + 1.
endfor
call GFD(1, stop_percent) for level nl.
call detailed placement.
```

Table 10.3. mPL5 algorithm.

Effects of Density Smoothing

Density-based constraints are used to approximate the pairwise cells nonoverlap constraints, significantly reducing the complexity of these from $O(n^2)$ to $O(n)$. This formulation is critical to the algorithm's scalability.

However, the bin-density function is not differentiable, which makes the problem (10.13) difficult to be solved by standard mathematical programming. Smooth approximation (10.15) of the density function is necessary. The ϵ in (10.15) plays an important role in the smoothing process. It not only makes the smoothing process well defined but also controls the smoothness of the smoothed density function. Figure 10.7 shows the density function of a one-dimensional placement and the smoothed density functions under $\epsilon = 10, 1, 0.1$. We see that the smaller ϵ, the more global the smoothing of the density function. In experiments, we observe that the larger the ϵ the slower the convergence of the GFD algorithm. The reason is that the larger the ϵ the less smoothness of the density function causes forces to act on cells more locally, leading to slower spreading. Hence, the choice of smaller ϵ gives faster convergence, however, at the expense of worse wirelength quality. Clearly, ϵ provides a knob for trading between speed and quality of the placement algorithm. In experiments, $\epsilon = 1$ is found to be a good trade-off.

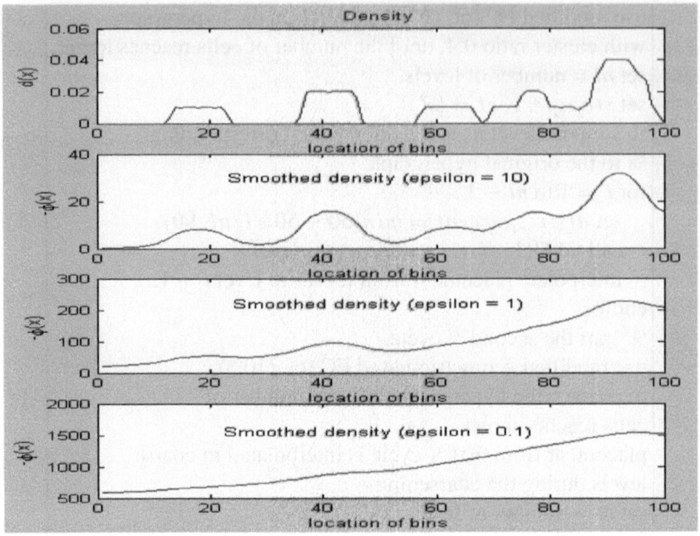

Fig. 10.7. Smoothness of the density function under different epsilon (ϵ) = 10, 1, 0.1. The definitive version was published in Proceedings of the 2006 Conference on Asia South Pacific Design Automation (ASP-DAC'06), pp. 188–194 http://doi.acm.org/10.1145/1118299.1118353.

Weighting of Forces

The gradients of the constraints with respect to cell center locations are approximated by (10.25), which not only speeds up the computations but also leads to faster convergence of the algorithm. The approximations (10.25) are viewed as cells spreading forces in [28]. Recently, a more stable implementation [59] of [28] shows that weighting of the forces acting on cell by the corresponding cell area gives a more stable spreading of the cells. The intuition is that each unit area of the cell should be getting the forces. Similar ideas were developed independently and used in our approximations (10.25). A more general form to approximate the gradient of the constraint with respect to cell v_k is given by:

$$\nabla_{x_k}\psi_{ij} = \frac{\psi_{i,j+1} - \psi_{i,j}}{h_x} * w(\text{area}(v_k)), \quad \nabla_{y_k}\psi_{ij} = \frac{\psi_{i+1,j} - \psi_{i,j}}{h_y} * w(\text{area}(v_k)) \tag{10.34}$$

if the center of cell v_k is inside B_{ij} and zero otherwise, where $w(x)$ is a monotone increasing function and area(v_k) denotes the area of cell v_k. In our experiments, $w(x) = cx^{0.8}$ is used. The weighted approximation (10.34) significantly improves the quality of placement, especially for the circuits with huge variations in cell area. Also, the value of the scalar $c > 0$ in $w(x)$ provides a trade-off between speed and quality: the larger c, the faster the convergence or spreading of cells.

Gradual Legalization and Fixing of Large Cells

The weighting of forces (cf. (10.34)) does improve the placement quality; however, it may cause instability for the spreading of large cells when most of the cells are well spread over the placement region. Equation (10.34) shows that larger cells[3] have larger forces acting on them, which may cause large perturbations of the large cells. It is therefore a good idea to fix the large cells once they are well placed. Fixing the large cells means that those large cells are first legalized [26] (no overlapping between those large cells) then their positions are fixed during the subsequent placement of smaller cells.

In the multilevel framework, the placement solutions at coarse levels provide good intermediate steps for fixing the relatively large cells. Similar ideas of fixing the large cells during coarse level placement has been used in [22]. Fixing large cells at coarser levels not only makes the GFD algorithm more stable but also accelerates the convergence, as it creates more connections to fixed cells.

Moreover, earlier legalization of the large cells also helps detailed placement [20]. In Figure 10.8, the left placement plot shows the global placement obtained without fixing the large cells during placement. The small amount of overlapping between large cells that global placement allows could create great difficulty in detailed placement. The right-hand side placement plot in Figure 10.8 shows that huge perturbations of the largest cells are produced after legalization which then causes huge perturbations during legalization of standard cells. Those huge perturbations created after legalization consistently lead to significantly increased wirelength of the placement.

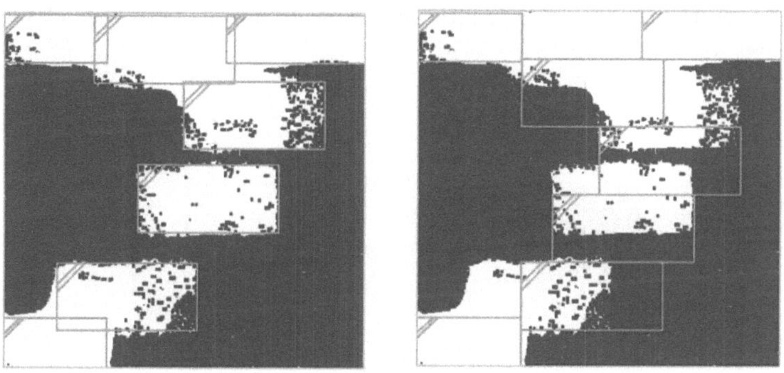

Fig. 10.8. Global placement is shown in the left figure and the placement after legalization of macros is shown in the right figure.

---

[3] In this section, the word "cell" is used generically for macros and standard cells.

In our experiments, macros with area 100× larger than the average cell area are legalized relative to one another and set fixed at all levels (excluding the coarsest level) just before the GFD algorithm is applied.

Wirelength Weighting

Proper weighting for spreading forces is crucial; otherwise, they may create significant degradation of placement quality. From the GFD algorithm (cf. Table 10.1) and the iterative scheme (10.29), the scaling factor α for the forces

$$\sum_{i,j} \lambda_{ij} \nabla \psi_{ij}$$

is increased while cells are not spreading. The τ in the iterative scheme (10.29) is decreased when the scheme does not converge. In the case when τ is too small and force scalar α is too large, the term $\nabla W(x^k, y^k)$ in (10.29), which corresponds to the wirelength objective, is diminishing. This causes the degradation of the placement quality. Hence weighting for the term $\nabla W(x^k, y^k)$ in (10.29) is introduced, giving the following iterative scheme:

$$\begin{cases} \begin{pmatrix} x^{k+1} \\ y^{k+1} \end{pmatrix} = \begin{pmatrix} x^{k} \\ y^{k} \end{pmatrix} - \tau(\beta \nabla W(x^k, y^k) + \sum_{i,j} \lambda_{ij} \nabla \psi_{ij}(x^k, y^k)) \\ (x^0, y^0) \qquad \text{is a given initial placement,} \end{cases} \tag{10.35}$$

where $\beta > 0$ is increased at the outer iterations of the GFD algorithm (cf. Table 10.4). Experiments show that the wirelength weighting significantly improves the placement wirelength in many testcases, at the expense of increased run time. The increased weighting of the wirelength increases attractions between cells and hence slows down cell spreading.

Fig. 10.9. Global placement with macros legalized and fixed after coarse levels placement.

Pin-to-Pin Half-Perimeter Wirelength Minimization

Half-perimeter wirelength in (10.2) is measured in terms of center locations of the cells. That is, each cell's pins are assumed located at the center of the cell. In real designs, pins of a cell are usually located on the boundary of the cell. Therefore, the center-to-center half-perimeter wirelength (10.2) may not a good approximation of the pin-to-pin wirelength (10.1). Because each net connects a unique set of pins, it is more costly to evaluate and hence to minimize (10.1). However, if (10.1) is approximated by log–sum–exp function (10.10), the complexity of minimizing center-to-center half-perimeter wirelength is the same as that of minimizing pin-to-pin half-perimeter wirelength, thanks to the equality $\exp(x_i + p^e_{x_i}) = \exp(p^e_{x_i})\exp(x_i)$, where $\exp(p^e_{x_i})$ is a constant that can be computed before the optimization steps. The same technique applies to the y-direction.

Whitespace Handling by Filler Cells

In the formulation (10.13), we have assumed the placement region has zero whitespace, that is, $K = 1$. In practie, there is usually $10 - 40\%$ whitespace which is reserved for post placement purpose. Also, cells cannot be packed too closely. Sufficient space around cells for routing the wires is necessary. However, setting K is significantly less than one would cause trouble to the GFD algorithm due to the following reasons.

First, Figure 10.10 shows a placement problem with total cells area much less than the placement region. One can easily see that the density equality cannot be satisfied, because the cells cannot be broken into pieces. Second, due to the fact that the forces are driven by the density constraint, cells are spread to occupy the whole placement region, causing excessive separation between cells and hence degradation of wirelength. Figure 10.11 shows the global placement on a testcase with 60% whitespace. The cells are evidently placed evenly throughout the region and very likely too far apart from each other to produce a result with competitive wirelength.

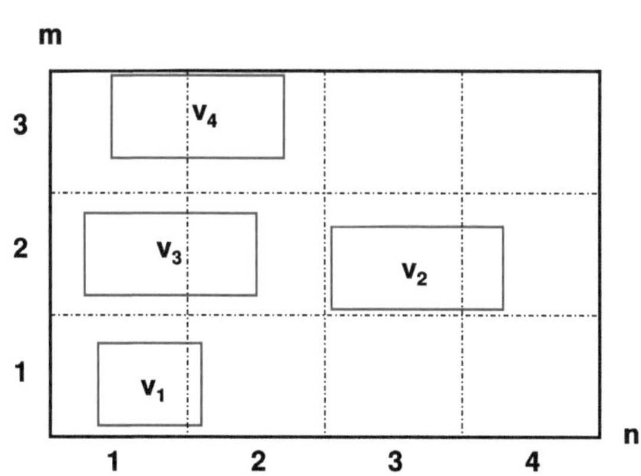

Fig. 10.10. A small example of circuit with low utilization.

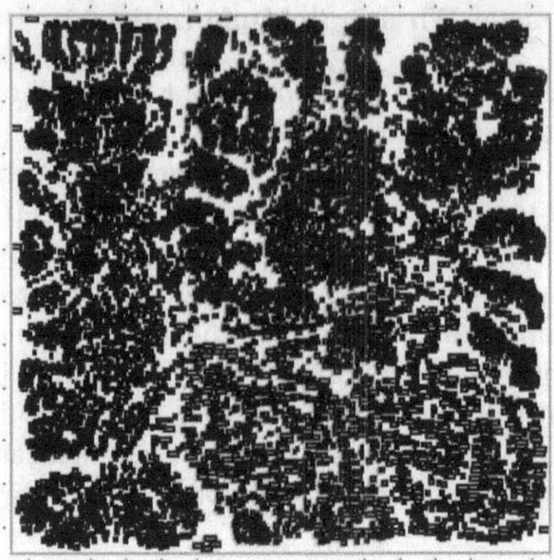

Fig. 10.11. Global placement on a testcase (#cells = 12506) with utilization = 0.4. The half-perimeter wirelength is 2.2024×10^6.

Fig. 10.12. Global placement on a testcase (#cells = 12506) with utilization = 0.4, with placement region shrunken from left and right boundary such that the utilization = 1 during the placement. The half-perimeter wirelength is 1.9250×10^6.

One simple fix is to shrink the placement region, from either directions of the chip boundary, so that the utilization ratio is close to one. Figure 10.12 shows the global placement of the same testcase as used in Figure 10.11. During placement, the placement region is shrunk by the same amount from both left and right boundary such that the utilization is one in the center subregion. This artificial region shrinkage significantly improves the wirelength, near 13%. However, it is easy to see that shrinking the region or trim away the excess whitespace has certainly limited the solution space. Also, there are several other possible directions to shrink the region. Picking the best from the placement solutions produced by different ways of region shrinking is not runtime efficient.

A reasonable way to handle whitespace is to reformulate (10.13) as an inequality-constrained minimization problem:

$$\begin{aligned} \min \quad & W(x, y) \\ \text{s.t.} \quad & D_{ij} <= 1, \quad 1 \le i \le m, 1 \le j \le n \end{aligned} \tag{10.36}$$

However, the Uzawa algorithm [4] is developed for solving an equality-constrained optimization problem. The GFD algorithm (cf. Table 10.1), based on the Uzawa algorithm, cannot be applied to solve (10.36). Whitespace handling by adding extra dummy cells or filler cells to the netlist is proposed in [7] for min-cut based placement algorithms. The idea gives a way to transform (10.36) into an equality constrained minimization problem. A set of artificial cells, named "dummy" or "filler" cells, with total area equal to the total amount of whitespace, is added to the netlist. Hence, the utilization is increased to one, and the problem becomes an equality-constrained minimization problem which can be solved by the GFD algorithm.

As there are no nets connecting dummy cells, their movements are driven by density only (cf. (10.29)). Initial placement of dummy cells is based on the density distribution of the initial placement of the cells from the original netlist. Recursive, top-down, four-way spatial partitioning of the dummy cells is used to distribute them throughout the placement region. The number of dummy cells assigned to each partition is proportional to the whitespace available in the region. The idea is to place dummy cells to low density regions initially. In addition to using forces to drive the dummy cells, frequent redistribution of dummy cells is carried out during intermediate stage of placement. This is to avoid dummy cells occupying the regions where the cells can move in to get a better placement wirelength. Figure 10.13 shows the global

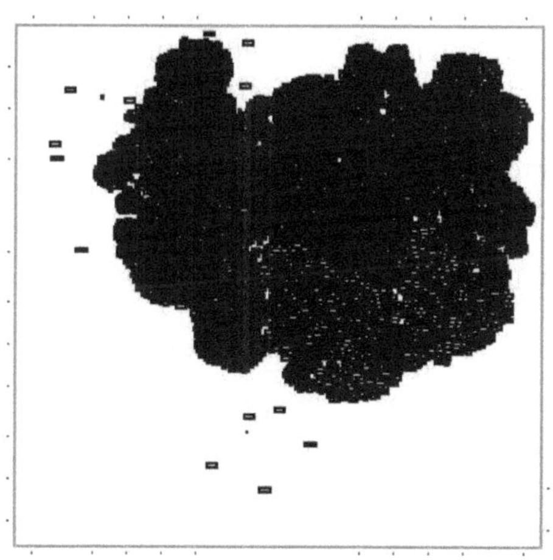

Fig. 10.13. Global placement on a testcase (#cells = 12506) with utilization = 0.4, with dummy cells added such that the utilization = 1 during the placement. The half-perimeter wirelength is 1.7203×10^6.

placement by using dummy cells to handle whitespace. It improves the wirelength by 12% over the method of region shrinking.

Stopping Criterion of GFD

The stopping criterion in the GFD algorithm (cf. Table 10.1), measured in terms of the percentage of nonempty bins, is not robust enough to guarantee a decent global placement, especially in the case where there are many large cells. Since each large cell automatically covers a significant number of bins, total overlap between small cells can be large, even when every bin has nonzero density. Also, GFD is not able to achieve the stopping criterion if the circuit has sufficiently low utilization (see Figure 10.10). A better stopping criterion is proposed in this section.

The new stopping criterion for a global placement is measured in terms of the average bin overflow. The amount of overflow for a bin is the amount of bin density exceeding the bin capacity. The average overflow for a set of cells V_c is defined as:

$$\mathrm{OVL}(V_c) \equiv \sum_{i=1}^{m} \sum_{j=1}^{n} \max(D(V_c)_{ij} - 1, 0) h_x h_y / \mathrm{area}(V_c) \tag{10.37}$$

where $D(V_c)_{ij}$ is the density of ijth bin and the density function is measured in terms of the area of the cells in V_c, area(V_c) the total area of the cells in V_c. Under sufficiently small size of bins and small overflow, the global placement is guaranteed little overlapping between cells and can be legalized without much increase in wirelength.

Minimum Perturbation Formulation

It is expected that the GFD algorithm is in slow convergence when the cells are well distributed. Slow reduction of overflow (cf. (10.37)) happens during the last few steps of the GFD algorithm, because the forces acting on cells diminish as the density in each bin becomes more even. To accelerate convergence and reduce computation, (10.23) is reformulated as

$$\begin{aligned} \min \quad & \sum_{v_i \in M} \{(x_i - x_i^o)^2 + (y_i - y_i^o)^2\} \\ \text{s.t.} \quad & \psi_{ij} = \bar{K}_\epsilon, \quad 1 \le i \le m, 1 \le j \le n \end{aligned} \tag{10.38}$$

where M is the set of movable cells and (x_i^o, y_i^o) is the initial center location of cell v_i. The idea is to minimize the displacement from a given placement subject to the density constraint. The objective in (10.38) is easier and faster to minimize than that in (10.10). It is more runtime efficient to switch to solve (10.38) when progress slows in the GFD algorithm. The GFD algorithm can be terminated earlier, and the placement solution is a good starting point for (10.38). It also gives a smoother transition from global placement to legalization and detailed placement. Experiments show that it is a good trade-off between quality and speed.

Enhanced GFD Algorithm

In this section, an enhanced GFD algorithm combining the techniques described in the previous sections is presented. Table 10.4 shows the new enhancement GFD algorithm, named EGFD.

In Table 10.4, P is the set of pads, F is the set of fixed cells, M is the set of movable cells not including dummy cells, and H is the set of dummy cells added to the netlist (cf. Sect. 10.5.3). EGFD takes in a percentage of overflow (cf. (10.37)) as stopping criterion. The algorithm terminates when both overflow of movable cells and overflow of total cells (excluding pads and dummy cells) is less than a target overflow.

Similar to the GFD algorithm (Table 10.1), EGFD uses the unconstrained minimizer of the quadratic wirelength (10.9) as the starting point, if an initial placement

```
EGFD(stop_percent_ovl)
set P = the set of pads.
set F = the set of fixed cells inside the placement region.
set M = the set of movable cells.
set H = the set of dummy cells added to the netlist.
set η = 100.
set μ = 1.5.
set α = √(max{|P|+|F|,1}) / (h_x h_y log|M|).
set β = 1.
set κ = stop_percent_ovl/100.
set λ = 0.
if initial placement not given
   use the unconstrained minimizer of the quadratic
   wirelength objective as an initial solution.
endif
Hierarchical distribution of dummy cells.
set old_std_ovl = curr_std_ovl = OVL(M).
for j = 1 to |M|
   λ = λ − α(ψ − K_ε).
   solve the iterative scheme (10.35).
   set old_std_ovl = curr_std_ovl.
   set curr_std_ovl = OVL(M).
   if curr_ovl ≤ κ and OVL(M ∪ F) ≤ κ
      break.
   endif
   if curr_std_ovl ≥ old_std_ovl
      α = μα.
   endif
   β = β + η.
   redistribution of dummy cells for every 3% reduction of OVL(M).
endfor
```

Table 10.4. Enhanced GFD algorithm.

use Best Choice (cf. (10.6)) to coarsen the hypergraph,
with cluster ratio 0.25, until the number of cells reaches target
set nl = number of levels created.
set $stop\_percent\_ovl = 15$
% suppose level nl is the finest level corresponding
% to the original hypergraph.
for $i = 1$ to $nl - 1$
  call EGFD($stop\_percent\_ovl + 5$) for level i.
  interpolate placement from level i to level $i + 1$.
  legalize and fix macros with area > 4X the average area at level i.
endfor
call EGFD($stop\_percent\_ovl + 5$) for level nl.
call EGFD($stop\_percent\_ovl$) by minimizing (10.38) for level nl.
call detailed placement.

Table 10.5. mPL6 algorithm.

is not given. In EGFD, the scaling factor α for forces is guided by the overflow of the movable cells instead of the number of nonzero density bins as in GFD. Also, there is a scaling factor β for the wirelength term in the iterative scheme (10.35). In addition, dummy cells are added to the netlist, and the dummy cells are hierarchically redistributed after every 3% reduction in the overflow of the movable cells. EGFD proceeds to (10.38) once the overflow is less than 5% away from the $stop\_percent\_ovl$ (the percentage for stopping criterion).

The multilevel algorithm using EGFD as relaxation is named mPL6 and is shown in Table 10.5. It uses Best Choice clustering [5] with cluster ratio 0.25. With Best Choice clustering, one V-cycle with cluster ratio 0.25 is usually enough to get a decent placement. mPL6 uses one V-cycle instead of the two V-cycles used in mPL5. Also, relatively large macros are fixed after their placement at coarse levels, as described in Sect. 10.5.3.

10.6 Legalization and Detailed Placement

Following global placement, the next step is to remove the overlap between macros and standard cells and further optimize the wirelength while keeping all constraints strictly satisfied.

A number of algorithms have been proposed for mixed-size placement, and they can be divided into two classes. The first class of algorithms removes the overlap between placeable objects during global placement, leaving detailed placement with only the task of further wirelength reduction. The examples in this class include Capo [1,49], mPG-ms [22], Dragon2005 [57] and PolarBear [24]. Capo combines a recursive min-cut-based placer with a fixed-outline floorplanner, Parquet [6]. Macros are first shredded into pieces and placed by the standard cell placer. The locations of macros are subsequently derived by reassembling the component pieces, and residual overlap is removed by the floorplanner. The second pass places standard cells with all macros fixed. A top-down "correct-by-construction" approach used in Capo [1,49] invoke Parquet at intermediate levels, with clustering of both small macros and standard cells used to reduce floorplanning run time. Another algorithm in this class is mPG-ms [22], which uses simulated annealing to gradually legalize macros and fix them in the intermediate levels of the multilevel optimization. Dragon2005 [57] is a two-pass simulated annealing-based placer. Standard cells and macros are placed together in the first pass. In the second pass, the macros are held fixed, and the standard cells are placed again. To further reduce wirelength, it shifts cells when swapping cells from different rows during detailed placement. PolarBear [24] combines recursive min-cut with fast wirelength-blind look-ahead legalization for every placement subproblem generated by the partitioning. The look-ahead step provides a fall-back legalized placements for subregions as they are defined and allows PolarBear to aggressively reduce wirelength even on low-whitespace test cases. Limited wirelength-aware look-ahead floorplanning on simplified subproblems has also been incorporated into Capo [49].

The second class of algorithms may leave overlap between macros and cells after global placement. Most analytical placers, including Kraftwerk [28], BonnPlace [11, 60, 61], Aplace [40], FDP [59], mPL5 [21], UPlace [63], and some min-cut-based placers, such as Fengshui [8, 36], belong to this category. In this case, the detailed placement is expected both to remove overlap and to reduce wirelength. BonnPlace uses a quadratic programming-based approach coupled with quadri-section [60]. To legalize macros, a bottom-up branch-and-bound search with linear programming (LP) is proposed. Standard cells are evened out between placement regions with a min-cost–max-flow formulation. Further wirelength reduction is achieved by solving an LP formulation on each row. Fengshui [8, 36] uses a greedy scheme that considers simultaneously perturbation of macros and wirelength minimization for legalization. Windows spanning multiple rows for cell permutation are used for wirelength reduction. Domino [27] iteratively improves wirelength by shredding cells into uniform pieces and solving a min-cost–max-flow formulation. UPlace [63] applies zone refinement for both legalization and wirelength reduction purposes. The objective it considers combines wirelength and zone height.

Most macro legalization schemes used in the second class suffer from two limitations. First, they may not produce a legal placement in the end. Second, they may cause a large perturbation to the global placement during legalization, resulting in longer wirelength. Figures 10.14 and 10.15 show an example of applying Fengshui's legalization scheme on a global placement generated by mPL6, the global placer presented in Sect. 10.5 (but without gradual macro fixing). The legalized wirelength is increased by more than 10% compared to the global placement wirelength.

In this section, we present a three-step approach, named XDP, for mixed-size detailed placement. First, a combination of constraint graph and linear programming is used to legalize macros. Then, an enhanced scan-based method is used to legalize the standard cells. Finally, a sliding-window-based cell swapping is applied to further reduce wirelength.

The remainder of this section is organized as follows: Sect. 10.6.1 describes the macro-legalization step, Sect. 10.6.2 describes the cell legalization step, and Sect. 10.6.3 presents the wirelength reduction step.

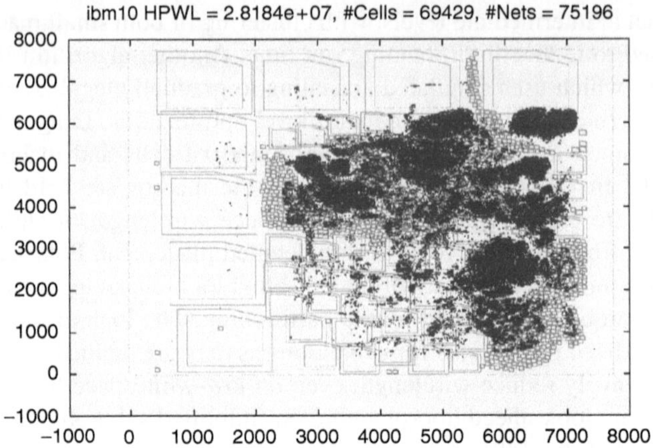

Fig. 10.14. An example global placement generated by mPL6 [19].

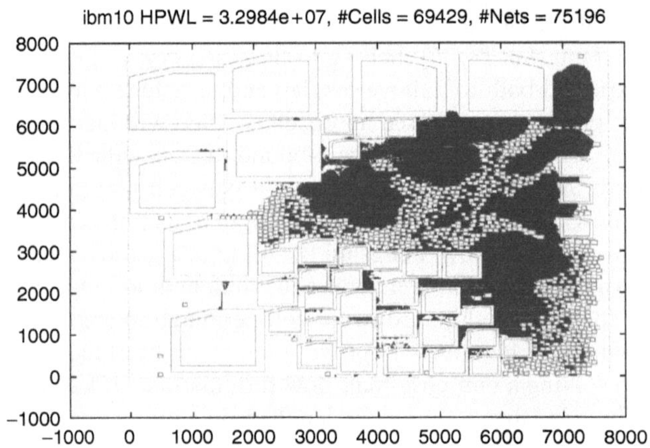

Fig. 10.15. Legalization by applying Fengshui's greedy method. The wirelength increases by more than 10%.

10.6.1 Macro Legalization

The first step of our algorithm removes the overlap between macros in the global placement, which can be formulated as the following problem:

Given a set of rectangular blocks, $M = \{m_1, m_2, \ldots, m_n\}$, pack the blocks within a rectangular region R without overlap. The objective is to minimize the perturbation, i.e., total movement of the blocks from their original locations.

Before a detailed description, we introduce some notation.

Let m_i be the ith macro. Its center coordinate in global placement is (x_i, y_i). Its width and height is w_i and h_i, respectively. The coordinate of m_i after macro legalization is denoted as (x_i', y_i').

Let the lower left corner of the placement region R be $(0, 0)$, the top right corner be (W, H).

Let G_h be a directed acyclic graph (DAG). For each macro m_i, v_{h_i} is the corresponding node in G_h. G_h has a source node v_{h_s} and a sink node v_{h_t}.

To represent the constraint that m_i should be on the left of m_j, a directed edge from v_{h_i} to v_{h_j} is inserted into G_h. The edge weight is set to be $(w_i + w_j)/2$. Our graph definition is similar to those widely used in floorplaning [43].

For each node in G_h, we calculate two values, $L(v_{h_i})$ and $R(v_{h_i})$, using (10.39).

$$\begin{aligned} L(v_{h_s}) &= 0 \\ L(v_{h_j}) &= \max(L(v_{h_i}) + \text{weight}(e_{ij}))\ \forall e_{ij} \in G_h \\ R(v_{h_t}) &= \max(L(v_{h_t}), W) \\ R(v_{h_i}) &= \min(R(v_{h_j}) - \text{weight}(e_{ij}))\ \forall e_{ij} \in G_h \end{aligned} \tag{10.39}$$

For each edge e_{ij} in G_h, we calculate slack(e_{ij}) using (10.40).

$$\text{slack}(e_{ij}) = R(v_{h_j}) - L(v_{h_i}) - \text{weight}(e_{ij})\ \forall e_{ij} \in G_h \tag{10.40}$$

It can be seen that the definition of slacks are analogous to those defined for timing analysis. For each node, we also calculate value disp(v_{h_i}) using (10.41). This is to model the potential displacement for each macro.

$$\text{disp}(v_{h_i}) = \begin{cases} L(v_{h_i}) - x_i & \text{if} \quad L(v_{h_i}) \geq x_i \\ x_i - R(v_{h_i}) & \text{if} \quad R(v_{h_i}) \leq x_i \\ 0 & \text{otherwise.} \end{cases} \tag{10.41}$$

In the end the total displacement of a constraint graph is defined using (10.42).

$$\text{disp}\,(G_h) = \sum_{v \in G_h} \text{disp}(v) \tag{10.42}$$

Similarly, we can define G_v and the corresponding values for the vertical direction.

Initial Constraint Graph Generation

Given a global placement, we examine each pair of macros m_i and m_j, and create a constraint edge between them. The edge can be either horizontal or vertical, depending on the relative locations of m_i and m_j. Figure 10.16 gives three relative locations that we consider. The type of the constraint edges is such that the macros are given the most flexibility in the constraint graphs. The edge weights are assigned accordingly.

Constraint Graph Adjustment

After the constraint graph construction, we traverse each graph and calculate the longest path.[4] If the longest path exceeds the chip dimension, some of the edges

[4] In case the graph thus constructed has cycles, we first derive a sequence-pair representation of the macros, and construct the constraint graphs according to the representation.

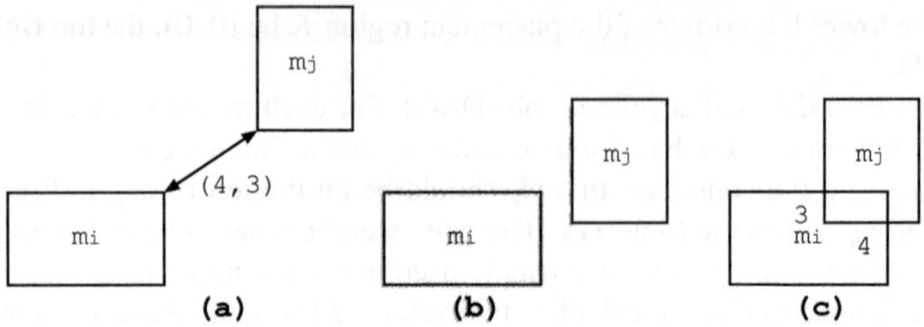

Fig. 10.16. Three types of relative macro locations which are used to determine the constraint type between each pair of macros. Constraint edge weight is assigned accordingly.

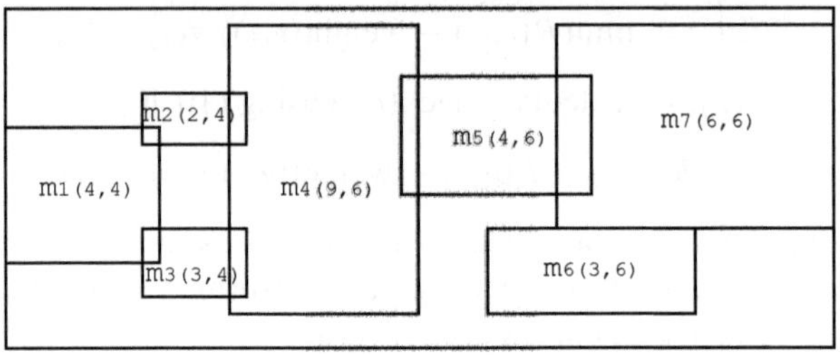

Fig. 10.17. An example of macros with overlap.

need to be adjusted to reduce the longest path. By adjustment we mean change an edge's direction from horizontal to vertical while keeping its head and tail,[5] or vice versa. In the following discussion, we assume the longest path in G_h exceeds the chip width, while the path in G_v is within the chip height.[6]

Formally, we need to solve the following subproblem:

Given G_h with $L(v_{h_t})$ greater than W, select a subset of constraint edges in G_h to move the G_v, so that the $L(v_{h_t})$ after adjustment is reduced, subject to the constraint that $L(v_{v_t})$ after adjustment should not be greater than H. The objective is to minimize disp(G_v) after the adjustment.

This problem needs to be addressed since identifying the right set of edges for adjustment may not be trivial under certain circumstances. Figure 10.17 presents a global placement of macros with dimensions. The dimension of the placement region is 25×10. Figure 10.18a presents the G_h corresponding to Figure 10.17. Edges in the critical path are highlighted with weights. Since macro 2 and 3 have the same width, we have two converging paths with the same length. Figure 10.18b gives the corresponding G_v. A straightforward method that examines one edge at a time will not pick e_{12}, e_{13}, e_{24}, or e_{34} for adjustment, since the final longest path will not

---

[5] We also investigated the alternative of swapping the head and tail of the constraint edge, depending on the global placement. Overall, we do not observe significant improvement in the final quality.

[6] In case the longest path in G_v exceeds the chip height already, we temporarily lift the chip height to be the same as the longest path in G_v.

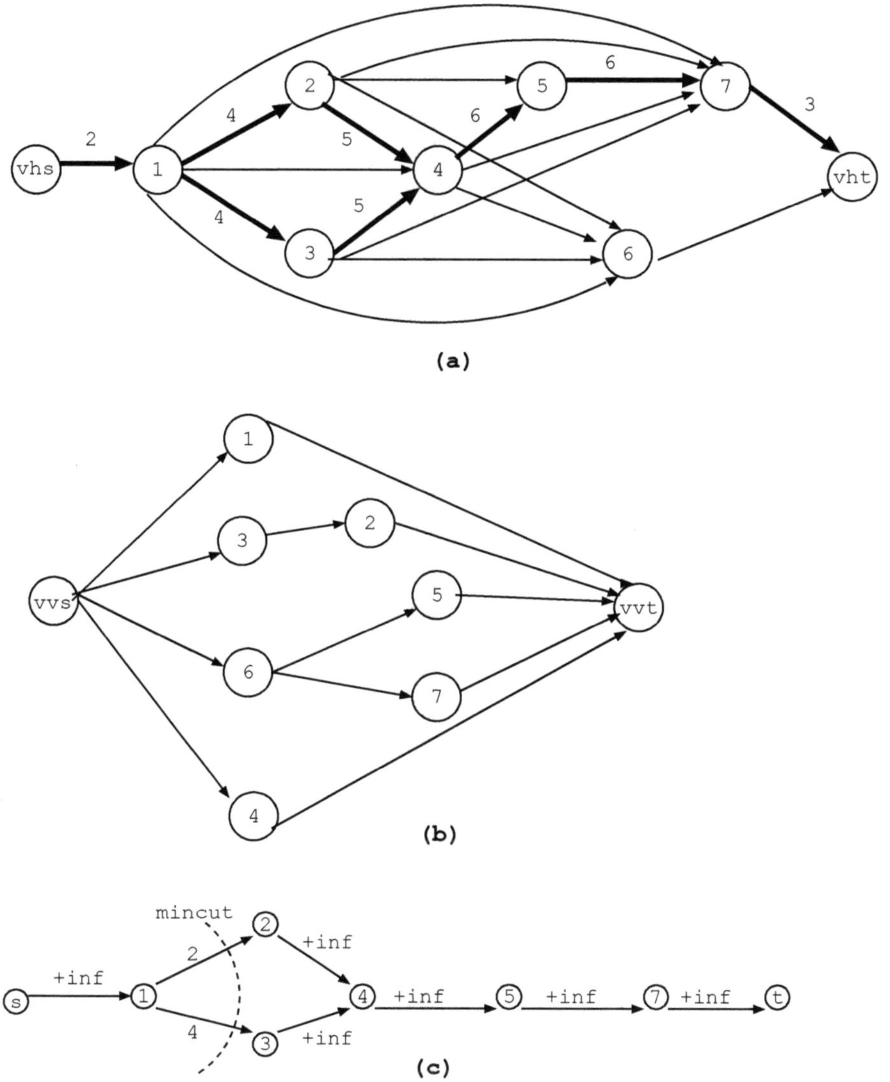

Fig. 10.18. (**a**) Constraint graph G_h. (**b**) Constraint graph G_v. (**c**) Corresponding G_c with edge capacity assigned. The min-cut identifies the set of edges which will be transformed from horizontal to vertical.

change. This leaves us with only the choice of e_{45} or e_{57}. However, adjusting either of them will make the longest path on the Y direction exceed the chip height.

To solve this problem, we extract a subgraph of G_h, consisting of edges with zero slack. This graph is similar to that used for timing optimization in logic synthesis [41, 44, 54, 62]. We name this subgraph the zero-slack network of G_h. According to this network, another DAG, G_c, will be constructed. Each edge and node in the zero-slack network have a corresponding counterpart in G_c. For an edge e_{ij} in the network, if adjusting causes the longest path in the Y direction to exceed the chip height, the corresponding edge capacity in G_c will be set to $+\infty$. Otherwise, the capacity is set using (10.43).

$$\max\left(y_i - R(v_j) + \frac{h_i+h_j}{2}, 0\right) + \max\left(L(v_i) + \frac{h_i+h_j}{2} - y_j, 0\right) \tag{10.43}$$

The first component is the potential perturbation on m_i's x coordinate because of the constraint edge under consideration. The second component is the potential perturbation on m_j's x coordinate because of the constraint edge adjustment. To reduce the complexity, we use $L(v_i)$ and $R(v_j)$ before the adjustment, rather than those values after the adjustment. All edges incident on v_{h_s} or v_{h_t} will be assigned a capacity $+\infty$. It can be seen that the definition of edge capacity is set to encourage choosing edges with potentially large slack on the orthogonal direction. A min-cut is then calculated on G_c. For each edge in the cut, the corresponding edge in G_h will be adjusted. Compared to [47], instead of permuting the sequence pair and evaluating the impact of the constraint graphs, we operate directly on the graphs, giving us more flexibility and finer granularity in the operations. Furthermore, our basic operations are more targeted to meeting the packing constraints.

Figure 10.18(c) gives the G_c for G_h with edge capacity assigned. The solution for this instance is the min-cut formed by e_{12} and e_{13}. Adjusting this increases the longest path on the Y direction to 9, but is still within the chip height. Figure 10.19 gives the final constraint graphs after the adjustment.

The adjustment process iterates until the longest paths in both graphs are shorter than the chip dimension, indicating we have found a set of nonoverlapping constraints that can be satisfied. Empirically, it terminates after a few iterations. [7]

Macro Coordinate Determination

The constraint graphs and the subsequent adjustment are essentially used to find a set of nonoverlapping constraints that can be satisfied. Our next stage is to determine the exact locations of the macros so that the total perturbation to macros is minimized. This can be formulated as the following linear programming problem:

$$\begin{array}{lll}
\min & \sum_{i=1}^{n} \left(w_{x_i} \times \mathrm{d}x_i + w_{y_i} \times dy_i\right) & \\
\text{s.t.} & -\mathrm{d}x_i \le x_i' - x_i \le \mathrm{d}x_i & \\
& -\mathrm{d}y_i \le y_i' - y_i \le \mathrm{d}y_i & \\
& x_j' - x_i' \ge \frac{w_i+w_j}{2} & \text{if } \exists e_{ij} \in G_h \\
& y_j' - y_i' \ge \frac{y_i+y_j}{2} & \text{if } \exists e_{ij} \in G_v \\
& \frac{w_i}{2} \le x_i' \le W - \frac{w_i}{2} & \\
& \frac{h_i}{2} \le y_i' \le H - \frac{h_i}{2} &
\end{array} \tag{10.44}$$

[7] It is possible that the iterations may not find a feasible solution. In reality, we have not observed any instance of failure on the example we tested, even with only $2-3\%$ of white space.

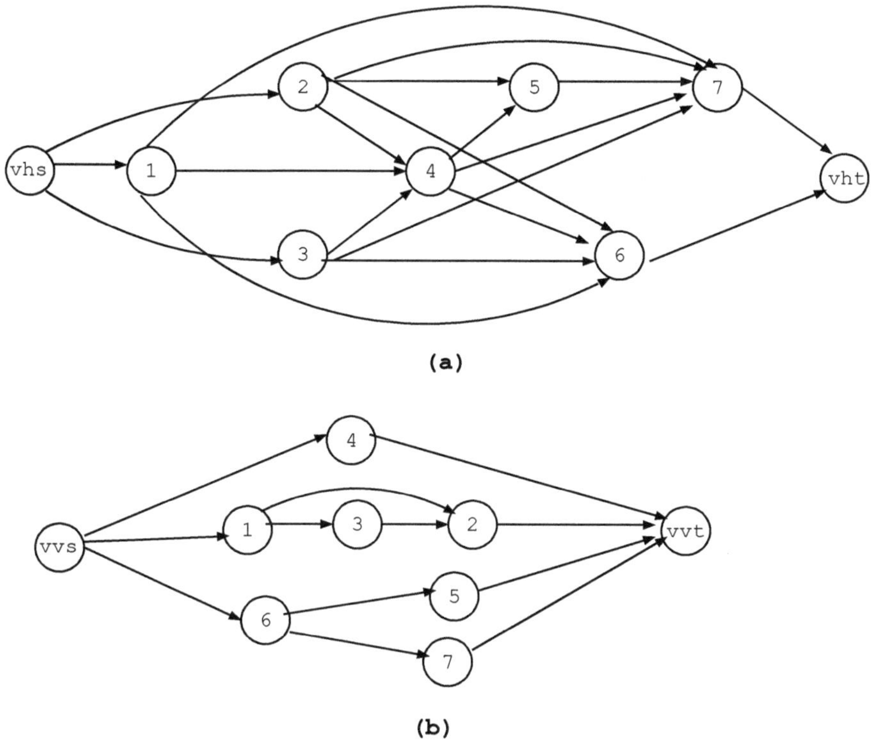

Fig. 10.19. (**a**) Constraint graph G_h after adjustment. (**b**) Constraint graph G_v after adjustment.

Here, the $\mathrm{d}x_i$ and $\mathrm{d}y_i$ are used to quantify the perturbation to m_i. Values w_{x_i} and w_{y_i} are positive weights that can be set to either one or the number of connections on each macro. The next two inequalities are derived from the edges in G_h and G_v. The last two constraints force the macros to stay with the chip region. Although the formulation is similar to that in [50, 60], we do not go through the bottom up branch and bound process, as proposed in [60] to exam both X and Y separation between each pair. Our constraint-graph-based method helps to prune the search space by following the relative order in the global placement. Furthermore, we only solve the LP after a legal packing of macros is guaranteed, while an LP may be tried for every possible combination of nonoverlapping constraints [60] in the worst case. The objective can also be enhanced to consider wirelength by the formulation of Mongrel [33], as in [56]. To solve the LP, we used a public domain interior-point LP solver, BPMPD [25].

10.6.2 Cell Legalization

Following macro legalization, the second step removes the overlap between standard cells. This step is to solve the following problem:

Given a placement where overlap only exists between cells, or cells and macros, remove the overlap between all objects and obtain a legal placement. The objective is still minimization of wirelength.

A greedy heuristic has been proposed for this purpose in [36], as an extension of [32] for mixed-size placement. A front-end contour designating the leftmost empty site on each row is maintained. Movable objects are traversed in ascending order of the x coordinate. The location of each object is determined by considering the combination of incident wirelength and displacement penalty. The front-end contour is updated after each object is placed. Although it gives a satisfactory result, this method cannot guarantee that all the macros can fit within the chip boundary when the legalization finishes. To mitigate this drawback, the global placement of Fengshui takes a conservative approach, packing the macros and cells very tightly to increase the chance of success during legalization [36]. Another alternative by APlace is to iteratively "squeeze" the cell locations and restart cell legalization until a legal soltion is obtained [38]. However, this strategy may not find a legal solution either.

We enhanced the method of [36] by introducing a back-end contour, which is initialized as the left contour of macros if they are packed to the right. Figure 10.20a illustrates the initialization of a back-end contour.

Before legalization, all the movable objects are sorted in ascending order of their left boundary. The placeable objects are examined one at a time. If the object is a cell, we scan each row and pick the site between the two contours that gives the shortest wirelength for the nets connected with it. The front-end on the target row is updated. If no site can be found for a cell, it will temporarily be put at its original location with

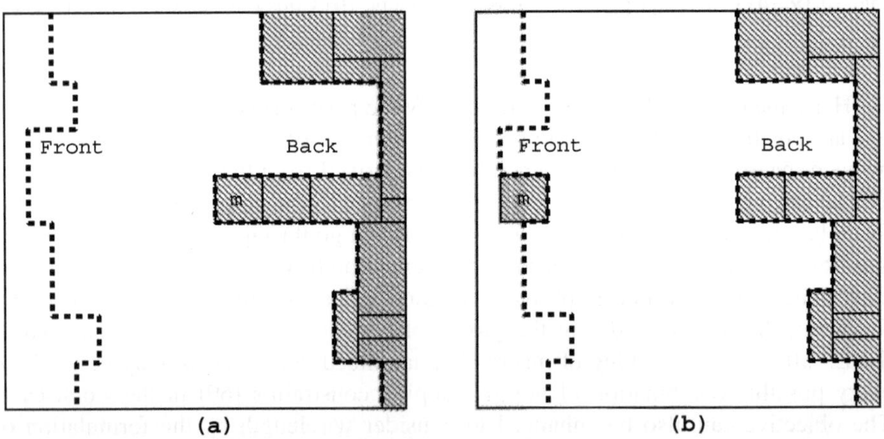

Fig. 10.20. (**a**) Front-end designates the leftmost site that can be occupied without overlap with already legalized objects. Back-end designates the rightmost site that can be occupied without overlapping with macros that have not been legalized yet. Back-end contour is initialized as the contour of macros if they are packed to the right boundary. (**b**) In addition to updating the front-end contour, the back-end contour of rows crossed by a macro will be updated after the macro is legalized.

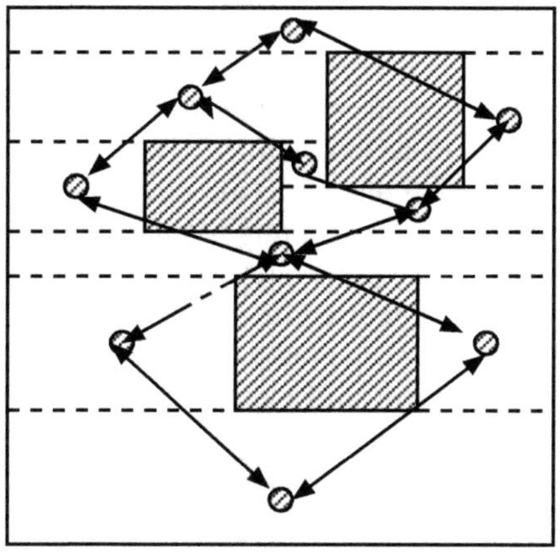

Fig. 10.21. Network flow based formulation to even out cells.

its physical dimension ignored. This will result in cell area overflow in certain regions of the chip, which will be dealt with in the additional step that follows. If the object is a macro, it will only be considered for movement between the interval determined by the two contours. This restriction guarantees legality of macros obtained from Sect. 10.6.1. An additional step after each macro legalization is to update the back-end contour of rows that the macro crosses, as shown in Figure 10.20(b).

Depending on the global placement, if cell area overflow remains in a certain part of the chip, we partition the chip into regions, and use the min-cost–max-flow formulation in [11, 60] to even out cells between different regions. Each region is represented as a node in a graph. A bidirectional edge is set up between each pair of adjacent regions, as illustrated by Figure 10.21. The node capacity is the difference between the region area and the total cell area in the region. The unit cost of an edge is the center-to-center distance between the two regions it connects. Since the edge cost is positive, the final solution has no cycles. A dynamic programming-based method is used to select the cells to move between regions. The occurrence of this situation depends partly on the global placement.

10.6.3 Further Wirelength Reduction

After a legal placement is obtained, the last step of our algorithm is to further reduce the wirelength. Here, we use a window spanning a single row or multiple rows and slide it across the chip. We enumerate all the possible configurations and pick the one giving the shortest wirelength of nets connected with the cells. After each such permutation is selected, the window is slid by half its width. This process is iterated until no further wirelength reduction is possible. This is the same process as that described in [36].

10.7 Numerical Results

In this section, we evaluate our placer mPL6 using four sets of circuits, including ISPD05, ISPD06, PEKO05, PEKO06. Tables 10.6 and 10.7 give the experimental results of our placer on ISPD05 and PEKO05 circuits, respectively. These benchmarks are used to evaluate placer's capability of minimizing the half-perimeter wirelength subject to nonoverlapping constraint. For PEKO05, the optimal placement solution is known, hence it can be used to evaluate the performance of a placer. In Table 10.7, we can see our placer can produce solutions which are around 30% away from optimal. Also, the run time of mPL6 is reasonably fast and highly scalable.

In Tables 10.8 and 10.9, we present the results of mPL6 on ISPD06 and PEKO06 circuits, respectively. These benchmarks are used to evaluate placer's capability of handling additional density constraints. In Table 10.8, we can see mPL6 is able to produce a placement solution with very low scaled bin overflow.[8] It demonstrates its ability to address the density constraint. The huge number of scaled bin overflow for newblue1 is due to the presence of many large movable macros, where they occupy more area than a bin can hold. For PEKO06, the optimal placement solution is known. In this benchmark, mPL6's HPWL is around 40% away from optimal HPWL. The degradation in the wirelength quality is reasonable, as it pays more effort on reducing the scale bin overflow. Note that the average overflow of the

circuit	GP		DP	
	WL	runtime	WL	runtime
adaptec1	8.00E+07	2109	77911923	768
adaptec2	9.36E+07	2210	91963774	771
adaptec3	2.15E+08	7804	2.14E+08	1572
adaptec4	1.97E+08	7280	1.94E+08	1528
bigblue1	1.02E+08	2696	96787268	917
bigblue2	1.56E+08	7440	1.52E+08	2846
bigblue4	8.79E+08	22963	8.29E+08	7536
bigblue3	3.59E+08	9996	3.44E+08	3493

Table 10.6. Experiment results on ISPD05 suite.

circuit	WL	WL / OPT WL	runtime
adaptec1	2.54E+07	1.27	4499
adaptec2	3.30E+07	1.32	4181
adaptec3	5.87E+07	1.43	9945
adaptec4	5.07E+07	1.29	7743
bigblue1	2.53E+07	1.21	4996
bigblue2	5.67E+07	1.34	8798
bigblue3	1.47E+08	1.55	17238
bigblue4	2.25E+08	1.31	38583

Table 10.7. Experiment results on PEKO05 suite.

[8] Detailed description of how the scaled overflow is computed can be found in [46].

circuit	WL	overflow	runtime
adaptec5	4.25E+08	1.42	8265.22
newblue1	6.69E+07	0.18	2251.67
newblue2	1.98E+08	1.72	6088.53
newblue3	2.84E+08	1.14	9695.73
newblue4	2.94E+08	1.77	5814.56
newblue5	5.31E+08	1.88	12348.67
newblue6	5.10E+08	1.62	12035.16
newblue7	1.07E+09	1.18	28384.81

Table 10.8. Experiment results on ISPD06 suite.

circuit	WL	WL / OPT WL	Overflow	runtime
adaptec5	1.08E+08	1.32	16	18478
newblue1	2.91E+07	1.42	151	6908
newblue2	4.49E+07	1.37	31	9174
newblue3	1.01E+08	1.38	21	15251
newblue4	6.83E+07	1.39	15	17361
newblue5	1.39E+08	1.36	15	25163
newblue6	1.29E+08	1.42	18	20153
newblue7	3.29E+08	1.59	26	63670

Table 10.9. Experiment results on PEKO06 suite.

optimal placement solution is around 10. That means we need to reduce the overflow at the expense of a wirelength increase.

To conclude, mPL6 can produce high quality placement solutions in a reasonable runtime. And it is highly scalable. It is also able to handle additional density constraint with little degradation in wirelength.

Acknowledgement. Financial supports from Semiconductor Research Consortium Contract 2003-TJ-1019, National Science Foundation grants ACI-0072112, CCF-0430077, and MSPA-MCS: 0528583 are gratefully acknowledged.

References

1. S.N. Adya, S. Chaturvedi, D.A. Papa J.A. Roy, and I.L. Markov. Unification of partitioning, floorplanning and placement. In *Proceedings of the International Conference on Computer Aided Design*, pages 550–557, Nov 2004
2. C.R. Anderson and C. Elion. Accelerated solutions of nonlinear equations using stabilized runge–kutta methods. Report, UCLA CAM, Apr 2004
3. C. Alpert, J.-H. Huang, and A.B. Kahng. Multilevel circuit partitioning. In *Proceedings of the Design Automation Conference*, pages 627–632, 1997
4. K. Arrow, L. Huriwicz, and H. Uzawa. *Studies in Nonlinear Programming*. Stanford University Press, 1958
5. C. Alpert, A.B. Kahng, G. Nam, S. Reda, and P. Villarrubia. A semi-persistent clustering technique for vlsi circuit placement. In *Proceedings of the International Symposium on Physical Design*, pages 200–207, Apr 2005

6. S.N. Adya and I.L. Markov. Consistent placement of macro-blocks using floorplanning and standard-cell placement. In *Proceedings of the International Symposium on Physical Design*, pages 12–17, Apr 2002
7. S.N. Adya, I.L. Markov, and P. G. Villarrubia. On whitespace and stability in mixed-size placement. In *Proceedings of the International Conference on Computer Aided Design*, pages 311–318, Nov 2003
8. A.R. Agnihotri, S. Ono, and P.H. Madden. Recursive bisection placement: Feng shui 5.0 implementation details. In *Proceedings of the International Symposium on Physical Design*, pages 230–232, Apr 2005
9. D.P. Bertsekas. *Constrained Optimization and Lagrange Multiplier Methods*. Academic Press, New York, 1982
10. W.L. Briggs, S.F. McCormick, and V.E. Henson. *A Multigrid Tutorial*. SIAM, Philadelphia, second edition, 2000
11. Ulrich Brenner, Anna Pauli, and Jens Vygen. Almost optimum placement legalization by minimum cost flow and dynamic programming. In *Proceedings of the International Symposium on Physical Design*, pages 2–9, April 2004
12. A. Brandt and D. Ron. *Multigrid Solvers and Multilevel Optimization Strategies*, chapter 1 of *Multilevel Optimization and VLSICAD*. Kluwer Academic Publishers, Boston, 2002
13. A. Brandt. Algebraic multigrid theory: The symmetric case. *Appl. Math. Comp.*, 19: 23–56, 1986
14. A. Brandt. Multiscale scientific computation: Review 2001. In T. Barth, R. Haimes, and T. Chan, editors, *Multiscale and Multiresolution Methods*. Springer Verlag, 2001
15. T.F. Chan, J. Cong, T. Kong, J. Shinnerl, and K. Sze. An enhanced multilevel algorithm for circuit placement. In *Proceedings of the International Conference on Computer Aided Design*, pages 299–306, San Jose, CA, Nov 2003
16. T.F. Chan, J. Cong, T. Kong, and J. Shinnerl. Multilevel optimization for large-scale circuit placement. In *Proceedings of the International Conference on Computer Aided Design*, pages 171–176, San Jose, CA, Nov 2000
17. T.F. Chan, J. Cong, T. Kong, and J. Shinnerl. *Multilevel Circuit Placement*, chapter 4 of *Multilevel Optimization in VLSICAD*. Kluwer Academic Publishers, Boston, 2003
18. R. Chan, T. Chan, M.K. Ng, and A. Yip. Cosine transform preconditioner for high resolution image reconstruction. *Linear Algebra and its Applications*, 316:89–104, 2000
19. T.F. Chan, J. Cong, M. Romesis, J.R. Shinnerl, K. Sze, and M. Xie. mPL6: a robust multilevel mixed-size placement engine. In *Proceedings of the International Symposium on Physical Design*, pages 227–229, Apr 2005
20. T. Chan, J. Cong, J. Shinnerl, K. Sze, and M. Xie. Enhanced robustness in multilevel mixed-size placement. In *SRC TECHCON*, Oct 2005
21. T. Chan, J. Cong, and K. Sze. Multilevel generalized force-directed method for circuit placement. In *Proceedings of the International Symposium on Physical Design*, pages 185–192, Apr 2005
22. C.C. Chang, J. Cong, and X. Yuan. Multi-level placement for large-scale mixed-size ic designs. In *Proceedings of the Asia South Pacific Design Automation Conference*, pages 325–330, 2003
23. J. Cong and S.K. Lim. Edge separability-based circuit clustering with application to multi-level circuit partitioning. *IEEE Tran. on Computer-Aided Design of Integrated Circuits and Systems*, 23(3):346–357, 2004
24. Jason Cong, Michail Romesis, and Joseph Shinnerl. Robust mixed-size placement under tight white-space constraints. In *Proceedings of the International Conference on Computer Aided Design*, pages 165–173, November 2005

25. Meszaros Csaba. Fast cholesky factorization for interior point methods of linear programming. *Computers and Mathematics with Applications*, 31:49–51, 1996
26. J. Cong and M. Xie. A robust detailed placement for mixed-size ic designs. In *Proceedings of the Asia South Pacific Design Automation Conference*, pages 188–194, Jan 2006
27. K. Doll, F.M. Johannes, and K.J. Antreich. Iterative placement improvement by network flow methods. *IEEE Transactions on Computer-Aided Design*, 13(10), October 1994
28. Hans Eisenmann and Frank M. Johannes. Generic global placement and floorplanning. In *Proceedings of the Design Automation Conference*, pages 269–274, 1998
29. L.C. Evans. *Partial Differential Equations*. American Mathematical Society, 2002
30. `http://momonga.t.u-tokyo.ac.jp/~ooura/fft.html`
31. A.V. Fiacco and G. P. McCormick. *Nonlinear Programming: Sequential Unconstrained Minimization Techniques*. John Wiley and Sons, Inc., New York, London, Sydney and Toronto, 1968
32. Dwight Hill. Method and system for high speed detailed placement of cells within an integrated circuit design. *US Patent No. 6,370,673*, 2002
33. S.-W. Hur and J. Lillis. Mongrel: Hybrid techniques for standard-cell placement. In *Proceedings of the International Conference on Computer Aided Design*, pages 165–170, San Jose, CA, Nov 2000
34. B. Hu and M. Marek-Sadowska. Fine granularity clustering for large scale placement problems. *IEEE Tran. on Computer-Aided Design of Integrated Circuits and Systems*, 23(4):527–536, 2004
35. G. Karypis, R. Aggarwal, V. Kumar, and S. Shekhar. Multilevel hypergraph partitioning: Application in vlsi domain. In *Proceedings of the Design Automation Conference*, pages 526–529, 1997
36. Ateen Khatkhate, Chen Li, Ameya R. Agnihotri, Mehmet C. Yildiz, Satoshi Ono, Cheng-Kok Koh, and Patrick H. Madden. Recursive bisection based mixed block placement. In *Proceedings of the International Symposium on Physical Design*, pages 84–89, April 2004
37. A. Kennings and I.L. Markov. Analytical minimization of half-perimeter wirelength. In *Proceedings of the Asia South Pacific Design Automation Conference*, pages 179–184, Jan 2000
38. Andrew Kahng, Sherief Reda, and Qinke Wang. Architecture and details of a high quality, large-scale analytical placer. In *Proceedings of the International Conference on Computer Aided Design*, pages 891–899, Nov 2005
39. J.M. Kleinhans, G. Sigl, F.M. Johannes, and K.J. Antreich. Gordian: Vlsi placement by quadratic programming and slicing optimization. *IEEE Trans. on Computer-Aided Design*, CAD-10:356–365, 1991
40. A.B. Kahng and Q. Wang. Implementation and extensibility of an analytic placer. In *Proceedings of the International Symposium on Physical Design*, pages 18–25, 2004
41. Singh K, A. Wang, R. Brayton, and A. Sangiovanni-Vincentelli. Timing optimization of combinatorial logic. In *Proceedings of the International Conference on Computer Aided Design*, pages 282–285, Nov 1988
42. C.Li and C.-K. Koh. On improving recursive bipartitioning-based placement. Report tr-ece-03-14, Purdue University ECE, 2003
43. H. Murata, K. Fujiyoshi, S. Nakatake, and Y. Kajitani. Rectangle-packing-based module placement. In *Proceedings of the International Conference on Computer Aided Design*, pages 472–479, 1995
44. G.D. Micheli. Performance-oriented synthesis of large-scale domino cmos circuits. *IEEE Trans. on Computer-Aided Design of Integrated Circuits and Systems*, 6:751–765, 1987

45. K.W. Morton and D.F. Mayers. *Numerical Solution of Partial Differential Equations*. Cambridge University Press, 1994
46. Gi-Joon Nam. Ispd 2006 placement contest: Benchmark suite and results. In *Proceedings of the International Symposium on Physical Design*, pages 167–167, 2006
47. Sudip Nag and Kamal Chaudhary. Post-placement residual-overlap removal with minimal movement. In *Proceedings of the Design Automation and Test in Europe*, pages 581–586, 1999
48. W. Naylor, R. Donelly, and L. Sha. Non-linear optimization system and method for wire length and delay optimization for an automatic electric circuit placer. *US Patent 6301693*, Oct 2001
49. A.N. Ng, I.L. Markov, R. Aggarwal, and V. Ramachandran. Solving hard instances of floorplacement. In *Proceedings of the International Symposium on Physical Design*, pages 170–177, New York, NY, USA, 2006. ACM Press
50. R. Okuda, T. Sato, H. Onodera, and K. Tamaru. An efficient algorithm for layout compaction problem with symmetry constraints. In *Proceedings of the International Conference on Computer Aided Design*, pages 148–153, November 1989
51. L.I. Rudin, S.J. Osher, and E. Fatermi. Nonlinear total variation based noise removal algorithms. *Physica D*, 60:259–268, 1992
52. J. Ruge and K. Stüben. Algebraic multigrid. In S.F. McCormick, editor, *Multigrid Methods*, pages 73–80. SIAM, Philadelphia, 1987
53. H.V. Sorensen and C.S. Burrus. Fast dft and convolution algorithms. In S.K. Mitra and J.F. Kaiser, editors, *Handbook for Digital Signal Processing*. John Wiley and Sons, New York, 1993
54. K.J. Singh. *Performance Optimization for Digital Circuits*. PhD thesis, Computer Science Department, University of California Berkeley, 1992
55. U. Trottenberg, C.W. Oosterlee, and A. Schüller. *Multigrid*. Academic Press, London, 2000
56. Xiaoping Tang, Ruiqi Tian, and Martin D.F. Wong. Optimal redistribution of white space for wire length minimization. In *Proceedings of the Asia South Pacific Design Automation Conference*, pages 412–417, January 2005
57. Taraneh Taghavi, Xiaojian Yang, and Bo-Kyung Choi. Dragon 2005: Large-scale mized-size placement tool. In *Proceedings of the International Symposium on Physical Design*, April 2005
58. K.P. Vorwerk and A. Kennings. An improved mulit-level framework for force-directed placement. In *Proceedings of the Design Automation and Test in Europe*, volume 2, pages 240–245, 2005
59. K.P. Vorwerk, A. Kennings, and A. Vannelli. Engineering details of a stable force-directed placer. In *Proceedings of the International Conference on Computer Aided Design*, pages 573–580, Nov 2004
60. Jens Vygen. Algorithms for large-scale flat placement. In *Proceedings of the Design Automation Conference*, pages 746–751, 1997
61. J. Vygen. Algorithms for detailed placement of standard cells. In *Proceedings of the Design Automation and Test in Europe*, pages 321–324, 1998
62. Songjie Xu. *Synthesis for Hign-Density and High-Performance FPGA*. PhD thesis, Computer Science Department, University of California, Los Angeles, 2000
63. Bo Yao, Hongyu Chen, Chung-Kuan Cheng, Nan-Chi Chou, Lung-Tien Liu, and Peter Suaris. Unified quadratic programming approach for mixed mode placement. In *Proc. Int. Symposium on Physical Design*, April 2005

11

NTUplace3: An Analytical Placer for Large-Scale Mixed-Size Designs

Tung-Chieh Chen[1], Zhe-Wei Jiang[1], Tien-Chang Hsu[1], Hsin-Chen Chen[2], and Yao-Wen Chang[1,2]
[1]Graduate Institute of Electronics Engineering
[2]Department of Electrical Engineering
National Taiwan University, Taipei 106, Taiwan
{donnie, crazying, tchsu, indark}@eda.ee.ntu.edu.tw; ywchang@cc.ee.ntu.edu.tw

11.1 Introduction

This chapter is focused on NTUplace3 [6], a large-scale mixed-size analytical placer that can handle modern placement considerations such as wirelength, preplaced blocks, and density. Like many modern placers, NTUplace3 consists of three major stages: global placement, legalization, and detailed placement. Global placement evenly distributes blocks and finds the best position for each block to minimize the target cost (e.g., wirelength). Then, legalization removes all overlaps among blocks and places standard cells row by row. Detailed placement further refines the solution.

The global placement of NTUplace3 is based on the multilevel framework which applies a two-stage technique of bottom-up coarsening followed by top-down uncoarsening. The coarsening stage iteratively clusters blocks based on connectivity/block size to reduce the problem size until the problem size is below a given threshold. Then, an initial placement is computed. In the uncoarsening stage, it iteratively declusters the blocks and refine the block positions to reduce the wirelength. The declustering process continues until the final placement is found.

During the uncoarsening stage, NTUplace3 applies the analytical model for the global placement. The objective function is based on the log–sum–exp wirelength model proposed by Naylor et al. [18]. To handle preplaced blocks, NTUplace3 applies a two-stage smoothing technique, Gaussian smoothing followed by level smoothing, to facilitate block spreading during global placement. The density is controlled mainly by cell spreading during global placement and cell sliding during detailed placement. We further use the conjugate gradient method with dynamic step-size control to speed up the global placement and apply macro shifting to find better macro positions.

During legalization, we remove the overlaps and place all standard cells row by row using a priority-based scheme based on block sizes and locations. We also incorporate a look-ahead legalization scheme into global placement to facilitate the legalization process. During detailed placement, we adopt cell matching and cell swapping to minimize the wirelength and cell sliding to optimize the density. We shall detail the techniques and evaluate them in the following.

11.2 Analytical Placement Model

Circuit placement can be formulated as a hypergraph $H = (V, E)$ placement problem. Let vertices $V = \{v_1, v_2, ..., v_n\}$ represent blocks and hyperedges $E = \{e_1, e_2, ..., e_m\}$ represent nets. Let x_i and y_i be the respective x and y coordinates of the center of the block v_i, and a_i be the area of the block v_i. The circuit may contain some *preplaced blocks* which have fixed x and y coordinates and are not movable. We intend to determine the optimal positions of movable blocks so that the total wirelength is minimized and there is no overlap among blocks.

To evenly distribute the blocks, we divide the placement region into uniform nonoverlapping bin grids. Consequently, the global placement problem can be formulated as a constrained minimization problem as follows:

$$\begin{aligned} &\min\ W(x, y) \\ &\text{s.t.}\ \ D_b(x, y) \le M_b, \ \text{for each bin b} \end{aligned} \tag{11.1}$$

where $W(x, y)$ is the wirelength function, $D_b(x, y)$ is the potential function that is the total area of movable blocks in bin b, and M_b is the maximum area of movable blocks in bin b. M_b can be computed by $M_b = t_{\text{density}}(w_b h_b - P_b)$, where t_{density} is a user-specified target density value for each bin, w_b (h_b) is the width (height) of bin b, and P_b is the *base potential* that equals the preplaced block area in bin b. Note that M_b is a fixed value as long as all preplaced block positions are given and the bin size is determined. Figure 11.1 gives the notation used in this chapter.

The wirelength $W(x, y)$ is defined as the total half-perimeter wirelength (HPWL) given by

$$W(x, y) = \sum_{\text{net } e} \left(\max_{v_i, v_j \in e} |x_i - x_j| + \max_{v_i, v_j \in e} |y_i - y_j| \right) \tag{11.2}$$

Since $W(x, y)$ is nonconvex, it is hard to minimize it directly. Thus, several smooth wirelength approximation functions are proposed in the literature, such as quadratic wirelength [8, 16],

$$\sum_{e \in E} \left(\sum_{v_i, v_j \in e, i<j} w_{ij}(x_i - x_j)^2 + \sum_{v_i, v_j \in e, i<j} w_{ij}(y_i - y_j)^2 \right) \tag{11.3}$$

L_p-norm wirelength [3, 14],

$$\sum_{e \in E} \left(\left(\sum_{v_k \in e} x_k^p \right)^{\frac{1}{p}} - \left(\sum_{v_k \in e} x_k^{-p} \right)^{-\frac{1}{p}} + \left(\sum_{v_k \in e} y_k^p \right)^{\frac{1}{p}} - \left(\sum_{v_k \in e} y_k^{-p} \right)^{-\frac{1}{p}} \right) \tag{11.4}$$

x_i, y_i	center coordinate of block v_i
w_i, h_i	width and height of block v_i
w_b, h_b	width and height of bin b
M_b	maximum area of movable blocks in bin b
D_b	potential (area of movable blocks) in bin b
P_b	base potential (preplaced block area) in bin b
t_{density}	target placement density

Fig. 11.1. Notation used in this chapter.

and log–sum–exp wirelength [2, 13, 18],

$$\gamma \sum_{e \in E} \left(\log \sum_{v_k \in e} \exp\left(\frac{x_k}{\gamma}\right) + \log \sum_{v_k \in e} \exp\left(\frac{-x_k}{\gamma}\right) + \right.$$

$$\left. \log \sum_{v_k \in e} \exp\left(\frac{y_k}{\gamma}\right) + \log \sum_{v_k \in e} \exp\left(\frac{-y_k}{\gamma}\right) \right) \quad (11.5)$$

The log–sum–exp wirelength model, proposed in [18], achieves the best results among these three models [3]. When γ is small, log–sum–exp wirelength gives a good approximation to the HPWL [18]. However, due to the intrinsic precision limitation of a computer, we can only choose a reasonably small γ, say 1% length of the chip width, so that it will not cause any arithmetic overflow.

The function $D_b(x, y)$ can be expressed as

$$D_b(x, y) = \sum_{v \in V}^{n} P_x(b, v) P_y(b, v) \quad (11.6)$$

where P_x and P_y are the overlap functions between bin b and block v along the x and y directions. Since density $D_b(x, y)$ is neither smooth nor differentiable, mPL [3] uses inverse Laplace transformation to smooth the density, while APlace [13] uses bell-shaped functions p_x and p_y for each block to smooth the density P_x and P_y, respectively. In [13], the bell-shaped potential function p_x is defined by:

$$p_x(b, v) = \begin{cases} 1 - a d_x^2, & 0 \le d_x \le w_v/2 + w_b \\ b(d_x - 2w_b - 2w_g)^2, & w_v/2 + w_b \le d_x \le w_v/2 + 2w_b \\ 0, & w_v/2 + 2w_b \le d_x \end{cases} \quad (11.7)$$

where

$$\begin{aligned} a &= 4/((w_v + 2w_b)(w_v + 4w_b)) \\ b &= 2/(w_b(w_v + 4w_b)) \end{aligned} \quad (11.8)$$

Here, w_b is the bin width, w_v is the block width, and d_x is the x direction difference between the block v and the center of the bin b. The range of block's potential is $w_v + 2w_b$ in the x direction. The smooth y-potential function $p_y(b, v)$ can be defined similarly.

By doing so, the nonsmooth function $D_b(x, y)$ can be replaced by the smooth one, $\hat{D}_b(x, y) = \sum_{v \in V}^{n} c_v p_x(b, v) p_y(b, v)$, where c_v is a normalization factor so that the total potential of a block equals its area.

A quadratic penalty method is used to solve (11.1), implying that we solve a sequence of unconstrained minimization problems of the form

$$\min \quad W(x, y) + \lambda \sum_b (\hat{D}_b(x, y) - M_b)^2 \quad (11.9)$$

with increasing λ's. The solution of the previous problem is used as the initial solution for the next one. We solve the unconstrained problem in (11.9) by the conjugate gradient (CG) method. We observe that CG with line search in [13] is not efficient enough since the line search spends most running time on the minimization process. Therefore, we further use CG with a dynamic step size to minimize (11.9). The dynamic step-size control leads to significantly better efficiency.

11.3 Core Techniques

We describe the underlying techniques used in the global placement, legalization, and detailed placement of NTUplace3 in this section.

11.3.1 Global Placement

As mentioned earlier, the global placement is based on the multilevel framework and the log–sum–exp wirelength model. A two-stage smoothing technique is used to handle preplaced blocks, and cell spreading is performed to optimize the density. We further use the conjugate gradient method with dynamic step-size control to speed up the global placement and apply macro shifting to find better macro positions.

Multilevel Framework

We use the multilevel framework for global placement to improve the scalability. Our algorithm is summarized in Figure 11.2. The multilevel framework applies a two-stage technique of bottom-up coarsening followed by top-down uncoarsening. Lines 1–4 give the coarsening stage. The initial placement is generated in line 5. Lines 6–22 give the uncoarsening stage. The details of each step are explained in the following.

During coarsening, we cluster blocks level by level to reduce the problem size, based on the first-choice (FC) clustering algorithm [3,15]. For the FC clustering algorithm, we examine the blocks in the circuit one by one, identify the blocks with the highest connectivity, and cluster the two blocks with the highest connectivity. After all blocks are processed once, we obtain a level of the clustered circuit. The FC clustering algorithm is then applied iteratively until the number of blocks in the resulting clustered circuit is less than a user-specified threshold.

After clustering, we solve an analytical placement problem by using the conjugate gradient method at each level of the uncoarsening stage. The conjugate gradient method requires an initial placement for the coarsest level, and this initial placement significantly affects the final placement quality and convergence speed. Therefore, we apply quadratic programming proposed by [16] and solve it by using an efficient solver [4].

The placement for the current level provides the initial placement for the next level. In each level, the bin grid size is set according to the number of clusters, the base potential P_b for each bin is computed, and the maximum area of movable blocks M_b is updated accordingly. Then, the value of λ is initialized according to the strength of wirelength and density gradients,

$$\lambda = \frac{\sum |\partial W(x, y)|}{\sum |\partial \hat{D}_b(x, y)|} \tag{11.10}$$

A conjugate gradient solver with dynamic step-size control is then used to solve the constrained minimization problem in (11.1) (in lines 10–17).

Macro shifting is then applied between uncoarsening levels to remove macro overlaps. After macro shifting, blocks are declustered, providing the initial placement for the next level.

We define the *overflow ratio* as the total overflow area in each bin over the area of total movable blocks as follows:

$$\text{overflow\_ratio} = \frac{\sum_{\text{Bin } b} \max\{D_b(x, y) - M_b, 0\}}{\sum \text{total movable area}} \tag{11.11}$$

where overflow_ratio ≥ 0.

Algorithm: **Multilevel Global Placement**
Input:
hypergraph H_0: mixed-size circuit
n_{max}: the maximum block number in the coarsest level
Output:
(x^*, y^*): optimal block positions

$level = 0$;
while $(BlockNumber(H_{level}) > n_{max})$
    $level{+}{+}$;
    $H_{level} = FirstChoiceClustering(H_{level-1})$;
Initialize block positions by $SolveQP(H_{level})$;
for $currentLevel = level$ **to** 0
    Initialize bin grid size $n_{bin} \propto \sqrt{n_x}$;
    Initialize base potential for each bin;
    Initialize $\lambda_0 = \frac{\sum |\partial W(x,y)|}{\sum |\partial \hat{D}_b(x,y)|}$; $m = 0$;
    do
        Solve min $W(x, y) + \lambda_m \sum (\hat{D}_b(x, y) - M_b)^2$;
        $m++$;
        $\lambda_m = 2\lambda_{m-1}$;
        if $(currentLevel == 0$ & $overflow\_ratio < 10\%)$
            Call $LookAheadLegalization()$ and save the best result;
        Compute $overflow\_ratio$;
    until (spreading enough or no further reduction in $overflow\_ratio$)
    if $(currentLevel == 0)$
        Restore the best look-ahead result;
    else
        Call $MacroShifting()$;
        Decluster and update block positions.

Fig. 11.2. Our global placement algorithm.

Our placer uses the overflow ratio to measure the evenness of block distribution instead of the *discrepancy* as in [13], where the discrepancy is defined as the maximum ratio of the actual total block area to the maximum allowable block area among all windows within the chip. The overflow ratio has a more global view since it considers all overflow areas in the placement region while discrepancy considers only the maximum density of a window in the placement region. The global placement stage stops when the overflow ratio is less than a user-specified target value, which is 0 by default.

Base Potential Smoothing

Preplaced blocks predefine the *base potential*, which significantly affects block spreading. Since the base potential P_b is not smooth, it incurs mountains that prevent movable blocks from passing through these regions. Therefore, we shall smooth the base potential to facilitate block spreading. We first use the Gaussian function to smooth the base potential change,

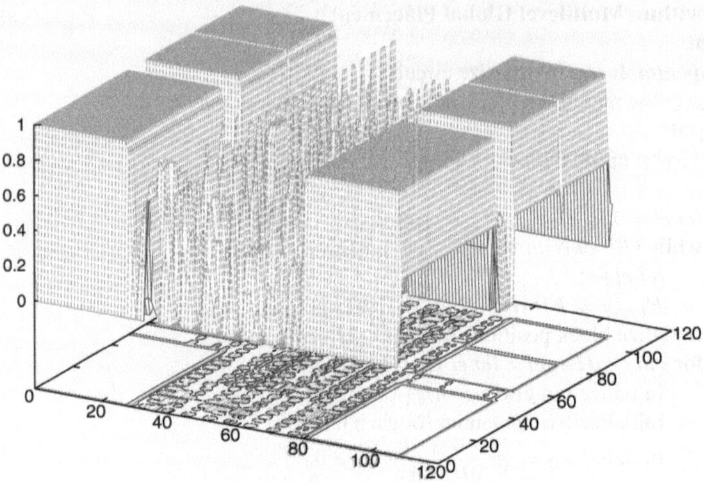

Fig. 11.3. The density profile of newblue2.

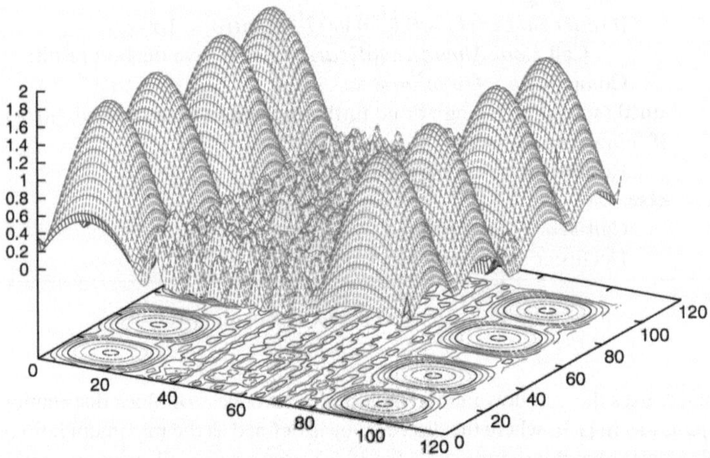

Fig. 11.4. Base potential using the bell-shaped function. The z-coordinate gives the value of $P_b/(w_b h_b)$. For a region with the potential level > 1.0, it means that the base potential in the region is larger than the bin area.

remove the rugged regions in the base potential, and then smooth the base potential level so that blocks can spread to the whole placement region.

The base potential of each block can be calculated by the bell-shaped function. However, we observe that the potential generated by the bell-shaped function has "valleys" among the adjacent regions of preplaced blocks. Figure 11.3 shows the density profile for the circuit newblue2, and Figure 11.4 illustrates the corresponding base potential generated by the bell-shaped function. The z-coordinate gives the value of $P_b/(w_b h_b)$. If a bin has $z > 1$, it means that the potential in the bin is larger than the bin area. There are several valleys in the bottom-left regions as shown in the figure, and these regions do not have any free space but their potentials are so low that a large number of blocks may spread to these regions. To avoid this problem, we calculate the exact density as the base potential, and then use the Gaussian function to smooth the base potential. The two-dimensional Gaussian has the form

$$G(x, y) = \frac{1}{2\pi\sigma^2} e^{-(x^2+y^2)/2\sigma^2} \tag{11.12}$$

where σ is the standard deviation of the distribution. Applying convolution to the Gaussian function G with the base potential P, $P'(x, y) = G(x, y) * P(x, y)$, we can obtain a smoother base potential P'. Gaussian smoothing works as a low-pass filter, which can smooth the local density change. The value σ defines the smoothing range; a larger σ leads to a smoother potential. In global placement, the smoothing range gradually decreases so that the smoothed potential approaches the exact density gradually. Figure 11.5 shows the resulting potential with σ being 0.25 times of the chip width.

After the Gaussian smoothing, we apply another landscape smoothing function [9, 12] to reduce the potential levels. The smoothing function $P''(x, y)$ is defined as follows:

$$P''(x, y) = \begin{cases} \overline{P'} + (P'(x, y) - \overline{P'})^\delta & \text{if } P'(x, y) \geq \overline{P'} \\ \overline{P'} - (\overline{P'} - P'(x, y))^\delta & \text{if } P'(x, y) \leq \overline{P'} \end{cases} \tag{11.13}$$

where $\delta \geq 1$. δ decreases from a large number (say 5) to 1, and a series of level-smoothed potentials are generated. Smoothing potential levels reduces "mountain" (high potential regions) heights so that blocks can spread to the whole placement area smoothly. Figure 11.6 shows the resulting level-smoothed potential of Figure 11.5 using $\delta = 2$.

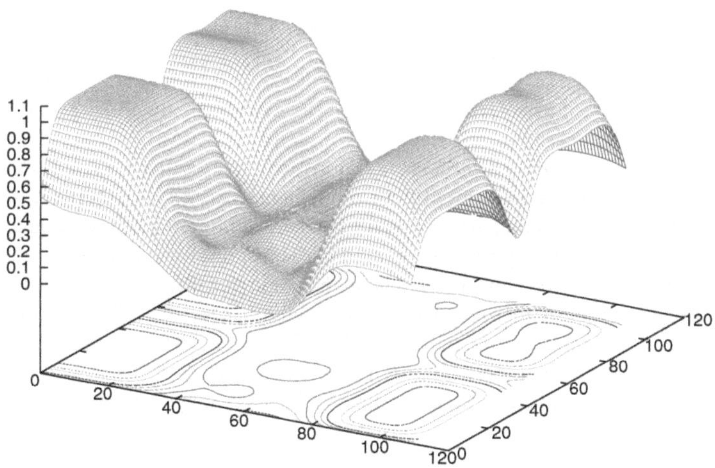

Fig. 11.5. Base potential using exact density and Gaussian smoothing results in a better smoothing potential.

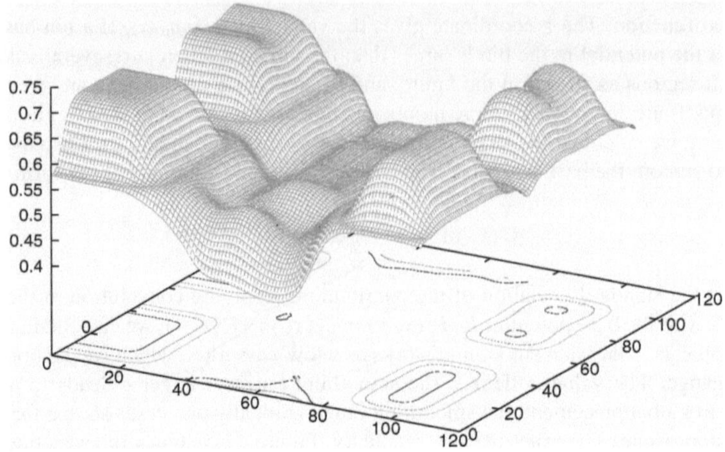

Fig. 11.6. Base potential of Figure 11.5 after level smoothing with $\delta = 2$. Note that the potential level ranges from 0.4 to 0.75, while the original potential level is between 0 and 1.1.

Conjugate Gradient Search with Dynamic Step Sizes

We use the conjugate gradient (CG) method to minimize (11.9). APlace uses the golden section line search to find the optimal step size, which spends most running time on the minimization process. Instead, our step size is computed by a more efficient and effective method. After computing the conjugate gradient direction d_k, the step size α_k is computed by $\alpha_k = s/||d_k||_2$, where s is a user-specified scaling factor. By doing so, we can limit the step size of block spreading since the total quadratic Euclidean movement is fixed,

$$\sum_{v_i \in V} (\Delta x_i^2 + \Delta y_i^2) = ||\alpha_k d_k||_2^2 = s^2 \tag{11.14}$$

where Δx_i and Δy_i are the respective amounts of the movement along the x and y directions for the block v_i in each iteration.

The value of s significantly affects the solution quality; a smaller s value leads to a better wirelength but a longer running time. In our implementation, we set s between 0.2 and 0.3 times of the bin width to obtain a good tradeoff between the running time and quality.

Figure 11.7 summarizes our conjugate gradient algorithm for minimizing the placement objective during global placement. The gradient and conjugate directions are initialized in line 1. The objective function is then iteratively optimized in lines 2–8 until no further improvement is found. To optimize the objective function $f(x_k)$, we first compute the gradient, the Polak–Ribiere parameter, and the conjugate directions in lines 3–5. Then, we can obtain the dynamic step size from equation (11.14) in line 6. Finally, we update the placement solution in line 7.

Macro Shifting

In the global placement stage, it is important to preserve legal macro positions since macros are much bigger than standard cells and illegal macro positions typically make legalization

Algorithm: Conjugate Gradient Algorithm with Dynamic Step-Size Control

Input:
$f(x)$: objective function
x_0: initial solution
s: step size

Output:
optimal x^*

Initialize $g_0 = 0$ and $d_0 = 0$;
do
  Compute gradient directions $g_k = \nabla f(x_k)$;
  Compute the Polak–Ribiere parameter $\beta_k = \frac{g_k^T(g_k - g_{k-1})}{||g_{k-1}||^2}$;
  Compute the conjugate directions $d_k = -g_k + \beta_k d_{k-1}$;
  Compute the step size $\alpha_k = s/||d_k||_2$;
  Update the solution $x_k = x_{k-1} + \alpha_k d_k$;
until $(f(x_k) > f(x_{k-1}))$

Fig. 11.7. Our conjugate gradient algorithm for the global placement optimization.

much more difficult. To avoid this, we apply macro shifting at each declustering level of the global placement stage. Macro shifting moves macros to the closest legal positions.

Integrating within the multilevel framework, only macros with sizes larger than the average cluster size of the current level are processed. Then, the legal macro positions provide a better initial solution for the next declustering level, and those macros are still allowed to spread at subsequent declustering levels.

11.3.2 Legalization

After global placement, legalization removes all overlaps and places standard cells in rows. Since the global placement gives the best positions for macros and standard cells without considering their overlaps, we shall remove the overlaps with minimal total displacement. We extend the standard-cell legalization method in [10] to solve the mixed-size legalization problem. The legalization order of macros and cells are determined by their x coordinates and sizes (widths and heights). Larger blocks get the priority for legalization. Therefore, we legalize macros earlier than standard cells. After the legalization order is determined, macros are placed to their nearest available positions and cells are packed into rows with the smallest wirelength. Despite its simplicity, we find this macro/cell legalization strategy works well on all benchmarks.

Recall that we performed block spreading during global placement. It is important to determine when to terminate the block spreading. If blocks do not spread enough, the wirelength may significantly be increased after legalization since blocks are over congested. If blocks spread too much, the wirelength before legalization may not be good even though the legalization step only increases wirelength a little. This situation becomes even worse when the density is also considered, since the placement objective is more complicated.

To improve the legalization quality, we use a look-ahead legalization technique during globe placement to make the subsequent legalization process easier. At the finest level of the multilevel placement, we apply legalization right after placement objective optimization in each iteration and record the best result with the minimum cost (wirelength and density penalty). Although the look-ahead legalization may take longer running time due to more iterations of legalization, we can ensure that blocks do not over spread and thus obtain a better legal placement. As a result, the look-ahead legalization significantly alleviates the difficulty in removing the macro and standard-cell overlaps during the later legalization stage, and eventually leads to a more robust placement result.

11.3.3 Detailed Placement

In the detailed placement stage, we work on the standard cells to further improve the placement quality. To preserve the prototype of the placement obtained from the legalization stage, we fix all macros and treat them as placement blockages in this stage. The objective of our detailed placement algorithm is to find a better position for each standard cell in the available free spaces.

The detailed placement stage consists of two stages: the wirelength minimization stage and the density optimization stage. In the wirelength minimization stage, we apply *cell matching* and *cell swapping* to reduce the total wirelength. In the density optimization stage, we apply the *cell sliding* technique to reduce the density overflow in congested regions. The flow of our detailed placement is summarized in Figure 11.8. In the following, we explain the cell-matching, cell-swapping, and cell-sliding algorithms.

Cell Matching

We extend the cell-matching algorithms used in our previous NTUplace versions [5, 11] to optimize the wirelength cost for a group of cells simultaneously. The cells in each matching step are chosen from a subregion in the placement region. Here we refer to the subregion as a *window*. Our algorithm formulates a weighted bipartite matching problem by matching the cells to the empty slots inside the window. To handle all cells inside the placement region $\mathcal{R}$, our algorithm divides $\mathcal{R}$ into an array of overlapped windows and iteratively rearranges the

Algorithm: **Detailed Placement**
Input:
  Legalized global placement
  Target density t_{density}
Output:
  Optimal block positions

1. **do** /**Wirelength minimization stage**/
2.   Cell Matching
3.   Cell Swapping
4. **until** no significant improvement in wirelength
5. **do**/**Density optimization stage**/
6.   Cell Sliding
7. **until** no significant improvement in solution quality

Fig. 11.8. Our detailed placement flow.

```
Algorithm: Cell Matching
Input:
   Legalized global placement
   Initial window size w
   Initial sweep direction d
Output:
   Optimal block positions

do
      Divide R into windows according to w
      for all_windows_in_R
            Select all cells inside the window
            Calculate the bipartite matching weights
            Solve the weighted bipartite matching problem
            Move the cells to their new position
      Perturb the window size
      Perturb the sweep direction
until no significant improvement in wirelength
```

Fig. 11.9. The cell-matching algorithm.

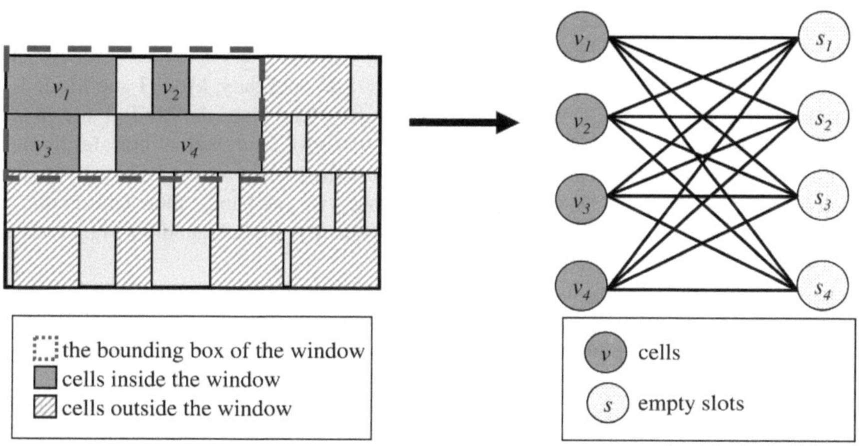

Fig. 11.10. Illustration of the cell-matching process.

cells inside each window to reduce the total wirelength. Figure 11.9 summarizes the flow of our cell-matching algorithm.

Figure 11.10 illustrates our cell-matching process. The dashed line denotes the boundary of the window. The set of cells $V_1 = \{v_1, v_2, v_3, v_4\}$ are selected, and the space occupied by those cells becomes a set of empty slots $V_2 = \{s_1, s_2, s_3, s_4\}$. With this set of cells and those empty slots, we formulate the cell placement as a weighted bipartite matching problem. To construct the bipartite graph $G = (V_1 \cup V_2, E)$, we add an edge $e(v_i, s_j)$ between every $v_i \in V_1$ and every $s_j \in V_2$. We assign the weight of edge $e(v_i, s_j)$, $c(v_i, s_j)$, by computing the wirelength cost of placing cell v_i in slot s_j. In order to keep the legality of our placement

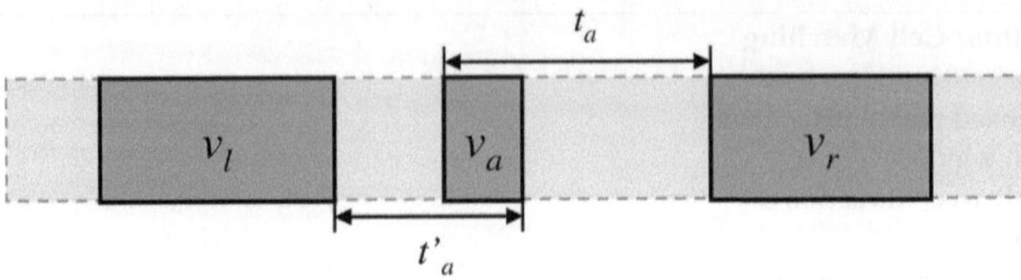

Fig. 11.11. The calculation of space of each slot. t_a is the slot space in Mode I, and t'_a is the slot space in Mode II.

solution, we remove edge $e(v_i, s_j)$ when w_i, the width of the cell v_i, is greater than t_j, the available space of slot s_j.

Now we further discuss the calculation of the space of each slot. In order to ensure the legality, our cell-matching algorithm has two modes: Mode I and Mode II. Figure 11.11 illustrates the definition of space in Mode I and Mode II. In Mode I, t_a, the space of slot s_a is calculated by

$$t_a = \left(x_r - \frac{w_r}{2}\right) - \left(x_a - \frac{w_a}{2}\right) \tag{11.15}$$

and in Mode II the space of slot s_a is given by:

$$t'_a = \left(x_a + \frac{w_r}{2}\right) - \left(x_l + \frac{w_l}{2}\right) \tag{11.16}$$

Here v_r is the cell right to v_a, and v_l is the cell left to v_a. $(x_i - \frac{w_i}{2})$ is the x-coordinate of the left boundary of cell v_i, and $(x_i + \frac{w_i}{2})$ is that of the right boundary. Mode I and Mode II are used alternately in our cell-matching algorithm. Figure 11.12 gives the Mode-I and Mode-II formulations for the circuit of Figure 11.10. With the above formulation, our algorithm can handle a circuit with different sizes of cells and ensure the legality of the resulting placement solution. Compared with the *Domino* [7] detailed placer, we can handle more cells at one time because we do not cut the cells into subcells. Consequently, we can handle a bigger window size and thus obtain a more global view.

In addition to exchanging the positions of the cells inside the window, the cell-matching algorithm has the ability to move the cells to other empty regions. If t_j, the available space of slot s_j, is large enough, our algorithm breaks s_j into several slots. We first find the maximum cell width $w_{\max}$ by

$$w_{\max} = \max_{v_i \in V}\{w_i\} \tag{11.17}$$

and then we can divide s_j into $\lfloor t_j / w_{\max} \rfloor$ slots. The excess $\lfloor t_j / w_{\max} \rfloor - 1$ empty slots can be considered in the bipartite matching formulation, and thus the cells $v \in V_1$ have the chance to move to sparer slots. Figure 11.13 illustrates this process with Mode I formulation. The set of cells inside the window is given by $V_1 = \{v_1, v_2, v_3\}$, and from the figure we get $w_{\max} = \max\{w_1, w_2, w_3\} = 3$. We can see that $t_2 = 6$ and $t_3 = 10$. Thus s_2 is split into $\{s_2, s_4\}$, and s_3 is split into $\{s_3, s_5, s_6\}$, where $V^+ = \{s_4, s_5, s_6\}$ is the set of spare slots. Now the formulation of bipartite matching becomes $G = (V_1 \cup V_2 \cup V^+, E)$.

Cell Swapping

The cell swapping technique selects k adjacent cells each time to find the best ordering by enumerating all possible orderings using the branch-and-bound method. Here, k is a user-specified

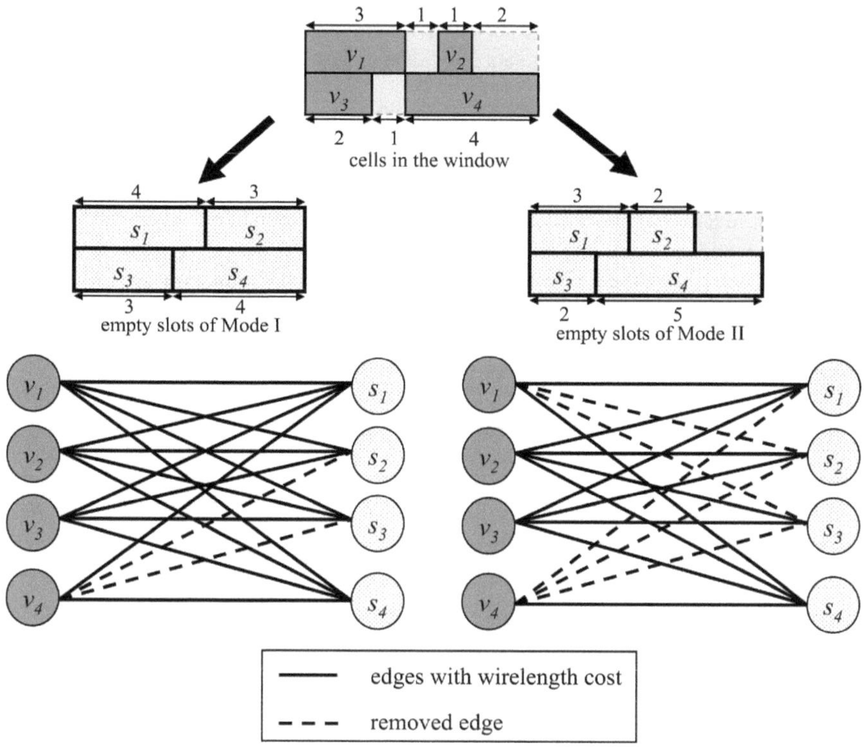

Fig. 11.12. Illustration of cell-matching formulation in Mode I and Mode II.

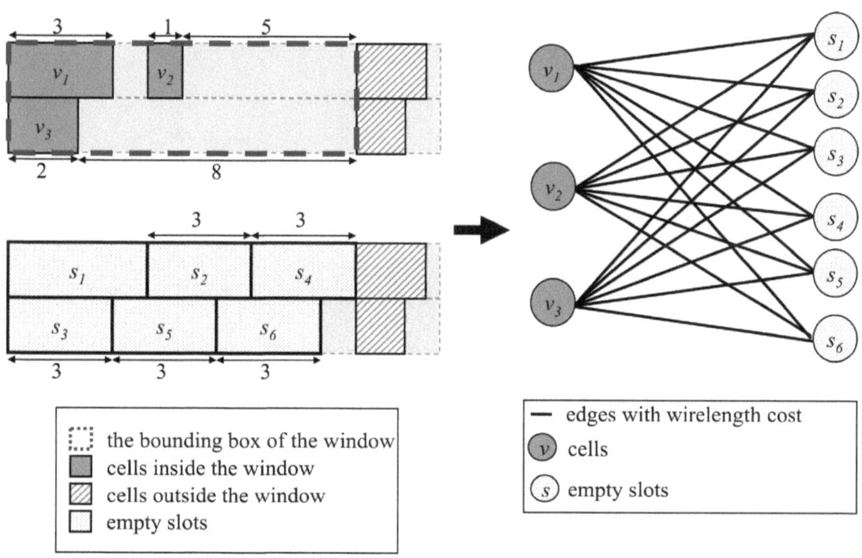

Fig. 11.13. Illustration of the cell-matching formulation with spare slots.

parameter. In our implementation, we set $k = 3$ for a good tradeoff between the running time and solution quality. This process repeats until all standard cells are processed.

Cell Sliding

The objective of cell sliding is to reduce the density overflow in the congested area. We divide the placement region into uniform nonoverlapping bins, and then iteratively reduce the densities of overflowed bins by sliding the cells horizontally from denser bins to sparser bins, with the cell order being preserved. Figure 11.14 illustrates the cell sliding process. Each iteration consists of two stages: left sliding and right sliding. In each stage, we calculate the density of each bin and then compute the area flow $f_{bb'}$ between bin b and its left or right neighboring bin b'. Here, $f_{bb'}$ denotes the desired amount of cell area to move from bin b to b'. Recall that we define D_b as the total movable cell area in bin b and M_b as the maximum allowable block area in bin b. If bin b has no area overflow or the area overflow ratio of b is smaller than b', that is $D_b \leq M_b$ or $D_b/M_b \leq D_{b'}/M_{b'}$, we set $f_{bb'} = 0$. Otherwise we calculate $f_{bb'}$ according to the capacity of b'. If bin b' has enough free space, we move the overflow area of bin b to b'. Otherwise, we evenly distribute the overflow area between b and b'. Therefore, $f_{bb'}$ is defined by

$$f_{bb'} = \begin{cases} D_b - M_b, & \text{if } (M_{b'} - D_{b'}) \geq (D_b - M_b) \\ \frac{D_b M_{b'} - D_{b'} M_b}{M_b + M_{b'}}, & \text{otherwise} \end{cases} \tag{11.18}$$

where the second condition of (11.18) is derived from

$$D_b - \left(M_b + \frac{(D_b - M_b + D_{b'} - M_{b'}) M_b}{M_b + M_{b'}} \right) = \frac{D_b M_{b'} - D_{b'} M_b}{M_b + M_{b'}} \tag{11.19}$$

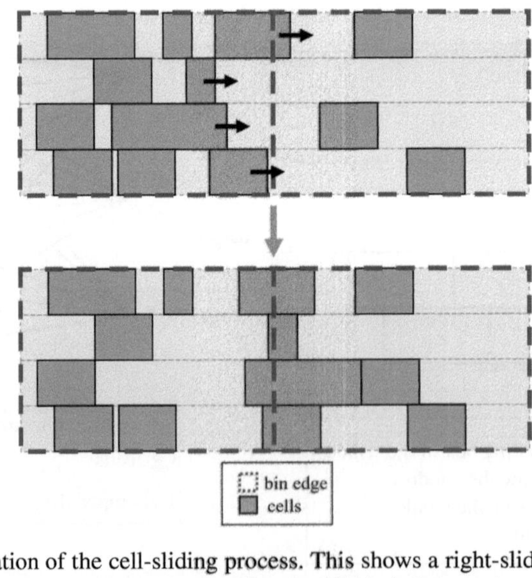

Fig. 11.14. Illustration of the cell-sliding process. This shows a right-sliding stage, where the cells are slid from left to right, and the density of each bin is balanced.

After the area flow $f_{bb'}$ is computed, we sequentially slide the cells across the boundary between b and b' until the amount of sliding area reaches $f_{bb'}$ or there is no more area for cell sliding. Then we update D_b and $D_{b'}$. In the right sliding stage, we start from the left-most bin of the placement region, and b' is right to b. In the left sliding stage, we start from the right-most bin, and b' is left to b, accordingly. We iteratively slide the cells from the area overflow regions to sparser regions until no significant improvement can be obtained (Figure 11.14).

11.4 Experimental Results

We conducted extensive experiments on the platform with an AMD Opteron 2.4 GHz CPU to examine the proposed techniques. We first show the effectiveness of the dynamic step-size control and the look-ahead legalization. Then, we give the HPWL and runtime breakdowns of our placer for the ISPD'05 and ISPD'06 benchmark suites. We also report the HPWLs and runtimes obtained from the L_p-norm wire model for both benchmark suites. Finally, we evaluate our results based on the PEKO-MS benchmarks which have known optimal wirelengths.

11.4.1 Dynamic Step-Size Control

To show the effectiveness of the dynamic step-size control, we performed experiments on adaptec1 with different step sizes. In Figure 11.16, the CPU times and HPWLs are plotted as functions of the step sizes. As shown in Figure 11.16, the CPU time decreases as the step size s becomes larger. In contrast, the HPWL decreases as the step size s gets smaller. The results show that the step size significantly affects the running time and the solution quality.

11.4.2 Look-Ahead Legalization

Table 11.1 lists the HPWLs after the legalization stage with (w/) and without (w/o) look-ahead legalization based on the ISPD-2005 benchmark suite. The table is divided into three parts. The first part gives the numbers of the applied legalizations (LG #), HPWLs, and the CPU times of the global placement with the look-ahead legalization. The second part gives the HPWLs and the CPU times of the global placement without the look-ahead legalization. The third part gives the HPWL and CPU-time ratios, computed by dividing the results without the look-ahead legalization by those with the look-ahead legalization. The results show that the look-ahead legalization can significantly reduce the wirelength by 24%, with even 2% reduction in the CPU time on average.

11.4.3 HPWL and Runtime Analysis

Table 11.2 gives the HPWLs and CPU times of the global placement (GP), the legalization (LG), and the detailed placement (DP) stages for the ISPD-2005 benchmark suite. On average, the legalization stage increases the wirelength by 7% while the detailed placement stage decreases the wirelength by 5%. For the CPU time, global placement spends 72% of the total runtime, which is much more than that of the legalization and the detailed placement stages.

Table 11.3 gives the HPWLs, DHPWLs (combined cost with wirelength and density), and the CPU times of each placement stage for the ISPD-2006 benchmark suite. Similar to the results for the ISPD-2005 benchmark suite, the legalization stage increases 7% wirelength

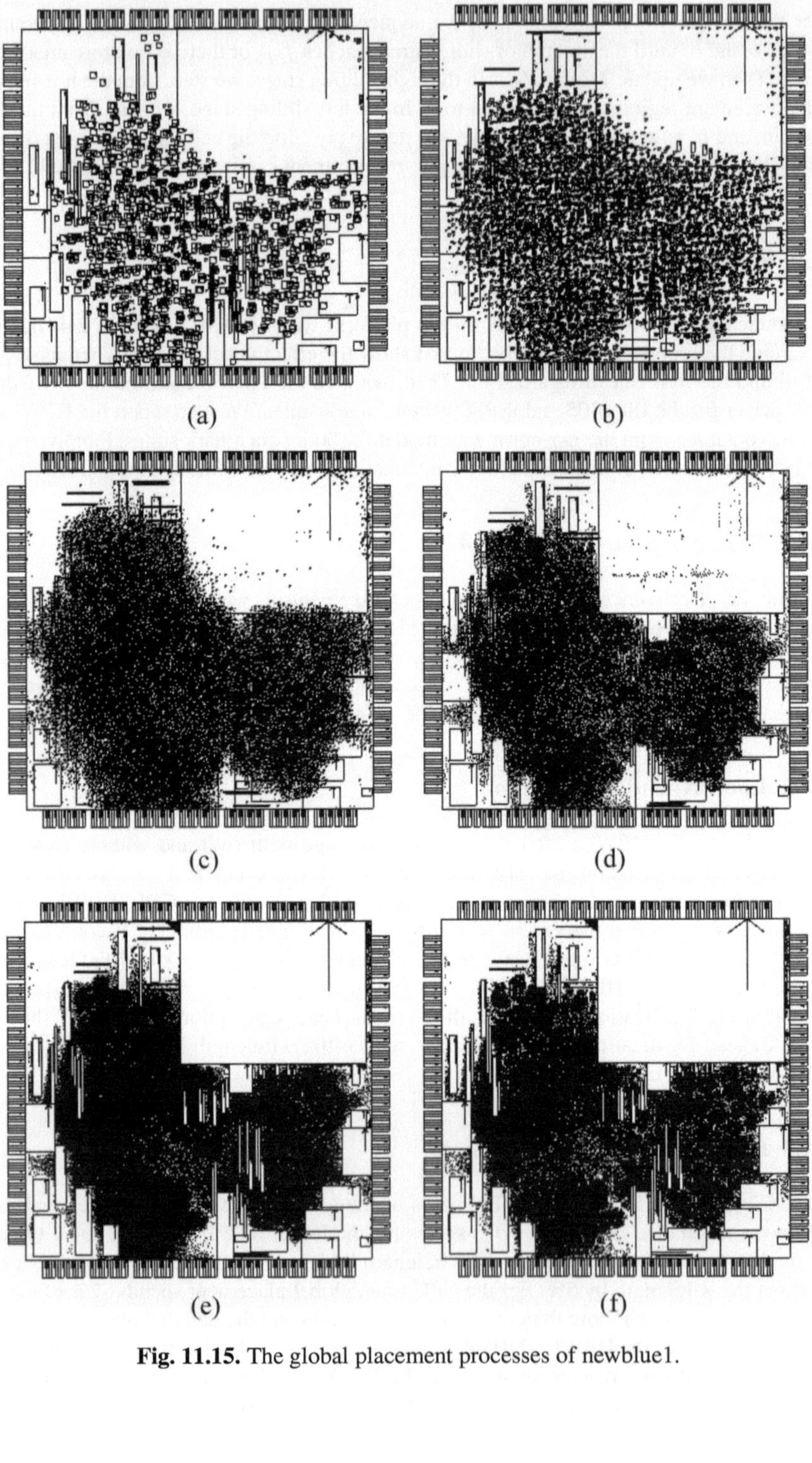

Fig. 11.15. The global placement processes of newblue1.

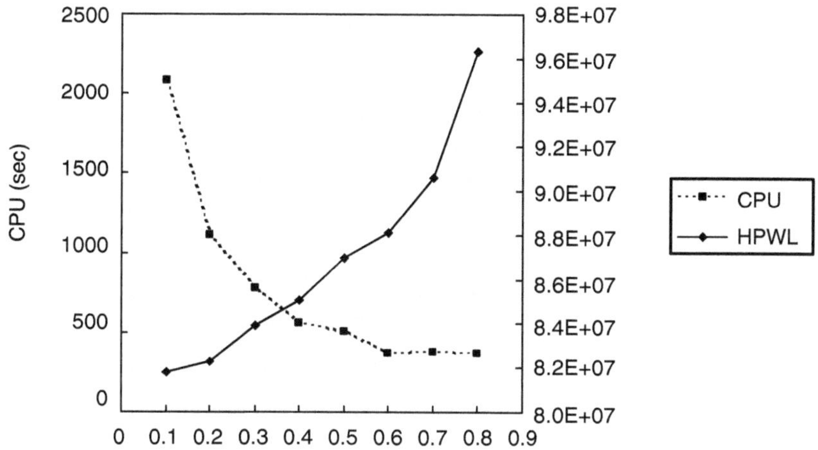

Fig. 11.16. The CPU times and HPWLs resulting from different step sizes based on the circuit adaptec1.

Table 11.1. The HPWLs and CPU times w/ and w/o the look-ahead legalization (LAL).

	w/ LAL			w/o LAL		ratio	
circuit	LG #	HPWL (× e6)	CPU (s)	HPWL (× e6)	CPU (s)	HPWL	CPU
adaptec1	3	83.94	810	86.93	814	1.04	1.01
adaptec2	1	93.09	887	110.50	1025	1.19	1.16
adaptec3	3	228.67	1803	244.31	1592	1.07	0.88
adaptec4	2	206.51	2150	219.06	2369	1.06	1.10
bigblue1	2	100.02	1499	102.24	1468	1.02	0.98
bigblue2	2	159.77	2856	167.84	2974	1.05	1.04
bigblue3	5	409.79	9362	999.92	8738	2.44	0.93
bigblue4	3	897.57	7995	941.07	8714	1.05	1.09
average						1.24	1.02

Table 11.2. HPWL and runtime results for the ISPD-2005 benchmark suite.

	HPWL (× e6)			CPU (s)			
circuit	GP	LG	DP	GP	LG	DP	total
adaptec1	82.10	83.92	80.93	608	105	80	803
adaptec2	90.39	93.07	89.85	640	36	135	824
adaptec3	207.09	227.12	214.20	1308	199	237	1767
adaptec4	196.63	203.28	193.74	1637	150	303	2114
bigblue1	92.19	100.54	97.28	1268	139	103	1523
bigblue2	152.24	161.04	152.20	2414	167	439	3047
bigblue3	331.35	382.72	348.48	2793	2122	723	5687
bigblue4	812.62	884.10	829.16	6127	2089	1957	10280
ratio	1.00	1.07	1.02	72%	14%	13%	100%

Table 11.3. HPWL and runtime results for the ISPD-2006 benchmark suite.

	HPWL (× e6)			DHPWL	CPU (s)			
circuit	GP	LG	DP	(× e6)	GP	LG	DP	total
adaptec5	373.84	402.38	375.05	448.58	8366	821	700	9971
newblue1	59.68	62.41	60.36	68.10	821	89	262	1194
newblue2	186.33	213.41	198.63	203.39	1961	1178	198	3380
newblue3	291.03	295.98	278.87	278.89	1107	53	690	1883
newblue4	274.52	287.98	271.01	301.19	5775	388	595	6812
newblue5	511.67	511.67	469.95	509.54	15601	1044	1159	17899
newblue6	474.83	511.02	482.19	521.65	13558	459	1105	15426
newblue7	1037.43	1126.88	1051.13	1099.66	23464	3006	1860	28734
average	1.00	1.07	1.01	1.08	79%	10%	11%	100%

Table 11.4. Wire-model comparisons based on the ISPD-2005 benchmark suite.

circuit	HPWL (× e6)	vs. LSE	CPU (s)	vs. LSE
adaptec1	80.90	1.00	1097	1.37
adaptec2	90.40	1.01	1043	1.27
adaptec3	216.06	1.01	3292	1.86
adaptec4	198.00	1.02	3542	1.68
bigblue1	97.15	1.00	2180	1.43
bigblue2	152.96	1.01	3962	1.30
bigblue3	353.03	1.01	12516	2.20
bigblue4	839.03	1.01	16521	1.61
average		1.01		1.59

Table 11.5. Wire-model comparisons based on the ISPD-2006 benchmark suite.

circuit	DHPWL (× e6)	vs. LSE	CPU (s)	vs. LSE
adaptec5	458.59	1.02	9971	2.11
newblue1	69.53	1.02	1194	1.02
newblue2	202.24	0.99	3380	1.23
newblue3	280.65	1.01	1883	1.13
newblue4	307.99	1.02	6812	1.88
newblue5	522.88	1.03	17899	1.00
newblue6	528.46	1.01	15426	1.59
newblue7	1103.20	1.00	28734	1.41
average		1.01		1.42

and the detailed placement stage decreases 6% wirelength on average. It should be noted that the detailed placement result incurs only 7% density penalty. Again, most CPU time was spent on global placement (79%).

11.4.4 Wire-Model Comparison

In Tables 11.4 and 11.5, we compare the log–sum–exp (LSE) and the L_p-norm wire models based on the ISPD-2005 and -2006 benchmark suites, respectively. As reported in Table 11.4,

Table 11.6. Results of the PEKO-MS-2005 benchmarks without density optimization.

	HPWL (× e6)				CPU (s)			
circuit	GP	LG	DP	LB	GP	LG	DP	total
adaptec1	23.35	32.82	25.44	20.05	187	2	357	560
adaptec2	37.49	47.90	32.91	24.96	445	2	570	1035
adaptec3	52.50	79.67	57.18	40.95	448	4	979	1463
adaptec4	49.37	74.19	52.16	39.39	352	4	1104	1497
bigblue1	22.13	32.38	25.95	20.85	193	2	340	554
bigblue2	48.84	71.68	53.19	42.25	467	5	1130	1646
bigblue3	118.29	181.73	131.92	94.39	6838	9	2440	9366
bigblue4	195 50	291 01	224 59	171.47	3268	20	3388	6802
average	1.23	1.78	1.31	1.00	38.37%	0.27%	58.08%	100%

Table 11.7. Results of the PEKO-MS-2006 benchmarks without density optimization.

	HPWL (× e6)				CPU (s)			
circuit	GP	LG	DP	LB	GP	LG	DP	Total
adaptec5	98.81	146.39	107.57	61.10	1327	7	2012	3346
newblue1	19.96	32.00	25.85	19.50	307	3	1991	2301
newblue2	40.53	66.56	47.62	27.30	1020	3	2157	3180
newblue3	88.43	121.62	94.14	30.30	1014	4	2017	3035
newblue4	55.96	82.83	61.81	43.60	486	6	1506	1998
newblue5	124.49	181.64	131.39	85.80	1349	11	2902	4262
newblue6	104.71	149.52	111.28	50.00	1091	11	8358	9460
newblue7	231.94	352.16	280.91	151.00	3824	19	12658	16501
average	1.58	2.34	1.77	1.00	26.15%	0.17%	73.68%	100%

the log–sum–exp wire model leads to smaller wirelength than that obtained by the L_p-norm one by about 1%, and better CPU times by about 59% for the ISPD-2005 benchmark suite.

For the ISPD-2006 benchmark suite, the objective is to optimize both wirelength and density, i.e., the density HPWL (DHPWL for short). The DHPWL is defined as follows [1,17]:

$$\text{DHPWL} = \text{HPWL} \times (1 + \text{density\_penalty}) \tag{11.20}$$

To compute *density_penalty*, we make the width and height of the bin grid equal to 10 circuit row height, and define *density_penalty* by

$$\text{density\_penalty} = (\text{overflow\_ratio} \times \text{bin\_area} \times \text{density\_target})^2 \tag{11.21}$$

where $overflow\_ratio$ is defined by (11.11).

As reported in Table 11.5, the log–sum–exp wire model again leads to smaller wirelenth than that obtained by the L_p-norm one by about 1%, and better CPU times by about 42% for the ISPD-2006 benchmark suite. It is clear that the log–sum–exp wire model is slightly more effective and significantly more efficient than the L_p-norm one, based on the ISPD-2005 and -2006 benchmark suites.

11.4.5 PEKO-MS Benchmarks

Tables 11.6 and 11.7 give the breakdowns of HPWLs and CPU times, based on the PEKO-MS benchmark suite with known wirelength lower bounds under wirelength optimization alone

Table 11.8. Results of the PEKO-MS-2005 benchmarks with density optimization.

	HPWL (× e6)					CPU (s)			
circuit	GP	LG	DP	LB	DHPWL	GP	LG	DP	total
adaptec1	19.99	33.28	26.16	20.05	27.84	200	2	407	609
adaptec2	31.56	56.85	40.21	24.96	44.70	423	2	460	885
adaptec3	48.67	92.33	64.75	40.95	72.73	536	3	806	1345
adaptec4	54.52	79.49	53.90	39.39	56.58	455	3	837	1295
bigblue1	22.21	32.54	25.77	20.85	26.28	226	2	261	489
bigblue2	48.77	75.63	56.60	42.25	57.99	947	4	985	1936
bigblue3	142.61	453.07	283.81	94.39	328.42	7870	7	2632	10509
bigblue4	225.67	336.29	238.19	171.47	242.73	6859	17	4047	10923
average	1.24	2.29	1.60	1.00		48.56%	0.23%	51.21%	100%

Table 11.9. Results of the PEKO-MS-2006 benchmarks with density optimization.

	HPWL (× e6)					CPU (s)			
circuit	GP	LG	DP	LB	DHPWL	GP	LG	DP	total
adaptec5	87.57	164.03	117.29	61.10	125.88	1990	7	1678	3675
newblue1	20.92	32.46	25.68	29.37	29.53	405	3	1700	2108
newblue2	45.05	68.80	47.39	27.30	48.58	1298	3	2165	3466
newblue3	99.15	132.99	98.85	30.30	104.01	1538	4	1995	3537
newblue4	53.44	85.86	65.59	43.60	67.46	1380	4	995	2379
newblue5	142.81	206.14	141.14	85.80	146.96	2816	10	1913	4739
newblue6	105.35	153.05	112.71	50.00	114.99	1284	9	6089	7372
newblue7	241.87	394.20	305.33	151.00	328.72	5987	18	13653	19658
average	1.65	2.52	1.85	1.00		39.95%	0.14%	59.91%	100%

(without considering the density cost). The columns "GP," "LG," and "DP" give the resulting wirelengths and the required CPU times from the respective global placement, legalization, and detailed placement stages. The column "LB" gives such wirelength lower bound for each circuit. As shown in Tables 11.6 and 11.7, our placer obtains about 1.31 and 1.77 times of the wirelength lower bounds on average for the PEKO-MS-2005 and PEKO-MS-2006 benchmark suites, respectively.

We also performed experiments for *both* wirelength and density optimization on the PEKO-MS benchmark suite. Tables 11.8 and 11.9 report the breakdowns of HPWLs and CPU times during the placement process, considering both wirelength and density costs, and the density HPWLs (DHPWLs) after the detailed placement stage. As reported in the tables, our placer obtains longer wirelengths than optimizing wirelength alone, which are about 1.6 and 1.85 times of the wirelength lower bounds on average for the PEKO-MS-2005 and PEKO-MS-2006 benchmark suites, respectively.

References

1. *ISPD 2006 Program.* http://www.ispd.cc/program.html.
2. T. Chan, J. Cong, J. Shinnerl, K. Sze, and M. Xie. mPL6: Enhanced multilevel mixed-size placement. In *Proceedings of ACM International Symposium on Physical Design*, pages 212–214, 2006.

3. T. Chan, J. Cong, and K. Sze. Multilevel generalized force-directed method for circuit placement. In *Proceedings of ACM International Symposium on Physical Design*, pages 185–192, April 2005. Best paper award at ISPD'2005
4. H. Chen, C.-K. Cheng, N.-C. Chou, A. B. Kahng, J. F. MacDonald, P. Suaris, B. Yao, and Z. Zhu. An algebraic multigrid solver for analytical placement with layout based clustering. In *Proceedings of ACM/IEEE Design Automation Conference*, pages 794–799, 2003
5. T.-C. Chen, T.-C. Hsu, Z.-W. Jiang, and Y.-W. Chang. NTUplace: a ratio partitioning based placement algorithm for large-scale mixed-size designs. In *Proceedings of ACM International Symposium on Physical Design*, pages 236–238, 2005
6. T.-C. Chen, Z.-W. Jiang, T.-C. Hsu, and Y.-W. Chang. A high-quality mixed-size analytical placer considering preplaced blocks and density constraints. In *Proceedings of IEEE/ACM International Conference on Computer-Aided Design*, 2006
7. K. Doll, F.M. Johannes, and K. Antreich. Iterative placement improvement by network flow methods. *IEEE Transations on Computer-Aided Design of Integrated Circuits and Systems*, 13:1189–1200, 1994
8. H. Eisenmann and F.M. Johannes. Generic global placement and floorplanning. In *Proceedings of ACM/IEEE Design Automation Conference*, pages 269–274, 1998
9. J. Gu and X. Huang. Efficient local search with search space smoothing: A case study of the traveling salesman problem (TSP). *IEEE Transaction on Systems, Man and Cybernetics*, 24(5):728–735, 1994
10. D. Hill. US patent 6,370,673: Method and system for high speed detailed placement of cells within an intergrated circuit design. 2002
11. Z.-W. Jiang, T.-C. Chen, T.-C. Hsu, H.-C. Chen, and Y.-W. Chang. NTUplace2: A hybrid placer using partitioning and analytical techniques. In *Proceedings of ACM International Symposium on Physical Design*, pages 215–217, 2006
12. A. B. Kahng, S. Reda, and Q. Wang. APlace: A general analytic placement framework. In *Proceedings of ACM International Symposium on Physical Design*, pages 233–235, 2005
13. A.B. Kahng and Q. Wang. Implementation and extensibility of an analytic placer. *IEEE Transations on Computer-Aided Design of Integrated Circuits and Systems*, 24(5), May 2005
14. A.B. Kahng and Q. Wang. A faster implementation of APlace. In *Proceedings of ACM International Symposium on Physical Design*, pages 218–220, 2006
15. G. Karypis and V. Kumar. Multilevel k-way hypergraph partitioning. In *Proceedings of ACM/IEEE Design Automation Conference*, pages 343–348, 1999
16. M. Kleinhans, G. Sigl, F.M. Johannes, and K. J. Antreich. Gordian: VLSI placement by quadratic programming and slicing optimization. *IEEE Transations on Computer-Aided Design of Integrated Circuits and Systems*, 10(3):356–365, 1991
17. G.-J. Nam, C.J. Aplert, and P.G. Villarrubia. The ISPD 2006 placement contest and benchmark suite. In *Slides presented at ISPD'06*, 2006
18. W.C. Naylor, R. Donelly, and L. Sha. US patent 6,301,693: Non-linear optimization system and method for wire length and dealy optimization for an automatic electric circuit placer. 2001

12

Conclusion and Challenges

Placement is one of the most important steps in the RTL-to-GDSII synthesis process, as it directly optimize the interconnects, which have become the bottleneck in circuit and system performance in the nanometer process technologies.

This book highlights the most dominant placement algorithms and implementation techniques up to year 2006, as demonstrated in the 2005 and 2006 ISPD placement contests. Given the exponential increase of the placement problem sizes, hierarchical or multilevel methods are typically needed for scalability. The hierarchical placement method traces back thirty years ago when min-cut based placement algorithms were first introduced, but have been refined a great deal in recent years as in Capo (Chap. 5). The multilevel placement method is much more recent, to a large extent promoted by the authors of the mPL placer in a sequence of publications since 2000. It has been adopted by a number of other analytical placers in recent years, including APlace (Chap. 7), FastPlace (Chap. 8), mFAR (Chap. 9), and NTUPlace3 (Chap. 11). Nevertheless, it might be surprising and puzzling to see that the flat placement tool Kraftwerk (Chap. 3) exhibits excellent scalability. The secret is that it uses a simplified quadratic wire length formulation, which can be solved by an efficient linear system solver using a multigrid (multilevel) method. Therefore, it uses the multilevel method implicitly to achieve the scalability.

Although the contest benchmarks provide numerical comparisons of various algorithms presented in this book, the editors would like to warn the reader that relying solely on a few benchmarking results to judge the merits of the underlying placement algorithms may not be totally reliable. Many implementation details, such as the choice of data structures, memory accessing patterns and various heuristics used by tie-breaking or placement refinements, may affect the final placement results. It is possible that a promising idea or algorithm does not achieve its full potential due to suboptimal implementations. Therefore, the reader needs to go beyond the numerical results to form deeper understanding of the scalability and optimization capability of various algorithms used in the placement contests. We sincerely hope that this book, with detailed algorithm and implementation description of each placement tool, helps the reader to achieve such level of understanding.

As we pointed in Preface of the book, the primary objective in both placement contests was wire length minimization (with some consideration of routability in the second contest). While we believe wire length minimization is very important, as the weighted wire length minimization provides a general framework for performance and routability optimization in placement, we also would like to encourage the researcher to apply and extend the wire length optimization formulation to address other placement constraints and/or objectives. Here are some examples for further improvements.

Congestion Control for Routability. Wire length minimization directly translates into *average* congestion minimization. This global objective function, however, may not be enough to mitigate *local* congestion. Modern circuits tend to have abundant white space. Management of available white space remains as essential consideration in congestion mitigation. Over-spreading cells (e.g., to achieve globally uniform white space distribution) might be good for routability but only at the cost of significant wire length degradation. More effective congestion prediction and reduction techniques are required.

Timing-aware Placement for Timing Closure. Placement is part of the physical synthesis process whose goal is to achieve timing closure (i.e. achieve the performance target determined in design specification). In fact, placement is positioned at the driver seat of the physical synthesis flow. Not only it can affect solution quality metrics (timing, routability, etc.) of physical synthesis, but also virtually all physical synthesis optimizations must communicate with the placer to maintain the legality of solutions. Placement and other physical optimization operations, such as buffering or gate sizing, may compete for the available chip area for timing optimization. How to seamlessly integrate placement with various physical optimization operations for timing closure is still an open problem.

Mixed-size Placement for System-on-Chip (SoC) designs. In order to reduce the design turn-around time, hierarchical designs and design reuse are practiced whenever possible, resulting in a lot more movable/fixed macro blocks in today's SoC designs. Simply, the capability of handling these large macros is a must in modern placement algorithm. In many cases, macros are placed manually by designers as demonstrated in placement contest benchmark circuits. Since those macros may have a huge influence on placement solutions, more automated and effective ways to handle those blocks at different stages of placement (for example, during floorplanning, global placement and legalization etc.) are needed.

3-Dimensional Placement for 3D ICs. The recent advances in packaging technology allow wafers to be stacked and connected together using through-silicon vias so that more functional blocks and circuit elements can be integrated into a single package. This new technology, called 3D IC design, introduces a new optimization dimension for placement. Straightforward extension of 2D placement formulation into 3D versions would not be enough because several technology issues must be addressed seamlessly in 3D placement. Thermal distribution, thermal/signal via placement, signal propagation delays between different wafer layers, and the possible need of supporting 3D macros, are some of the most important concerns. It remains to be seen what is the most efficient and effective placement engine to support such new 3D technology.

Other Placement Constraints and Objectives. In addition to wire lengths, congestion and timing, modern circuits have to address various other issues and constraints. For example, aggressive power minimization needs to be supported where the design may have multiple voltage islands. Thermal constraints, power/ground issues with IR drops, clocking, crosstalk and signal integrity, DFM (Design for Manufacturability) etc., all have strong implications to circuit placement and need to be addressed in the near future.

Given these challenges, we believe that circuit placement is a still an open and active research topic. More exciting development and progress is waiting for us.

Index

MIX
Papier aus verantwortungsvollen Quellen
Paper from responsible sources
FSC® C105338

If you have any concerns about our products,
you can contact us on
ProductSafety@springernature.com

In case Publisher is established outside the EU,
the EU authorized representative is:
Springer Nature Customer Service Center GmbH
Europaplatz 3, 69115 Heidelberg, Germany

Printed by Libri Plureos GmbH
in Hamburg, Germany